T0302285

Materials Issues in Art and Archaeology VIII

MATERIALS RESEARCH SOCIETY
SYMPOSIUM PROCEEDINGS VOLUME 1047

Materials Issues in Art and Archaeology VIII

Symposium held November 26–28, 2007, Boston, Massachusetts, U.S.A.

EDITORS:

Pamela B. Vandiver
University of Arizona
Tucson, Arizona, U.S.A.

Blythe McCarthy
Freer Gallery of Art/Arthur M. Sackler Gallery
Smithsonian Institution
Washington, D.C., U.S.A.

Robert H. Tykot
University of South Florida
Tampa, Florida, U.S.A.

Jose Luis Ruvalcaba-Sil
Instituto de Fisica
Universidad Nacional Autonoma de Mexico
Mexico City, Mexico

Francesca Casadio
The Art Institute of Chicago
Chicago, Illinois, U.S.A.

Materials Research Society
Warrendale, Pennsylvania

CAMBRIDGE
UNIVERSITY PRESS

University Printing House, Cambridge CB2 8BS, United Kingdom

One Liberty Plaza, 20th Floor, New York, NY 10006, USA

477 Williamstown Road, Port Melbourne, VIC 3207, Australia

314-321, 3rd Floor, Plot 3, Splendor Forum, Jasola District Centre, New Delhi - 110025, India

79 Anson Road, #06-04/06, Singapore 079906

Cambridge University Press is part of the University of Cambridge.

It furthers the University's mission by disseminating knowledge in the pursuit of education, learning and research at the highest international levels of excellence.

www.cambridge.org
Information on this title: www.cambridge.org/9781558999886

Materials Research Society
506 Keystone Drive, Warrendale, PA 15086
http://www.mrs.org

First published 2009
First paperback edition 2012

Single article reprints from this publication are available through University Microfilms Inc., 300 North Zeeb Road, Ann Arbor, MI 48106

CODEN: MRSPDH

A catalogue record for this publication is available from the British Library

ISBN 978-1-558-99988-6 Hardback
ISBN 978-1-107-40860-9 Paperback

CONTENTS

*Invited Paper
\#Best in Symposium

*Invited Paper
#Best in Symposium

RECONSTRUCTION OF PAST TECHNOLOGIES

*Invited Paper
^Special Student Award

METHODOLOGY AND INSTRUMENTATION

INTERDISCIPLINARY OR
CROSS-DISCIPLINARY CONTRIBUTIONS

*Invited Paper

DEDICATION

Edward V. Sayre   **Frederick R. Matson**

This volume is dedicated to two pioneers who were close friends, Edward V. Sayre and
Frederick R. Matson. Each passed away last year, 2007. Both scholars provided many new insights
into the understanding and preservation of artistic and cultural heritage. Both were dedicated to
achieving answers to archaeological, artistic and conservation questions through scientific inquiry
and interdisciplinary teamwork. Ed Sayre initiated our first Materials Issues in Art and Archaeology
conference at the Materials Research Society in 1988. He had at least three careers, first in the
Chemistry Department at Brookhaven National Laboratory and teaching at the Institute of Fine Arts,
New York University, then as head of the Conservation Laboratory of the Boston Museum of Fine
Arts, and lastly as the senior physical scientist at the Conservation Analytical Lab, Smithsonian
Institution. Ed Sayre's accomplishments are cited in the volume, *Patterns and Process: A Festschrift
in Honor of Dr. Edward V. Sayre*, edited by Lambertus van Zelst (Smithsonian Center for Materials
Research and Education, Smithsonian Institution,* Suitland, Maryland, 2003). In the entertaining first
chapter, Professor Fred Matson recollects the first collaborative effort that considered use of neutron
activation analysis for the chemical characterization of well-provenanced ceramics. Fred Matson's
accomplishments were documented in 2008 by an exhibition entitled "The Ceramic Legacy of
Frederick R. Matson" at the Matson Museum of Anthropology, Department of Anthropology,
Pennsylvania State University, that was curated by Clare McHale Milner and Tracy Peshoff.
Matson's archaeological studies of ceramic provenance using petrography, of ceramic processing
and firing, of early glass technology and his ethnographic study of village potters in Southwest Asia
and Egypt are legendary, as is the book he edited, *Ceramics and Man* (Wenner-Gren Foundation and
Aldine Publishers, 1965)—a book that began the study of "ceramic ecology." Both men were
generous with their wisdom, shared wonderful but pointed, heuristic stories, and they mentored many
thankful, enlightened students, including several authors in this volume.

*The Center is now called the Museum Conservation Institute, and the volume is available at no
cost from Ann N'Gadi, Information Officer, at ngadia@si.edu.

PREFACE

"Materials Issues in Art and Archaeology VIII" is the result of Symposium Y, held November 26–28 at the 2007 MRS Fall Meeting in Boston, Massachusetts. This volume is the eighth in a series, and represents our goal of presenting cutting-edge and interdisciplinary research used to characterize: (1) cultural materials and intangible cultural properties; (2) the technologies of selection, production and use through which materials are transformed into objects and artifacts; (3) the science underlying their deterioration, preservation and conservation; and (4) the development of sensors and tools for non-destructive, in-situ examination of artifacts as well as innovative technology for their characterization. Studies were solicited that use the methods and techniques of materials research to understand degradation and design strategies to promote long-term preservation of material culture and cultural heritage, e.g., works of art, culturally significant artifacts, and archaeological sites and complexes and their environments. Preserving cultural heritage extends beyond artifact preservation. It includes developing a critical understanding of how our predecessors used technology and craft to solve problems of survival and organization and to make the symbols or representations of what was important to them. It discloses patterns of technology-transfer from one field to another and allows us to gain insights into artists' intentions and processes. It provides evaluation tools and skills such that preservation expectations are based on performance criteria and life histories of the constituent materials.

The call for papers solicited contributions that included a wide range of topics, as follows:

- Materials science applied to promote understanding and long-term preservation of cultural heritage
- Analytical studies of art objects and archaeological artifacts; of particular interest are developments in non-destructive techniques and replicative studies
- Production, microstructure, and performance parallels between ancient materials and processes, and modern technologies, especially surface and nanoscale interactions
- Application of new instrumental techniques to the study and environmental monitoring of art and archaeological materials and to materials for display, storage and transport, including development of new instrumentation and protocols
- Nanotechnology and its impact on conservation treatments and diagnostics

This symposium gave evidence of the proliferation of successful multi-disciplinary collaborations among researchers in museums, universities and laboratories, as well as the interdisciplinary maturity of many researchers whose research has spanned fields of endeavor at a world-class level of expertise and experimentation.

We thank and congratulate the presenters for producing strong contributions to the understanding and safeguarding of our diverse, yet global heritage. Some contributions in this series have reported world-class research, and that research has later been reported in top-quality journals. Other contributions have been the first professional contributions by students or researchers who, for one reason or another, were not able to publish outside their local specialty or country. Thus, this conference always has had a didactic and heuristic function. Of the first-time, pre-doctoral student authors, the editors (without Vandiver) voted to recognize Israel Favela with a special student award at this conference.

We welcome contributions to the next symposium that will be proposed for the 2010 MRS Fall Meeting in Boston. If you intend to submit an abstract and would like to be considered as an invited speaker, or if your topic and research have the potential of high impact in the popular press, please contact Vandiver directly (Vandiver@mse.arizona.edu) and the group of editors will consider your request. We are interested in organizing a workshop for the next meeting. Please contact us if you are interested in contributing to this workshop with a demonstration, or if you have recommendations or suggestions for another successful meeting in Boston. We have thought about presenting a workshop at MIT focusing on the topic of coatings and coverings, colored and decorative, or protective and invisible, including foils, lacquers and others, and would be interested in comments or suggestions on this topic.

Of the thirty-four papers that were presented, twenty-nine are included in this volume, one of which is from a previous meeting. The invited speakers were allotted double time at the conference and double space in the proceedings, and their contributions are noted in the "Contents" with asterisks. Award winning papers are also noted by special symbols. Many of these papers are connected by current research themes and problem orientations in technical art history, archaeological science and conservation science. However, the significance of the papers may be elusive to a non-specialist. To educate students and non-specialists and excite them about our field, MRS spends considerable effort to involve the lay press. Based on the abstracts that conference attendees submitted and our knowledge of the field, we constructed 10 lay abstracts of the conference highlights for the press that answered the question of "Why the press might be interested?" You can read the articles in this volume and make comparisons with our expectations. For the next symposium, we invite you to contextualize your problem and formulate its potential impact on non-specialists by composing and submitting your own lay abstract directly to P. Vandiver.

In the Area of Technical Art History:
Jennifer Mass and students from the Winterthur Museum and Sculpture Garden: A valuable painting has been entrusted to analysis at the Cornell High Energy Synchrotron Source. The confocal XRF technique was tested on a synthesized painting with many paint layers, and it was able to non-destructively see into the painting and to differentiate the compositions of the various paint layers. This test study was presented at the last MRS symposium four years ago. The present paper solves a problem of attribution or collaboration between two famous northern Renaissance painters by using this and other analytical techniques to reconstruct the processes of fabrication of the oak panel and the sequence of application of the paint layers.
Katherine Eremin of the Strauss Conservation Center, Harvard University, and Jennifer Giaccai, Walters Art Gallery, each present a paper on Thai painting: No one has ever studied the materials and techniques of Thai painting. Rumor has it that gum from the tamarind tree was used as a medium for the pigments. The harsh tropical conditions in Thailand have prevented preservation of painted works of art much beyond 400 years, unlike the paintings of China and Japan, a few of which have lasted 1300 to 1400 years. Eremin has found a previously unidentified green pigment, a hydrated copper citrate. She synthesized the pigment according to a seventeenth century Venetian recipe, and then she tested the stability and effects of various binders to determine whether the green color was the result of weathering or an intentional colorant. Eremin's analyses show a change in the palette of pigments that documents the opening of Thailand to pigments imported from further to the west. Giaccai adds further data to track the changes in pigment type and use over time using banner paintings from the Walters collection.
Marc Walton of the Getty Conservation Institute and Getty Villa: Attic Greek ceramic vase painting in the 6th to 4th centuries BCE consisted of red and black slips, and a rare coral red color, compared in Greek poetry to the color of red wine. Walton analyzes the composition and microstructure using some new, high magnification equipment (FIB/STEM) to reconstruct, for the first time, a special process for the production of the coral red color.

In the Area of Conservation Science and Engineering:
George Scherer et al., Princeton University: As salts in groundwater rise and precipitate in outdoor stone sculpture, gravestones, and building stones among others, the pores in those stones enlarge and cracks cause the stone to weather and break apart. Scherer presents results of modeling and experimentation that demonstrate both the mechanism and a process to reduce the pressure in the pores below the breaking strength of the stone.

Rui Chen, and her mentor, Prof. Paul Whitmore, Art Conservation Research Center of Carnegie Mellon: Materials proposed for use in the display of museum objects often outgas and may deteriorate objects unless tested. The authors propose and have made and tested a nano-particled, silver-film sensor that will serve as a proxy, or singing canary, to warn of corrosion or deterioration processes from the proposed material. This sensor works because the silver film is composed of particles in the range of 30 nanometers that react more rapidly than the objects they protect.

In the Areas of Archaeological Science and Reconstructions of Past Technologies:

Charles Kolb of the National Endowment for the Humanities: This is the first presentation differentiating various ceramic traditions and technologies of northern Afghanistan that were excavated at archaeological sites active in the 4th to 2nd millennia BCE as the Silk Road was being established as a trade route. These sites were excavated in the 1960s prior to the period of war that has engulfed Afghanistan. Such studies are important to establish our knowledge of Afghan culture and to rebuild Afghan cultural heritage.

Tomoko Katayama of Kyoto University: Five hundred to seven hundred year-old clothes from Mongolia were dated and found to contain gold, copper, iron and lead. The textiles were analyzed nondestructively using synchrotron radiation. The analyses add to our knowledge of the manufacture, use and degradation of textiles by members of the culture of ancient Mongolia in the time of the Mongolian Empire.

Lesley Frame of the University of Arizona: The experimental research involves the first application of modern metallurgical analysis for determination of solidification rates and modeling of process temperature to ancient metallurgy. The study shows that in the second and third millennia BCE at the site of Godin in northwestern Iran, arsenical-bronze and tin-bronze knives were made with a limited range of processing variability to produce functional, well-engineered blades; whereas, pins, bracelets and other decorative objects appreciated for their beauty were made with a wide range of heat treatments. Their use did not mandate as precise processing control, and their variability also indicates the probability of wider knowledge of the processing of bracelets.

Andrew Shortland, Cranfield University; Katherine Eremin and James Armstrong, Harvard University, et al.: Nuzi was a second millennium BCE provincial capital city that is near the Kirkuk oil-fields in northern Iraq. It was excavated in the 1930s and the artifact collection is housed at the Semitic Museum of Harvard University. This study of glass, metal, faience and ceramic technologies is significant because it aims to understand the inter-relationships among these technologies. For instance, the first brass alloys are found at Nuzi, but glass beads are colored blue-green by pure copper, indicating separate streams of raw materials for these two crafts. In addition, analyses of many now-deteriorated objects have established the chemical traces of their original colors and helped to reconstruct their original technologies.

Chandra Reedy, University of Delaware: Just as UNESCO has legislation to designate and preserve World Heritage sites, its new program aims to preserve intangible cultural heritage, such as music, dance, theater, festivals and craft knowledge. Reedy's study documents the modern practices of making Buddhist ritual objects in eastern Tibet, Amdo region (where the present Dalai Lama was born and now part of Sichuan, China), such that we can better understand and preserve both the tangible craft technologies and intangible aspects of decision-based and ritual-based behaviors. Based on study of traditional practices of making ritual objects, barley dough sculptures (*torma*) and unfired votive clay objects (*tsha-tshas*) often covered with a thin foil of gold, we better document and preserve the intangible, knowledge-based elements of cultural heritage. This paper makes a theoretical breakthrough and presents both ethnographic and analytical data.

The MRS science museum exhibition, Strange Matter, is still circulating throughout North America, and we recommend that readers see it or visit the MRS website (http://www.strangematterexhibit.com/). In particular, the beginning of the exhibition draws parallels between materials engineering in the Bronze Age and the present, and then proceeds to modern wonder of materials, analysis and underlying science.

We thank the MRS staff for its support. They labor for months to make the meeting progress with flawless organization and order. We appreciate their support of this symposium which probably is the only venue that offers interaction among conservation scientists, archaeological scientists and materials scientists. We wish to thank the participants, the session chairs, the reviewers and our home institutions for their contributions to making this symposium and proceedings a success. We acknowledge the considerable time and effort spent in the review process by Jennifer Giaccai and Michelle Taube.

Pamela B. Vandiver
Blythe McCarthy
Robert H. Tykot
Jose Luis Ruvalcaba-Sil
Francesca Casadio

August 2008

MATERIALS RESEARCH SOCIETY SYMPOSIUM PROCEEDINGS

MATERIALS RESEARCH SOCIETY SYMPOSIUM PROCEEDINGS

Prior Materials Research Society Symposium Proceedings available by contacting Materials Research Society

Technical Art History

Mater. Res. Soc. Symp. Proc. Vol. 1047 © 2008 Materials Research Society

Collaboration or Appropriation? Examining a 17th c. Panel by David Teniers the Younger and Jan Brueghel the Younger Using Confocal X-ray Fluorescence Microscopy

Jennifer L. Mass[1], Arthur R. Woll[2], Noelle Ocon[3], Christina Bisulca[4], Tomasz Wazny[5], Carol B. Griggs[5], and Matt Cushman[6]

[1]Conservation Department, Winterthur Museum, 5105 Kennett Pike, Winterthur, DE, 19735
[2]Cornell High Energy Synchrotron Source, Cornell University, 200L Wilson Laboratory, Ithaca, NY, 14853
[3]Conservation Department, North Carolina Museum of Art, 2110 Blue Ridge Road, Raleigh, NC, 27607
[4]Conservation and Scientific Research, Freer Gallery of Art, Smithsonian Institution, Washington, DC, 20560
[5]Cornell Tree-Ring Laboratory, Cornell University, B48 Goldwin Smith Hall, Ithaca, NY, 14853
[6]Williamstown Art Conservation Center, 225 South Street, Williamstown, MA, 01267

ABSTRACT

This paper presents the results of a multidisciplinary investigation of *The Armorer's Shop* (North Carolina Museum of Art), a 17th century painting on panel attributed to David Teniers the Younger of Flanders. The study was motivated by x-radiographic observations suggesting an atypical panel construction and by the discovery that the armor depicted in this painting is nearly identical to that of several other works, all but one of which are attributed to Jan Brueghel the Younger, a contemporary Flemish master and relative of Teniers. Stylistic analysis strongly supports the hypothesis that Teniers painted the background, figures and objects depicted around the armor, and that Brueghel completed the armor itself. A broad range of materials analysis techniques, including cross-section microanalysis, dendrochronology, and confocal x-ray fluorescence microscopy (CXRF), were used to establish whether the panel construction and palette composition are consistent with this hypothesis. Dendrochronology shows that the panel was fabricated from three distinct wood planks, and suggests that the smallest of these, the armor plank, was painted approximately twenty years before the other two. CXRF demonstrates that this plank was painted before being attached to the other two. To the authors' knowledge, this is the first report of a painting being re-used in this way, and the first evidence of collaboration between these two painters.

INTRODUCTION

The Armorer's Shop, by David Teniers the Younger (figure 1) depicts in its foreground a seated armorer and a richly detailed pile of parade armor. A 1946 article identifies most of this armor, some of which is still exhibited today in Vienna, Brussels, and Krakow [1]. The middle ground depicts several workers at a forge, above which hangs a dragon, a symbol for alchemy. The painting is signed *D. Teniers* on the log upon which the armorer sits. During a visual examination of the painting, the panel was found to have a highly unusual construction.

Figure 1. *The Armorer's Shop,* attributed to David Teniers the Younger, 56.5×80.7 cm^2, oil on panel, NCMA. The long solid lines indicate the interface between the armor-containing plank and the remainder of the panel. The dashed line indicates a third join discovered during dendrochronological analysis. The short lines indicate the locations of confocal XRF scans across the interface. The numbered arrows show the locations of samples used for conventional microanalysis. The unnumbered arrow highlights the particular CXRF scan shown in Figs. 4-5.

Typically when a large panel is constructed from multiple planks, the planks are assembled such that the joins are all parallel to the grain. In *The Armorer's Shop*, the parade armor in the lower left corner is painted on its own distinct plank, which is attached to the remainder of the painting along two perpendicular joins indicated by the solid white lines in figure 1. There is increased paint loss along the right-hand, vertical join because it is perpendicular to the grain. Both joins are visible as raised ridges in raking light, and are corroborated by the transmission x-radiograph in figure 2.

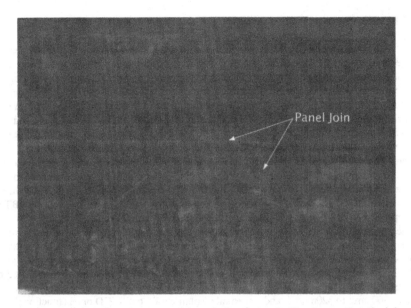

Figure 2. Transmission x-ray radiograph of the painting shown in Fig. 1, revealing the joins between the rectangular portion and remainder of the painting. The light cross-hatch pattern throughout the radiograph arises from a mahogany cradle. The radiograph reveals that the helmet on the left-hand table of the painting was originally part of a full suit of armor.

This atypical panel construction prompted further study and analysis of the painting, and eventually led to the discovery that the armor pile in figure 1 is nearly identical to armor depicted in six other paintings, all but one of which are attributed to Jan Brueghel the Younger. Although collaboration among two or more painters was common practice among Northern European painters in the 17th century [2], this painting, constructed by the incorporation of an already-painted single plank into a larger scene, is unique. Moreover, it is unclear whether the term 'collaboration' should be applied to this work at all.

This paper presents a study of the construction, composition, and palette of *The Armorer's Shop*, to determine the origin and history of this uniquely constructed painting. It presents a discussion of the visual, holistic and art historical evidence pertaining to the hypothesis that Jan Brueghel the Younger painted the armor pile, and that Teniers painted the remainder of the work. This is followed by the results of the conventional microanalysis of the painting, including Raman spectroscopy and x-ray fluorescence spectroscopy (XRF). Results obtained using synchrotron-based 3D, or confocal XRF (CXRF) are presented that unambiguously reveal the chronology of construction, and these are followed by results of a dendrochronological analysis of the panel components. In accord with the CXRF results, the dendrochronology data suggests that the plank containing the armor was painted approximately twenty years before being joined with two other planks to complete the work.

EXPERIMENT

X-ray radiography (figure 2) was performed using a Picker Hotshot unit at an operating voltage of 20 kV and a tube current of 3 mA, with an exposure time of 90 seconds. Fourier transform infrared microspectroscopy (FTIR) was performed using a Thermo-Nicolet Magna 560 IR spectrometer with a Nicolet Nic-Plan microscope in transmission mode. For each sample (mounted on a diamond half-cell) 128 scans were acquired from 4000 cm^{-1} to 650 cm^{-1} at a spectral resolution of 4 cm^{-1}. Spectra were collected with Omnic E.S.P. 6.1a software and analyzed using the Infrared and Raman Users Group (IRUG) database and commercial polymer and organic chemical libraries. Non-destructive, qualitative energy dispersive XRF was performed using a Bruker ArtTAX μXRF spectrometer with a molybdenum tube operated at 50 kV, 600 microamps, and 200s collection time. A polycapillary focusing optic was used to achieve an approximately 70 micron incident beam size, and Intax version 4.5.18.1 software was used to interpret spectra. Scanning electron microscopy was conducted with a Topcon ABT 60 SEM operated at 20 kV, a 22 mm stage height and a 20 degree sample tilt. Paint layer thicknesses from SEM images were calibrated and measured using ImageProPlus software (Media Cybernetics). Energy dispersive x-ray spectra were collected using an EDAX x-ray detector, an Evex pulse processor and multi-channel analyzer, and Evex Nanoanalysis software. Raman spectroscopy was performed on a Renishaw inVia Raman spectrometer using a 785 nm diode laser, a 50x objective, 1200 l/mm grating, a laser power of 3 mW at the sample, a spectral range of 100 cm^{-1} to 3200 cm^{-1}, and a spectral resolution of 1 cm^{-1}/CCD pixel (functional resolution of 3 cm^{-1}). Data was also collected using a JY Horiba LabRAM Aramis Raman spectrometer with a 50x objective, 785 and 633 nm lasers, a laser power of 8 mW at the sample, a 1200 l/mm grating, a spectral range of 200 cm^{-1} to 1600 cm^{-1}, and a spectral resolution of 1 cm^{-1}/CCD pixel.

The CXRF experiments were carried out at CHESS station D1, using monochromatic radiation at 18 keV, selected using a 1% bandpass multilayer monochromator. A single-bounce monocapillary, fabricated at CHESS[3], was used to provide a focused incident beam of approximately 5×10^9 photons/second into a 20 μm-diameter spot. A double-focusing polycapillary lens (X-ray Optical Systems) with an input acceptance angle of 25°, was used to collect x-ray fluorescence from the sample and direct it onto a Rontec Xflash silicon drift detector. The detector resolution is approximately 0.16 keV. The two optics define a 3D sample volume as described previously [4]. The energy-dependent depth resolution with the setup used for the scans presented here varied smoothly from 31 μm at 4.5 keV to approximately 15 μm at 16 keV. The painting was mounted on a large-area, high resolution 3D scanning stage equipped with an easel-style mount [5]. To increase the distance between the painting and polycapillary lens, the sample surface was oriented 32° from the incident beam, rather than 45°, as in prior experiments.

For dendrochronology, one end of each plank in the support panel was prepared, and the widths of all rings were measured to 0.05 mm precision. The outer growth rings of the opposite ends were also prepared and measured, both to be sure that each plank's outermost ring was counted and to determine whether any sapwood rings were present (sapwood consists of the outer rings next to the bark: these rings are generally removed due to low durability, leaving just the heartwood). In addition, all edges of the panel were examined for structural and wood anatomical features. The data from each plank were compared with several established Baltic,

German, and Dutch oak chronologies to determine the source of the wood and the outer ring dates of the planks, using standard dendrochronological statistical and visual techniques [6].

RESULTS and DISCUSSION

Visual & Stylistic Examination

After the initial observation of *The Armorer's Shop's* unusual construction, the painting was examined for corroborating evidence that the armor panel was painted either by a different painter or at a different time than the remainder of the painting. The armor is finely painted, contrasting with the irregular, coarsely painted brushstrokes of the surrounding forge. However, this difference alone does not constitute evidence of a second artist, since the depiction of armor would require much more detail compared to the rather rough elements in the rest of the composition. The difference could also indicate an increased degree of importance placed upon the pile of armor.

Compositional changes might also indicate that more than one artist contributed to the work. Infrared reflectography (data not shown) and x-radiography reveal possible attempts to integrate the pre-existing image of the armor with the surrounding composition. Figure 2 shows that the helmet on the table to the left was originally part of a full suit of armor. The question of whether a different painter might have painted the armor section was addressed by comparing *The Armorer's Shop* to other works by Teniers and his contemporaries. Remarkably, five paintings attributed to Jan Brueghel the Younger (1601-1678) have nearly identical depictions of armor as in *The Armorer's Shop* [7,8]. During an *in situ* examination of the *Allegory of Touch*, (Musée Calvet, Avignon, France) an overlay from *The Armorer's Shop* was used to precisely compare compositional and stylistic elements with those of the Avignon painting. Most of the armor pieces in the two paintings were found to match precisely in size, palette and execution, suggesting that the same artist painted both piles of armor. The biggest difference between the two is a full suit of armor at the far left of the *Allegory of Touch*, where *The Armorer's Shop* shows only a helmet on a table. However, as noted above, x-radiography revealed this helmet to have originally been part of a full armor suit until it was painted over with a table.

The correspondence between *The Armorer's Shop* and the armor in five paintings attributed to Brueghel the Younger suggest that *The Armorer's Shop's* armor was painted by Jan Brueghel the Younger. Teniers and Brueghel the Younger were both active in the Antwerp art guild in the early 17th century, and were also related through Teniers' marriage to Brueghel's half sister. Woolet [2] and Ertz make reference to collaborative projects between Teniers and Brueghel, yet no examples of such work have been firmly identified.

In general, the composition and execution of *The Armorer's Shop* fits quite well into Teniers' compositions of the 1640s. The figures, palette and spatial configuration are typical of Teniers' work. Moreover, both parts of the panel depict common elements found in other of Teniers' compositions, including the earthenware jug in the top left window and the tables in the bottom right and bottom left corners. Although several works of Teniers include depictions of armor, that of *The Armorer's Shop* stands out in both its execution style and quality. Other paintings that are confidently attributed to Teniers show his armor and other metallic objects having a rectangular quality to the contour and application of the highlight, not defining shape or curve. In comparison, Brueghel's armor paintings reveal that he applied highlights in a well-planned manner, always accentuating form as well as illustrating the reflective quality of the armor. These qualities apply to the armor in *The Armorer's Shop*, and suggest that Jan Brueghel the Younger contributed to this section of the work.

Conventional Microanalysis

Conventional microanalysis was carried out on both sections of the painting to identify any distinctions in their compositions and microstructures. A clear difference in the chemical composition of the two parts of the painting could support the hypothesis that the painting was executed at two different times or by different artists. Conventional energy dispersive XRF was applied to numerous locations in both sections of the painting to determine the elemental composition of the palette and ground. In addition, seven cross-sections were taken from the painting, the locations of which are indicated in figure 1.

Results from conventional XRF analysis suggest a traditional 17th century palette for both sections of the painting: vermilion and iron oxide reds, lead-tin and iron ochre yellows, lead white, azurite blue, umber browns, and flesh tones created by mixing lead white, vermilion, iron oxide red, and umber [9-11]. A translucent copper-containing green was used to represent the interior fabric of the armor, possibly verdigris or copper resinate (sampling for further analysis was not possible in these areas) [12]. Calcium could be found in each spectrum, suggesting the presence of a chalk ground as should be expected for 17th c. Northern European painting on panel [9,11,13,14]. Both lead and copper were found in virtually every spectrum regardless of the color of the presentation surface, suggesting that lead white and a copper-containing pigment such as azurite were used in a lower-lying paint layer such as an imprimatura [11,14,15]. No significant difference was observed between the palettes of the two sections. Fourier transform infrared spectroscopy and dispersive Raman spectroscopy were applied to five cross-sections from the painting to verify compounds inferred from conventional XRF. FTIR on all samples revealed characteristic absorbances for lead white in a drying oil. Strong bands were observed at 1407 and 1530 cm^{-1}, the carbonyl stretching bands for hydrocerussite and lead carboxylate soaps respectively. Bands at 2928 cm^{-1} and 2854 cm^{-1} (C-H stretching bands), at 3365 cm^{-1} (O-H stretching band), and 1711 cm^{-1} (carbonyl stretching band) were due to a drying oil binding medium. The major carbonyl band, at 1530 cm^{-1}, was due to the presence of lead carboxylates. The formation of such metal carboxylate salts are expected for a painting of this age prepared with a lead white-based palette [16]. Dispersive Raman spectroscopy of sample 18 revealed strong scattering bands at 1084 cm^{-1}, 708 cm^{-1}, and 278 cm^{-1}. All three of these bands match the reference spectrum for calcite, confirming the phase identification of the ground.

Cross-section samples were also studied with SEM-EDS to evaluate paint layer thicknesses and microstructures. Figure 3 shows a backscattered electron image of one of the cross-sections (sample 1). Above the Ca-rich ground layer, figure 3 clearly shows a 2-10 micron layer consisting primarily of lead, identified as lead white on the basis of FTIR results discussed above. The topmost layer shown in figure 3 contains significant Pb, Fe, Al, Si, Ca, and P, elements that suggest the presence of lead white, iron ocher, bone black, and calcite, consistent with the palette inferred from the conventional XRF data.

10 μm

Figure 3. Backscattered SEM image of sample 1 (see Fig. 1), taken from the grey area from the armor pile, showing the presence of a lead white-containing imprimatura layer. Magnification 2000X.

The lead-rich layer immediately above the ground in figure 3 is identified as the imprimatura layer that was suggested by the conventional XRF data described above. The cross-section samples all show a lead-rich region directly above the ground, which corroborates the application of a lead white-based imprimatura to the ground before additional paint layers were applied. In addition to lead, SEM-EDS analysis of samples 2, 15, and 18 directly above their ground layers all revealed the presence of small amounts of copper. This data, combined with the ubiquitous copper observed in the ED-XRF data and the occasional presence of dark blue particles directly above the ground observed by visible-light microscopy, suggests that the imprimatura layer is interspersed with small amounts of a copper-containing blue pigment such as azurite. An azurite-containing imprimatura would create a cool-toned gray ground, and the addition of copper-containing pigments to imprimatura layers to facilitate rapid drying is well known [11].

The identification of calcium and phosphorus together in the grey particles of the imprimatura layer in figure 3 is suggestive of bone black, and the identification of aluminum and silicon in the dark (low-Z) particles that have platey habit in this layer is suggestive of the clay minerals typically associated with iron ochres. The scumbled background of both the armor and forge sections appear to be prepared from a mixture of pigments that includes lead white, bone black, iron ochers and calcium carbonate. The identification of copper in the imprimatura layer on both sections of the painting is notable. The composition, number and thickness of paint layers in *The Armorer's Shop* are consistent with 17th century northern European painting practice, but do not reveal any substantive differences between the armor and forge sections outlined in figure 1.

Confocal X-Ray Fluorescence

To obtain direct evidence for the chronology of the paint layers throughout the painting, a new, non-destructive technique, confocal x-ray fluorescence microscopy (CXRF) was applied. CXRF, or 3D scanning XRF, combines two x-ray focusing optics to resolve x-ray fluorescence from a particular, 3D volume in space [4,17-20]. By scanning the sample through this volume, the composition of a layered sample as a function of depth is determined.

Figure 4 shows intensity versus depth profiles of several fluorescence lines obtained from a single depth scan using CXRF, taken on the armor panel adjacent to the join between the armor and forge sections. The panel was translated by 6 μm between each point (moving the confocal volume deeper into the painting), and the data collection time at each point was 2 seconds. Each profile has been normalized by its maximum value to allow visual comparison of the relative peak positions. The data indicate a top layer (between 25 and 75 microns depth) consisting of Ca, Fe, Mn (not shown), Cu, and Pb. Within the layer, the Ca, Fe, and Cu peaks are approximately 5-10 microns to the left, or above the Pb peak. This implies an increasing Pb concentration immediately below the presentation surface, consistent with the observation of a lead-rich imprimatura layer in figure 4.

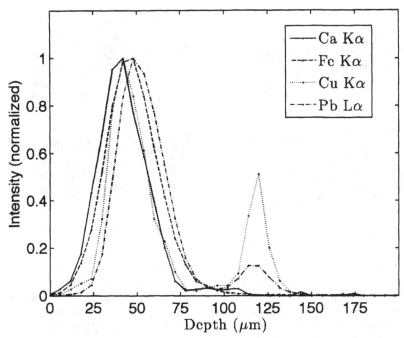

Figure 4. Intensity vs. depth of several fluorescence lines obtained from a single depth scan using CXRF (performed at the left side of the lower line in the inset). The intensities have been normalized with respect to their maxima for visual comparison. The top layer consists of Ca, Fe, and Cu-based pigments on a Pb-based imprimatura layer. A second Pb and Cu-containing layer, visible approximately 0.1 mm below the presentation layer, is indicative of a second, earlier imprimatura.

The most conspicuous feature of figure 4 is the presence of additional Pb Lα and Cu Kα fluorescence peaks towards the right of the figure, that appear to indicate the presence of an additional layer of pigment. Figure 4 represents just one of a series of 21 depth scans taken at 1mm lateral intervals, and spanning the interface between planks 1 and 2 (this scan is the horizontal line in figure 1 demarcated by a grey arrow). In Figs. 5a-d, the data from all of these scans are combined to form cross-sectional representations of elemental concentration across this interface as a function of depth. Each plot in figure 5 shows the 2D distribution of integrated intensity of a particular fluorescence peak.

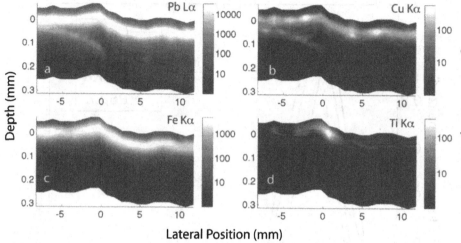

Figure 5. Virtual cross-section across the interface between the armor and forge sections of the painting (scan position is indicated by the unnumbered arrow in figure 1). The Pb Lα and Cu Kα cross-sections reveal the presence of an additional, buried imprimatura layer on the armor section of the painting. Iron is present only in the top layer. The presence of a titanium-based pigment used during restoration at the location of the join between the two panels is indicated in d.

Figures 5a-b clearly show that the extra Pb Lα and Cu Kα peaks towards the right-hand side of figure 4 correspond to a buried layer that slopes downward as it approaches the interface between the two planks, then abruptly stops. The location of the join, in this scan, is indicated both by the abrupt termination of this buried layer and by the sudden change in slope and height of the surface that make the join visible under raking light. This buried layer, likely an imprimatura on the earlier composition, was evidently applied to the armor plank before the plank was joined to the remainder of the panel. The separation between the top and buried layers is explained by the application of additional ground material to smooth any gaps caused by height differences at the interface after the planks were physically attached. Subsequently, new imprimatura and pigment, corresponding to the upper layer in Figs. 4-5, was applied to both sides of the join. A series of other scans across the interface between the planks containing the armor and forge, indicated by the horizontal lines in figure 1, showed the same key feature as in figure 5: a buried, lead- and copper-containing layer on the armor side of the join. In contrast, no buried layers were observed on the forge section, suggesting that this section was only painted after the join. Thus, the data show that the armor plank, and probably the armor itself was painted first, and that *The Armorer's Shop* composition was conceived to make use of this earlier image.

A notable feature of figure 5 is the fact that the height difference of the surface across the join is only about 0.05 mm, attesting to the exceptional craftsmanship that must have been employed to join these planks. In addition, figure 5d shows titanium-containing retouching paint from the most recent campaign of conservation treatment. The paint was used to in-paint a small area of loss caused by the slight movement of the wood planks over time.

Confocal XRF was also used to study the paint layer stratigraphy of many other areas of the painting beyond the joins, and indicated that in some regions of the composition the imprimatura serves as the presentation surface. This is consistent with 17th century Northern European painting practice, where an imprimatura layer was often used as a middle-toned ground from which the painter could build up highlights.

Dendrochronology

In preparation for dendrochronological analysis, a careful examination of the structure and tree-ring patterns on all sides of *The Armorer's Shop* was performed, and revealed two unexpected features of the overall construction. First, the panel is comprised not of two, but of three separate planks, as illustrated in figure 6. Visual inspection indicated that the armor plank (plank 1 in figure 6) was glued onto the bottom plank (plank 2), which was evidently planed down expressly for this purpose and which extends the full width of the painting. This insertion is visible in the left and bottom edges of the planks. To the authors' knowledge, such an insertion of one plank into another is unique to this panel [15]. The second unusual feature, also indicated in figure 6, is the orientation of the bottom two planks relative to their ring-growth direction. Generally, planks are placed with their inner rings at the outer edges of a painting panel due to the stronger structural support of the inner rings [21]. In this painting, only the top plank follows this convention, whereas both bottom planks have their *outer* rings at the *outer* edge of the panel. The orientation of plank 2 was evidently chosen to match that of plank 1 to minimize damage due to warping. These observations both indicate that *The Armor's Shop* panel was conceived to make use of the pre-existing painting of armor on plank 1.

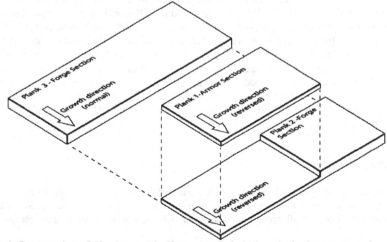

Figure 6. Construction of *The Armorer's Shop*, as revealed by optical microscopy and dendrochronology. Arrows indicate the growth direction of each of the wood planks. For the bottom two planks, this growth direction is reversed compared to conventional practice, in which planks are oriented so that their older, harder wood faces outward.

The orientation of the armor plank with respect to its growth direction also suggests that this portion of *The Armorer's Shop* was originally a study, rather than part of a larger, separate composition. In all of the paintings by Jan Brueghel the Younger that depict this parade armor, the armor appears in the foreground, or lower portion of the painting. If the armor plank in *The Armorer's Shop* had been part of a different, larger painting, one would thus still expect its lower edge to coincide with the lower edge of the painting, and therefore with the older, harder wood.

Based on modern and archaeological oak samples collected in the Baltic region [22] , the median number of sapwood rings in these oaks is 15, with 90% of samples having between 9 and 23 [23]. For 17th century paintings, the period of time between the felling and painting is typically 5 ± 3 years, again with ~90% certainty [24]. This range includes the time needed for splitting the timbers into boards, for transportation and sale, for seasoning the wood, and for preparation and finishing of the panels. Each of the three planks in *The Armorer's Shop* contains over 140 tree-rings, and none contain any sapwood. All were found to match sufficiently with other oaks from the Baltic region that their tree-ring sequences could be dated [9,11,15].

The outer rings of the two forge planks (planks 2 and 3 in figure 6) date to 1620 and 1624. Since these two planks were painted at the same time, the later date is used to estimate the painting date. The average of 15 sapwood rings (range of 9-23) plus 5 years between felling and painting (range of 2-8) gives a most likely date of 1644 or after for the completed work, with a statistically likely range of 1638-1652. In contrast, the outer ring of the armor plank dates to 1605, yielding a most likely painting date of 1625 and a probable range of 1619-1633. Thus, dendrochronological analysis suggests that the armor plank was painted approximately 20 years before the remainder of the painting.

In the case of the armor plank, the range of likely painting dates was substantially narrowed by comparing the ring sequence of this plank to those of planks from other panel paintings that have been inscribed with a date. The ring sequences of the armor plank from *The Armorer's Shop* along with those from two other paintings were compared: *The Art Cabinet of Crowned Prince Vladislav IV* (Vladislav IV became king of Poland in 1632) attributed to an unknown Antwerp painter with signature "Here(?) fecit 1626" [25] and *Lamentation* by an unknown Antwerp painter (after Hugo van der Goes), dendrochronologically and art-historically dated to *ca.* 1628 (The Royal Museums of Fine Arts of Belgium in Brussels, inv. No. 678. See [26], pp. 256-264). The degree of dendrochronological correlation among these three panels is high enough to conclude that the wood in these panels originated from trees in the same forest stand. The dates of execution for the other two paintings, 1626 and 1628, combined with a range of up to 5 years for seasoning of the wood, suggest that the armor plank was most likely painted between 1624 and 1630.

CONCLUSIONS

The wide variety of analysis techniques presented here each adds insight to the construction and history of *The Armorer's Shop*. Holistic examination alone suggests that 1) the armor is contained on a small, rectangular plank that was incorporated into a larger composition and 2) the armorer and forge sections of the work were painted by Teniers, while the armor plank was most likely painted by a different artist, possibly Jan Brueghel the Younger. Scientific analysis corroborates each of these inferences. Conventional analysis shows that all of the construction materials used throughout the painting, including altered regions identified by IR and X-ray radiography, are consistent with those known to be used in 17th century Flemish oil painting. Confocal XRF provides direct evidence that the armor panel was painted before the complete panel was constructed. Finally, dendrochronology shows that all three planks forming the panel were felled in the early 1600s, and that the armor plank was most likely painted 20 years before the other two planks. In view of Brueghel the Elder's death on January 13, 1625 [2], the range, 1624-1630, of likely dates for the completion of the armor plank strongly supports Brueghel the Younger as the artist of the armor pile.

There is evidence for collaboration between Teniers and Brueghel the Younger [2,8]; however, no examples of such a collaboration have been firmly identified. *The Armorer's Shop*, then, could represent one of the only known collaborations between these two important 17th century painters. In view of these results, in 2005 the attribution of *The Armorer's Shop* was changed to David Teniers the Younger and Jan Brueghel the Younger. The probable 20-year gap between the execution of the two parts of the painting raises the question as to whether Brueghel the Younger was aware that his painting was incorporated into a larger work. If not, it may be more appropriate to refer to *The Armorer's Shop* as an appropriation rather than a collaboration. It is hoped that such questions may be addressed by further research.

ACKNOWLEDGMENTS

The authors gratefully acknowledge helpful discussions and support of Dennis P. Weller, William P. Brown, Sol Gruner, Don Bilderback, Detlef Smilgies, and Sterling Cornaby. We also thank Ida Sinkevi, Ruth Gleisberg, Jottany Pagenel, Frank Zuccari, Mark Leonard, Tiarna Doherty, Anne Woollett, James Martin, Peter Kuniholm, Kristof Haneca, and Kathy Dedrick. Fran Adar, Catherine Matsen, and Ken Sutherland are gratefully acknowledged for their contributions to the Raman, FTIR, and cross-sectional analysis of the painting.

This work is supported by FAIC, the Kress Foundation, and the Mellon Foundation. Further support was provided by the NSF under DMR 0415838. CHESS is supported by the National Science Foundation and NIH-NIGMS via NSF grant DMR-0225180.

REFERENCES

1. S.V. Grancsay, *The Journal of the Walters Art Gallery* 9, 23-40 (1946).
2. A.T. Woollett, A. van Suchtelen, T. Doherty, M. Leonard, and J. Wadum, *Rubens and Brueghel: A Working Friendship*, (J. Paul Getty Museum, Waanders Publishers, Los Angeles, Mauritshaus, Den Haag, Zwolle, 2006) pp. 1-41.
3. D.H. Bilderback, *X-Ray Spectrometry* 32(3), 195 (2003).
4. A.R. Woll, J. Mass, C. Bisulca, R. Huang, D.H. Bilderback, S. Gruner, and N. Gao, *Applied Physics a-Materials Science & Processing* 83, 235 (2006). Also, A.R. Woll, S. Gruner, D. Bilderback, N. Gao, C. Bisulca, and J. L. Mass, *Materials Issues in Art and Archaeology VII*, edited by P. B. Vandiver, J. L. Mass, and A. Murray, (Mater. Res. Soc. Proc. 852 Pittsburgh, PA, 2005) pp. 281-290.

5. A. Woll, *CHESS 2005 News Magazine (http://www.chess.cornell.edu/pubs/2005/index.htm)*, 41 (2005).
6. P. Klein, in *Recent developments in the technical examination of early Netherlandish painting: Methodology, limitations & perspectives*, edited by M. Faries et al. (Cambridge, Harvard University Art Museums, 2003) pp. 65-81.
7. K. Ertz, *Jan Brueghel der Ältere. Die Gemälde mit kritischem Oeurvrekatalog* (DuMont Buchverlag, Köln, 1979).
8. K. Ertz, *Jan Breughel der Jüngere. Die Gemälde mit kritischem Oeuvrekatalog* (Luca Verlag, Freren, 1984).
9. C. Brown, *Rubens's Landscapes: Making and Meaning* (National Gallery Publications, London, 1996) pp. 99-104 and 116-121.
10. E. Van de Wetering, in *Preprints, Historical Painting Techniques, Materials and Studio Practice*, edited by A. Wallert et al. (The Getty Conservation Institute and University of Leiden, Leiden, Los Angeles, 1995).
11. A. Wallert, *Still Lifes : Techniques and Style : An Examination of Paintings from the Rijksmuseum*, (Waanders, Rijksmuseum, Zwolle, Amsterdam, 1999).
12. R. Woudhuysen-Keller, and P. Woudhuysen, in *Looking Through Paintings: The Study of Painting Techniques and Materials in Support of Art Historical Research*, edited by E. Hermens et al. (Archetype publications, London, Baarn, 1998) pp. 133-146.
13. D. Bomford , C. Brown, and A. Roy, *Art in the making: Rembrandt*, (National Gallery, Washington, D.C., 1998).
14. N. Van Hout, in *Looking Through Paintings: The Study of Painting Techniques and Materials in Support of Art Historical Research*, edited by E. Hermens et al. (Archetype publications, London, Baarn, 1998) pp. 195-225.
15. E. Hendriks, and J.C. Verspronck, in *Looking Through Paintings: The Study of Painting Techniques and Materials in Support of Art Historical Research*, edited by E. Hermens et al. (London, Baarn, Archetype publications, 1998) pp. 227-263.
16. J. Boon, in *Proceedings of the Sixth Infrared and Raman Users Group Conference*, edited by M. Picollo, (il prato, Florence, Italy, 2004) pp. 66-130.
17. K. Janssens, K. Proost, and G. Falkenberg, *Spectrochimica Acta Part B-Atomic Spectroscopy*, **59**(10-11), 1637-1645 (2004).
18. B. Kanngiesser, W. Malzer, and , I. Reiche, *Nuclear Instruments & Methods in Physics Research Section B-Beam Interactions with Materials and Atoms*, **211**(2), 259-264 (2003).
19. B. Kanngiesser, W. Malzer, A.F. Rodriguez, and I. Reiche, *Spectrochimica Acta Part B-Atomic Spectroscopy*, **60** (1), 41-47 (2005).
20. L. Vincze, B. Vekemans, F.E. Brenker, G. Falkenberg, K. Rickers, A. Somogyi, M. Kersten, and F. Adams, *Analytical Chemistry* **76** (22), 6786-6791 (2004).
21. J. Wadum, in *The Structure and Conservation of Panel Paintings*, edited by K. Dardes et al. (The Getty Conservation Institute, Los Angeles, 1998) pp. 149-177.
22. T. Wazny, in *Constructing Wooden Images*, edited by C. van de Velde et al. (VUB Press, Brussels, 2005) pp. 115-126.
23. T. Wazny, *Aufbau und Anwendung der Dendrochronologie für Eichenholz in Polen.* Dissertation, (Universität Hamburg, 1990).

24. J. Bauch, and D. Eckstein, *Studies in Conservation* 15, 45-50 (1970).
25. The Royal Castle in Warsaw, inv. No. ZKW 2123
26. C. Stroo, P. Syfer-D'Olne, A. DuBois, and R. Slachmuylders, *The Flemish Primitives. II. The Dirk Bouts, Petrus Christus, Hans Memling and Hugo van der Goes Groups. Catalogue of Early Netherlandish Painting in the Royal Museums of Fine Arts of Belgium*, (Brepols, Brussels, 1999).

Mater. Res. Soc. Symp. Proc. Vol. 1047 © 2008 Materials Research Society 1047-Y02-06

Trace Element Indicators of Fabrication Technology for Coral Red and Black Gloss Decoration on Greek Attic Pottery

Marc Walton, and Karen Trentelman
Getty Conservation Institute, 1200 Getty Center Drive, Suite 700, Los Angeles. CA 90049-1684

ABSTRACT

Laser ablation inductively coupled plasma time-of-flight mass spectrometry (LA-ICP-TOFMS) was used to study the trace element chemistry of coral red and black gloss slip decoration on Greek Attic pottery (6th century BC). The distribution of trace elements in the body fabric and glaze slips were found to be correlated suggesting the raw materials came from a single source. Furthermore, the so-called high calcium and magnesium (HCM) coral red was found to be a less refined material than black gloss, with trace element characteristics suggestive of a carbonate phase in the raw material. This carbonate component may have imparted refractory properties to the HCM coral red slip material during the three-stage oxidative-reductive-oxidative firing used to produce Attic pottery, allowing it to remain porous and re-oxidize during the final firing step, thus creating its final red color. The so-called low calcium and magnesium (LCM) coral red, on the other hand, was found to be more refined than the HCM coral red slip which suggests that two separate firings would have been needed to produce the red color of this material.

INTRODUCTION

Attic vase painting was already an aesthetically mature and technically sophisticated art during the 6th century B.C. (Figure 1). By this time, the ancient Greek artisans of Attica had learned to process coarse-grained clay to produce fine-grained slips. Also, manipulation of kiln temperatures and atmospheres was practiced with an extraordinary level of achievement. The basic operation of the Greek potters kiln has been known since the early work of Noble [1], who described a single three-stage firing process in which vessels were stepped through successive oxidation, reduction, and oxidation stages. The reductive stage was critical in that it was during this stage that the red-colored hematite (Fe_2O_3) produced in the first oxidation stage was transformed to black-colored magnetite (Fe_3O_4). Also during this stage, fine-grained clay slips that had been applied to areas intended to be black would fuse into a glassy mass locking the magnetite in place. In the final oxidation stage, air would re-enter the coarser-grained and porous ceramic body, re-oxidizing the magnetite back to hematite, producing a buff red color. However, the magnetite locked in the sintered slip material would not re-oxidize leaving a glossy surface treatment referred to as "black gloss". This complex process of firing allowed Greek potters to achieve the contrast between red and black that is now a well known hallmark of their figurative vase painting tradition.

Figure 1. J. Paul Getty Museum's *Kleophrates Krater* (87.AE.974), showing both black and red glosses Reproduced from [3].

In addition to 'black gloss', red-colored slips called 'coral red' were employed. The composition of coral red and black gloss slip decoration on Greek Attic pottery has been the subject of an ongoing study [2, 3]. As reported previously, the black gloss and coral red illite clay slips were found to have similar compositions with subtle, yet significant, variations. Most notably, the black gloss and coral red gloss slips were found to separate into three distinct groups with respect to the CaO-MgO content, as shown in Figure 2 (adapted from [3]). The black gloss samples form a single group, characterized by low CaO (<1%) and MgO between 1-2.5%, (labeled I in Figure 2). By contrast, the coral red glosses split into two groups: i) coral red with low CaO and low MgO content – low calcium-magnesium (LCM) coral red (labeled II in Figure 2); and ii) coral red with high CaO and MgO (>1.5%) content – high calcium-magnesium (HCM) coral red (labeled III in Figure 2). In addition to the above compositional differences, scanning transmission electron (STEM) imaging revealed different ceramic fabric textures between the black gloss and coral red (Figure 3, adapted from [3]). Both of the coral red groups showed open porosity with visible distinct clay particles (Figure 3, images b and c) whereas the black gloss (Figure 3, image a) appeared highly vitrified, with the original shape of the clay particles remaining only as fused relics. Size comparison between the visible clay platelets of coral red and the relic particles in the black gloss revealed that both materials originally consisted of similarly sized clay particles.

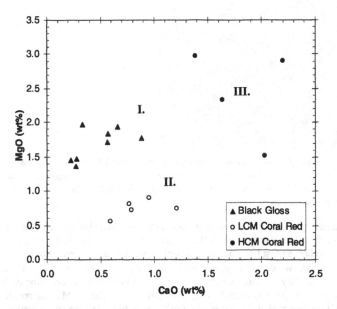

Figure 2. MgO-CaO bivariate plot showing the major compositional groups of black gloss (I.), LCM coral red (II), and HCM coral red (III).

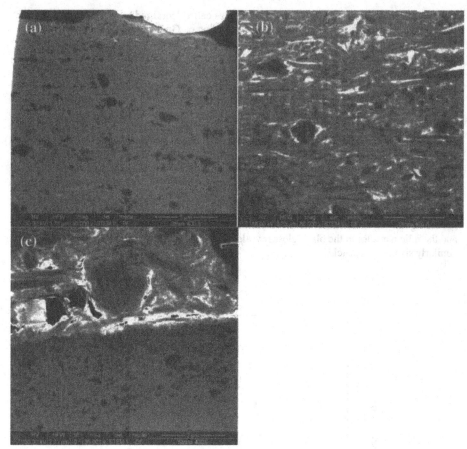

Figure 3. Scanning Transmission Electron images of black gloss and red. Image (a) shows sintered texture of black gloss. Image (b) shows LCM coral red. Image (c) shows HCM coral red on top of black gloss in JPGM. The light areas in images (b) and in the top part of (c) are open pore voids.

We have argued [3] that the similar chemistry and grain size of LCM coral red to black gloss would likely cause it to turn black if exposed to the three-stage firing process. Therefore, a different firing scheme must have been employed to produce a coral red slip from the LCM material, most likely two separate firings: an initial firing with a reduction step to produce the black gloss, followed by a re-application of gloss slip and another second firing under oxidative conditions to produce the red colored gloss. Alternatively, for the HCM coral red, the higher CaO and MgO content may have imparted a refractory property to the slip material, allowing it to be fired under the single three stage firing process without turning black. This interpretation of the data was based on the observation that coral red decoration sometimes appeared underneath black gloss when these two slip applications were superimposed [4]. Such a layered sequence of black over red would not be possible if both materials were composed of clay slip that easily vitrified when fired under reductive conditions.

In this paper we refine our considerations of the fabrication technology of coral red through the examination of trace elements as determined by laser ablation inductively coupled plasma time-of-flight mass spectrometry (LA-ICP-TOFMS). By characterizing the trace elements, the observed differences in the major element chemistry discussed above are clarified, in particular, whether the black gloss and the two coral red materials are geochemically related or came from different clay deposits. Also, as has been discussed by Kingery [5] and Tite et al. [6], the black gloss and coral red were refined through the process of levigation - the settling of Attic clay in large water vats. During levigation, coarser-grain and heavier particles sink, leaving a suspension of fine clay particles in solution. Comparing trace elements in the ceramic body to those in both the coral red and black glosses has further elucidated this settling process.

EXPERIMENTAL

A GBC Optimass 9500 Inductively Coupled Plasma Time of Flight Mass Spectrometer (GBC Scientific Equipment, Dandenong, Victoria, Australia) coupled to a New Wave Laser UP213 laser ablation system (New Wave Research, Inc., Freemont, California) was used for all analyses presented here. The laser was a Q-switched frequency-quadrupled Nd:YAG laser with an output of 213 nm, operated at a repetition rate of 20 Hz for a duration of 20 seconds. Analyses were performed using spot mode with an aperture spot size of 30 μm. The laser was automatically refocused over the course of the run to ensure equal laser power was applied to the material surface as the laser pit was produced. Ablation was performed in a helium sweep (0.400 L/min) admixed with Argon (0.800 L/min) prior to injection into the plasma torch. The plasma was operated under cool conditions (850 Watts), optimal for dry aerosol introduction, with a nebulizer flow (as determined by both He and Ar flow rates) of 1.2 L/min, plasma flow of 10.00 L/min and auxiliary flow of 0.500 L/min. The ICP-TOFMS voltages were optimized while monitoring mass ^{120}Sn during the ablation of the multi-elemental glass 616 from NIST (Sn ~1ppm; National Institutes of Standard and Technology, Gaithersburg, Maryland).

External calibration of the ICP-TOFMS response was performed by ablating a series of NIST SRM glasses: 610, 612, and 614. Data was reduced following the procedures of Gratuze et al. [7] for laser ablation generated aerosols using Si as a normalizing factor. Data accuracy was assessed by the ablative analysis of Corning Archaeological Glass D (Corning Museum) and by basalt supplied by the USGS (BHVO-1, United States Geological Survey, Colorado).

The current study group (Table 1) was composed of four un-mounted black gloss sherds (three of which had sufficient body ceramic attached to be analyzed with laser ablation) and one representative sample each of LCM and HCM coral red (which had previously been mounted in cross-section [2,3]). To allow direct comparison, the black gloss samples were only collected from vessels that also had coral red decoration. Only one example each of LCM and HCM coral red from the original study group was wider than 30 μm in section (the smallest laser spot size which produces sufficient aerosol for sensitivity to trace elements by ICP-TOFMS) and therefore suitable for this study.

Object (JPGM #)	Type	Date (B.C.)	Colors Sampled
76.AE.130.42	Mastoid Cup	6th century	black and body
97.AE.22.21	Phiale	500-470	black and body
86.AE.280	Kylix	500-470	black
97.AE.22.19	Phiale	500-470	black, body, and HCM coral red
76.AE.96.1	Phiale	500-471	LCM coral red

Table 1: List of samples included in this study.

The black gloss/ body ceramic samples were introduced to the laser ablation chamber without further preparation. The examples of LCM and HCM coral red, which were previously mounted and coated in carbon for analysis by SEM/EDX, were cleaned prior to LA-ICP-TOFMS analysis by polishing using 1 μm diamond grit (Buehler, Illinois). Each sample was analyzed six times with the highest and lowest values dropped before averaging.

RESULTS

The concentration of 54 calibrated masses was measured for each of the nine samples listed in Table 1. In Figure 4, the average body ceramic, black gloss, HCM and LCM coral red results are presented and reorganized into periodic groups (*alkali* and *alkali earth*, groups 1-2; *transition metals*, groups 3-12; *metalloid and non-metals*, groups 13-16; and *rare earth elements*, lanthanide/actinide series) to facilitate examination of elemental trends. The overall distribution pattern of all four types of material are quite similar, with the notable exception of the Mg and Ca concentrations, reflecting the fact, as discussed above [3], that these elements serve as differentiators between these materials. The four materials similarly show variation in the abundance of Sr, the significance of which will be discussed below. The patterns also deviate around the elements associated with metallurgical practices (Cu, Ag, Au and Pb). These few variations notwithstanding, the majority of the elements show nearly identical abundances, indicating the samples have similar geochemical characteristics, and likely were formed from a single clay deposit.

DISCUSSION

The compositional differences between samples as determined by Mg and Ca had been previously noted in discussion of the major element chemistry [3]. However, the variation in Sr abundance was not apparent in these studies as it is present in quantities below the detection limit of SEM/EDX analysis for this element. However, the LA-ICP-TOFMS data presented here now allows the relationship of Ca to Sr to be evaluated, as shown in Figure 5.

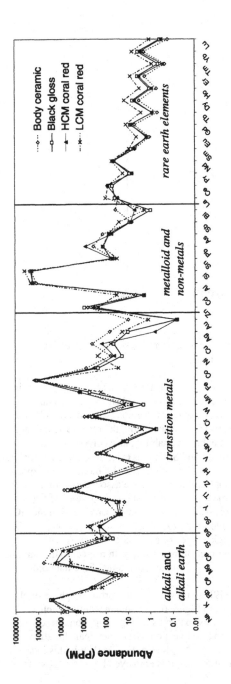

Figure 4: Average body ceramic, black gloss, HCM and LCM coral red abundances reorganized into periodic groups.

25

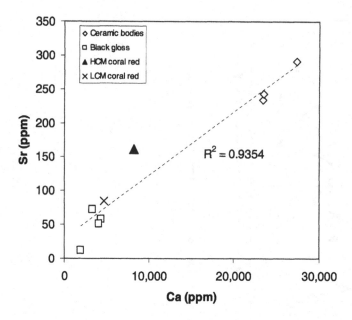

Figure 5. Ca-Sr correlation. The systematic decrease in concentrations shows selective removal of the mineral phases that hosts these elements.

Ca and Sr exhibit a strong correlation (R^2=0.94), demonstrating that Sr is hosted within the calcium fraction of the clay, as might be expected since Sr often substitutes for Ca due to their similar ionic radii. In the course of weathering, Ca and Sr are amongst the most mobile cations, and are readily leached from clay sediments [8]. Consequently, the relative abundance of Ca and Sr to other elements in the clay matrix will exhibit variability, as can be observed in Figure 5. As the groundwater carrying the leached Ca and Sr cations evaporates, it leaves behind $CaCO_3$ and $SrCO_3$, and these carbonates are often found interstratified with natural clay deposits. Therefore, it is likely that the Ca and Sr in the samples in this study are associated with a variable level of carbonate mineral phase present in each material: the highest concentrations of Ca and Sr are found in the body ceramic, with successively decreasing concentrations found in the HCM coral red, LCM coral red and the black gloss samples. The fact that the samples lie on a single correlation line links the samples together to a single source of carbonate. The systematic decrease of Ca and Sr concentrations of nearly an order of magnitude with respect to the ceramic body suggests this mineral phase may have been removed in the processing of the materials during levigation.

As discussed by Wronkiewcz and Condie [9], along with Sr, the carbonate phase of clay-bearing rocks also contains Mg and Mn. As shown in Figure 6, to evaluate whether such a carbonate phase was selectively removed from the clay materials during levigation, these elements (Sr, Mg, and Mn) are plotted against a modified version of the chemical alteration index (CAI, [10]). The CAI is calculated here as Al/(Al+Na+K+Ca+Fe+Ti). While this index is often used to interpret weathering patterns of clay minerals, it is employed here as a proxy for the

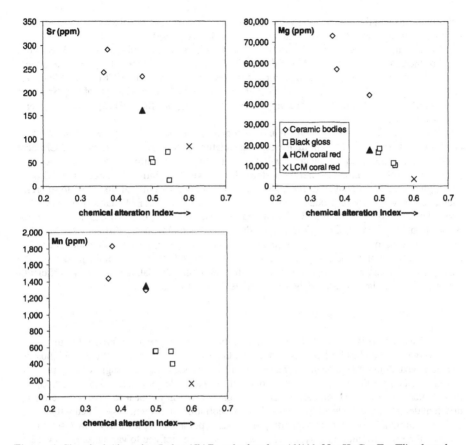

Figure 6. Chemical Alteration Index (CAI), calculated as Al/(Al+Na+K+Ca+Fe+Ti), plotted against trace elements typically found in carbonate phases of clay bearing rocks.

clay refinement process. Illitic clay contains the illite mineral (a potassium-aluminum sheet silicate) as its backbone together with variable associated minor minerals. As the illite clay is refined during levigation, the minor mineral components are lost leaving behind a pure and fine-particled illite fraction. CAI ratio follows this process: the greater the CAI ratio, the greater the loss of minor mineral components from the parent clay. A ratio approaching unity represents a highly weathered material with kaolin (aluminum sheet silicate) as an end member.

As shown in Figure 6, all three elements exhibit a negative correlation with increasing CAI ratio. For each element, the highest concentrations (corresponding to the lowest CAI ratios) are found in the ceramic bodies (open diamonds); the ceramic body therefore must represent the most unrefined material. HCM coral red (filled triangle) has a slightly higher CAI ratio and relatively moderate levels of trace elements when compared to the ceramic body. The black gloss samples (open squares) have lower concentrations of Sr, Mg, and Ca and a higher CAI ratio when compared to HCM coral red. Finally, LCM coral red (cross) has the lowest trace element concentrations and the highest CAI value; LCM coral red therefore represents the most refined material (least amount of carbonate) among the samples examined.

Assuming the ceramic body material is representative of unrefined clay, the observed trend of increasing CAI from HCM coral red, black gloss to LCM coral red suggests these slip materials were produced by levigating unrefined clay. Significantly, the HCM coral red appears to contain a carbonate phase, as indicated by the relatively high concentrations of Mg, Sr, and Mn. Similarly, LCM coral red, which appears to have undergone the most rigorous refinement is mostly devoid of carbonate.

These differences in refinement and carbonate concentration between HCM and LCM coral red gloss material allow an improved interpretation of coral red fabrication. As HCM coral red is fired, the entrained carbonates break down, emitting carbon dioxide, leaving behind pockets available for air to re-enter the material for re-oxidation during the three-stage firing. The high degree of refinement of the LCM coral red clay material may at first seem to suggest such pockets would not be produced, but this is inconsistent with STEM data which showed open porosity in both HCM and LCM coral red glosses. However, if, as postulated in our earlier publication, the LCM coral red gloss was produced in a second oxidative firing, it would not have been exposed to the same high temperatures (around 1000 °C) under which the black gloss (which shares a similar chemistry) was produced. One hypothesis to explain the porosity of the LCM material is that the lower temperatures in the subsequent oxidative firing are insufficient to sinter the refined clay material to the same degree as the black gloss.

CONCLUSIONS

The examination of the trace element distribution and degree of refinement in the samples in this study has furthered our understanding of the characteristics of coral red glosses. The observed similarity between the trace elements in the body and slips suggests the raw materials used in Attic vase painting were derived from a single geological source. Although these materials may have been derived from a single source, they appear to have been refined to different degrees, as evidenced by the varying concentration of elements associated with a carbonate mineral clay fraction, supporting previous hypotheses that levigation was used to produce the raw clay slip. Finally, the slips were found to be refined to various degrees with respect to the ceramic body clay. The elevated levels of carbonate minerals observed in HCM coral red were postulated as accounting for the observed porosity of this material, which in turn would have allowed it to re-oxidize in the final stage of a three-stage firing sequence to produce the red gloss. On the other hand, the LCM coral red material, with a bulk composition closely related to black gloss, was found to be more refined, and thus would have required a subsequent firing under only oxidative conditions to produce the red color.

ACKNOWLEDGMENTS

The authors would like to thank Beth Cohen for organizing the exhibition "Colors of Clay" at the J. Paul Getty Museum, which provided the impetus for this project. We would also like to thank Jerry Podany and Jeffrey Maish of the J. Paul Getty Museum Antiquities Conservation Department for providing samples, helpful discussions, and support.

REFERENCES

1. Noble, J.V. *The Techniques of Painted Attic Pottery*, (Watson-Guptill Publications, New York, 1965).

2. Walton, M., Doehne, E., Trentelman, K., Chiari, G., Maish, J., and Buxbaum, A., A Preliminary Investigation of Coral Red Glosses Found on Greek Attic Pottery. In *Colors of Clay Workshop Proceedings*, edited by K. Lapatin, (J. Paul Getty Museum, Los Angeles, 2008)

3. Walton, M., Doehne, E., Trentelman, K., Chiari, G., Maish, J., and Buxbaum, A. Characterization of Coral Red Slips on Greek Attic Pottery. *Archaeometry*, in press (2008).

4. Cohen, B. Coral-red Gloss: Potters, Painters, and Painter-potters. In *Colors of Clay* edited by B. Cohen, pp. 44-54 (J. Paul Getty Museum, Los Angeles, 2006)

5. Kingery, D. Attic Pottery Gloss Technology. *Archaeomaterials,* 5, 47-54 (1991).

6. Tite, M.S., Bimson, M. and Frestone, I.C. An Examination of the High Gloss Surface Finishes on Greek Attic and Roman Samian wares. *Archaeometry*, 24, 117-126 (1982).

7. Gratuze, B., Blet-Lemarquand, M., Barrandon, J.-N. Mass Spectrometry with Laser Sampling: A New Tool to Characterize Archaeological Samples. *Journal of Radioanalytical and Nuclear Chemistry*, 247(3), 645- 656 (2001).

8. Rollinson, H. *Using Geochemical Data: Evaluation, Presentation, Interpretation.* (Longman Group Limited Harlow, U.K., 1993).

9. Wronkiewicz, D., and Condie, K., Geochemistry of Archaean Shales from Witwatersand Supergroup, South Africa: Source-area Weathering and Provenance. *Geochemica et Cosmochemica Acta*, 51, 2401-2416 (1987).

10. Nesbitt, H., and Young, G. Prediction of Weathering Trends of Plutonic and Volcanic Rocks Based on Thermodynamic and Kinetic Considerations. *Geochemica et Cosmochemica Acta*, 48, 1523-1534 (1984).

Mater. Res. Soc. Symp. Proc. Vol. 1047 © 2008 Materials Research Society   1047-Y06-04

Materials and Techniques of Thai Painting

Katherine Eremin[1], Jens Stenger[1], Narayan Khandekar[1], Jo Fan Huang[2], Theodore Betley[3], Alan Aspuru-Guzik[3], Leslie Vogt[3], Ivan Kassal[3], and Scott Speakman[4]

[1]Straus Center for Conservation, Harvard University Art Museums, 32 Quincy Street, Cambridge, MA, 02138
[2]Philadelphia Museum of Art, Philadelphia, PA, 19130
[3]Department of Chemistry and Chemical Biology, Harvard University, Cambridge, MA, 02138
[4]Center for Materials Science and Engineering, Massachusetts Institute of Technology, Cambridge, MA, 02139

ABSTRACT

Samples from Thai manuscripts and banner paintings dated from the late 17th to the early 20th centuries were analyzed by X-ray fluorescence, scanning electron microscopy, X-ray diffraction, Raman spectroscopy and Fourier transform infrared spectroscopy to determine the materials used. This revealed a chronological change in palette with the introduction of imported pigments in the later 19th and 20th centuries. In the late 17th and 18th centuries, the main green used on the manuscripts was an organic copper salt with malachite used on the banner paintings and on only one manuscript. Both the organic copper salt and malachite were replaced by emerald green and mixtures of Prussian blue plus gamboge on both manuscripts and banner paintings and of Prussian blue with chrome yellow or zinc yellow (zinc potassium chromate) on some manuscripts. Similarly, indigo in the late 17th and 18th century manuscripts and banner paintings was replaced by Prussian blue and then synthetic ultramarine in the 19th century. Chrome yellow was used in addition to gamboge in one later 19th century manuscript.

The organic copper salt used in the 18th century was identified as a copper citrate phase. This has not been identified previously as an artist's material although its use has been postulated based on historic texts. Copper citrate was synthesized to provide reference material for comparison with the phase found in the Thai manuscripts. The color of these synthesized copper citrates varies from deep blue-green to light blue depending on hydration. The most hydrated form is closest in color but is not identical to the deep green phase found on the Thai manuscripts.

Lead white was the main white pigment in all but one manuscript, which contained huntite, a magnesium calcium carbonate. Huntite also occurred in mixtures with other pigments in two other manuscripts. In all the works studied, red lead, vermilion and red earths were used for red, orange and pink shades and red earth used in brown areas.

INTRODUCTION

The Harvard University Art Museums (HUAM) has an important collection of Southeast Asian manuscripts, which includes a large number of Thai manuscripts dating from the late 17th to 20th centuries. These were part of a bequest in 1984 from Philip Hofer, Curator of Printing and Graphic Arts at Harvard's Houghton Library, and a collector of both Western and Asian art works. The present study focused on twelve horizontal accordion-fold manuscripts with ink, color, and gold on khoi paper, made from the inner bark of the khoi tree (*Streblus aspera*). A page from an early (late 17th to early 18th century) and late (late 19th to early 20th century) are shown in Figures 1a and 1b. All but one of the HUAM manuscripts studied was two-sided. The manuscripts examined vary in subject, text and proposed date, from the late 17th century to the 20th century. Samples from four manuscripts and seven banner paintings from the Asian Art Museum in San Francisco, dating to the 18th and 19th centuries, were also examined for comparison

Figure 1a: illustration from a late 17th to early 18th century Thai manuscript, 1984.511 and 1b: illustration from a late 19th century Thai manuscript, 1984.517

METHODOLOGY

A non-destructive X-ray fluorescence (XRF) study was undertaken on the HUAM manuscripts to determine the elemental composition of all colors. Several pages were examined in each manuscript. Following this, micro-samples were removed from selected areas for further analysis. These samples and those provided by the Asian Art Museum were examined using Raman spectroscopy, Fourier transform infrared (FT-IR) spectrometry, scanning electron microscopy (SEM) and/or X-ray diffraction (XRD) to confirm the phases present and identify organic materials.

Reference samples of copper citrate were synthesized by dissolving citric acid and $Cu(NO_3)_2 \cdot 3 \cdot (H_2O)$ in deionized water. Ammonium or sodium hydroxide was added until the pH was stable at 4. This solution was then subjected to solution or hydrothermal synthetic protocols. The former involved refluxing for 4-8 h to precipitate a light green material which was collected on a sintered glass frit, washed with deionized water, and dried under dynamic vacuum on a Schlenk line or in a convection oven at 145°C for several hours. For the hydrothermal syntheses, the same solution was sealed in either a teflon-lined pressure vessel and heated to 140°C for 24h, or added to a thick-walled glass reactor, sealed with a teflon stopcock and heated in an oil bath held at 140°C for 24h. Both setups were allowed to cool to room temperature gradually (over 8h). The material was isolated by decanting the remaining solution phase, and washing the resultant product with water on filter paper. Further samples were produced following a medieval recipe for reacting verdigris with lemon juice to prepare a green pigment recipe (1).

The XRF system used was a Bruker Artax micro-XRF spectrometer with a Mo tube and a Silicon Drift (SDD) detector. Polycapillary optics give a spot size of approximately 70 microns. The operating conditions used were 40 kV and 400 mA with a He stream employed in some analyses to enhance detection of light elements (below calcium). All analyses were qualitative only.

The Raman spectrometer used was a Bruker Optics 'Senterra' dispersive Raman microscope with an Olympus BX51M microscope. The Raman spectrometer has 532 nm, 633 nm, and 785 nm lasers available. There are three gratings for the 785 nm laser, covering the 70-3283 wavenumber range, four gratings for the 633 nm laser covering 60-3532 wavenumbers, and three gratings for the 532 nm laser covering 65-3700 wavenumbers. The spectrometer resolution is ~ 3-5 wavenumbers (dependent on the wavenumber). The system uses an Andor 'iDus' CCD detector, operated at -55°C. There are five software-controlled settings for the power of each laser: 100%, 50%, 25%, 10% and 1%. Estimated actual power on the sample at the 100% setting is 8.5 milliwatts for the 532 nm laser, 10.9 mW for the 633 nm laser, and 37.5 mW for the 785 nm laser. The microscope has 20X, 50X and 100X objectives, with laser spot sizes approximately 5, 2, and 1 microns respectively. The microscope contains a joystick-controlled motorized stage and setting of the analysis area is accomplished with the aid of an attached video camera. The instrument is controlled via OPUS software version 5.5.

FT-IR spectrometric analyses were carried out using a Nicolet 510 instrument coupled to a Spectra-tech IR-plan infrared microscope with a 32x objective. The sample was compressed onto a diamond cell (2 mm x 2 mm) with a stainless steel roller and the sample area defined by double apertures contained in the microscope. An absorbance spectrum (4000-500 cm^{-1}) was measured (resolution setting 8 cm^{-1}) and subtracted against a blank background. The spectrum was compared with a database of artists' materials at the Straus Center for Conservation.

A Jeol JSM-6460 LV Scanning Electron Microscope with an Inca X-sight Oxford Instruments energy dispersive detector was used to examine some samples to determine the elemental composition. The samples were analyzed in low vacuum (35Pa) without carbon coating at a voltage of 20kV.

X-ray diffraction (XRD) of selected samples was undertaken on a Bruker D8 Multipurpose Diffractometer. Samples were collected over a 2theta range of 30° to 90° using CuKα radiation at 40kV, 40mA. The primary X-ray beam was collimated to approximately 0.5mm with a monocapillary and the diffracted X-rays detected with a GADDS detector.

RESULTS

The analytical results for the manuscripts and banner paintings indicated a varied palette which changed from the late 17th to 20th centuries. The results are presented by color, with the most complex discussed first.

Green

The green pigments proved the most varied in these manuscripts. Most manuscripts contained more than one green shade and in many of the 19th century manuscripts a light green was used for the underdrawing. A number of different green pigments and mixtures were identified in the project and are discussed below.

The only inorganic element detected by either XRF or SEM in the bright green pigments used in the three oldest manuscripts, 1984.501, 1984.511 and 1984.512 dating from the late 17th to early 19th centuries, and two possibly later 19th century manuscripts, 1984.506 and 1984.507, was copper, indicating a copper green. However, although samples gave good FTIR spectra, these did not correspond to any available reference spectra. This ruled out the common green copper pigments, such as malachite, verdigris, or other organic copper salts. XRD of these samples yielded spectra that matched hydrated copper citrate, as shown in Figure 2.

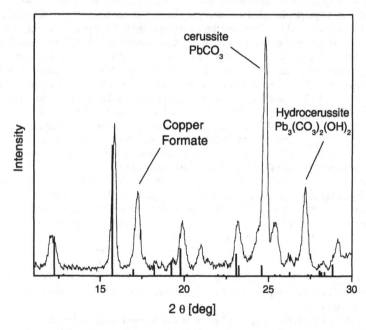

Figure 2: XRD pattern of the unknown green (black line) and the copper citrate reference (red bars). Other compounds present are copper formate, cerussite and and hydrocerussite.

Examination of recent literature showed that the FTIR spectra were similar to those from hydrated copper citrate $Cu_2C_6H_4O_7(H_2O)_2$ (2), illustrated in Figure 3. Copper citrate was also identified in a sample from the Asian Art Museum manuscript 1993.11.2, dated to 1725-1775.

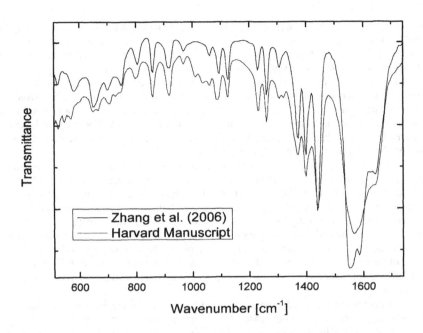

Figure 3: FTIR spectrum of copper citrate reference (2) and from manuscripts.

The crystal structure of the copper (II) citrate is unusual in forming a 3D open framework polymer with large void channels and Cu(II) in 2 different coordination environments with 5 fold coordination (2). Minor copper formate and/or copper chloride were also identified by FTIR, XRD and SEM in some of the green copper citrate samples, possibly as alteration products. Copper citrate salts are discussed by Scott who noted that these had not yet been found on any art works (1). The presence of copper citrate on these manuscripts hence represents the first known use of copper citrate as a pigment. The lack of significant associated phases and the prevalence of this phase in the earlier manuscripts indicate that copper citrate was the original pigment used, rather than an alteration product of another copper green pigment. However, malachite was found in the Asian Art Museum manuscript 2006.27.85, dated to 1781. So far, this is the only malachite identified in any of the manuscripts analyzed. Malachite was also found in samples from two of the banner paintings, 2006.27.122.7 from the 18th century and F2002.27.80.1 from the 19th century. Copper citrate was not found in any of the banner paintings examined in this study but recent work by Giaccai found copper citrate under gilding in an 18th century banner painting (3).

Scott replicated a mid to late 17th century Venetian recipe for reacting verdigris with lemon juice said to give a "most beautiful green color ... like an emerald" to produce copper citrate pentahydrate, $Cu_4(C_6H_5O_7)_2(H_2O)_5$ (1). His color measurements showed that the color spectrum for the copper citrate was shifted into the blue compared to verdigris itself (1). The

Venetian recipe states that care should be taken not to let the brush touch water and Scott notes that the copper citrate salts "have dubious stability". Water was easily lost from the copper citrate complex even at ambient temperatures, the gray green crystals becoming opaque and then light blue as water loss progressed, and that the dehydrated copper citrate slowly absorbed water from the air over time (2). The thermal behavior was attributed both to the long bond lengths of the two $Cu-O_{(H2O)}$ bonds which allows the weakly bonded water molecules to be removed readily and to the fact these loosely bound water molecules occur on the edge of the large unoccupied void channels, hence are able to move through them once released (2). The color of the crystals produced by these authors (2) appears to be different from that of the Thai manuscript samples or those produced by Scott (1), possibly due to differences in hydration.

The potential instability of the copper citrate raises the question of whether any alteration has occurred to the manuscript samples and hence whether the hydration state and color of the original phase differed significantly from that now present. Visual examination indicates that the bright green areas on the manuscripts contain scattered translucent bright green crystals in a more powdery and slightly lighter matrix. A number of copper citrate species are recorded in the literature with different degrees of hydration, eg. copper citrate pentahydrate, copper citrate sesquihydrate and copper citrate dihydrate, but thermal analysis is required to determine the degree of the hydration and hence differentiate these. There was no obvious difference between the FTIR or Raman spectra from the larger translucent crystals and the more powdery matrix on the manuscripts.

Synthesis of copper citrate was undertaken using both the medieval recipes and laboratory grade chemicals. In both cases, the precipitates formed were identified as poorly crystalline or polycrystalline copper citrate by XRD and FTIR. The XRD spectra were relatively weak, although identifiable, due to the lack of good crystallinity, which may explain the generally weak spectra from the manuscript samples.

The hydrothermal syntheses and reaction of verdigris with lemon juice produced a dark green material, whilst the solution syntheses produced a light, aquamarine material. Oven-drying any of these materials resulted in a distinct blue material (typical of tetrahedral $Cu(II)$) that absorbs water from the atmosphere over the period of a week to regain a green hue. The FTIR spectra from the different colored phases were extremely similar and would not allow them to be discriminated. The most hydrated form is closest in color but is not identical to the deep green phase found on the Thai manuscripts. It is possible that the color in the Thai samples may be modified by the presence of trace amounts of another metal. None was detected by either XRF or SEM measurement. However, most samples are contaminated to some extent by associated phases, such as kaolin or lead white, and metals present at very low levels may not have been detected by the analyses undertaken to date. Elemental analysis will be undertaken to determine the hydration state of the various synthesized phases and relate these to color. This will allow the Raman, FTIR and XRD spectra of different forms of copper citrate to be compared with each other and with data obtained from the manuscript samples. Direct measurement of the hydration state of the manuscript samples is hampered both by the limited sample available and contamination from associated phases.

In contrast to the earlier works, samples from bright green areas of HUAM manuscripts, 1984.508, 1984.510, 1984.516 and 984.517, dated from the 19th to early 20th centuries, contained emerald green, a copper arsenic acetate. Samples from dark green areas of these manuscripts contained a mixture of Prussian blue and gamboge. The same mixture, Prussian blue and gamboge, was identified in samples from the underdrawing of the 19th century manuscripts

1984.506, 1984.516 and 984.517. Emerald green was also found on two 19th century banner paintings from the Asian Art Museum, 2006.27.76 and F2002.27.73, and in all green areas of HUAM manuscript 1984.433, dated to the early 20th century, where it was often used over carbon black to give a darker shade.

Optical examination of samples from bright green areas of HUAM manuscript 1984.521, dated to the 19th century, showed these consisted of red, white and yellow-green particles. Raman and FTIR spectroscopy indicated these were a currently unidentified red, barite and a mixture of zinc yellow and Prussian blue respectively. SEM, FTIR and Raman analysis indicated that the form of zinc yellow used was zinc potassium chromate rather than pure zinc chromate. The former has the approximate composition $K_2O.4ZnCrO_4.3H_2O$ although this can vary considerably due to variations in $ZnO:Cr_2O_3$ ratio (4). In contrast, yellow-green areas and the green underdrawing contained a mixture of Prussian blue and gamboge.

Three different greens can be distinguished visually on another 19th century HUAM manuscript 1984.524, and appear to consist of a mixture lead chromate and Prussian blue, with ultramarine (the blue pigment used in this manuscript) mixed in to control the shade. The bright green contained significant barite whilst the yellow-green and dark green contained significant gypsum.

Blue

Three distinct blue pigments were identified in the manuscripts and banner paintings. Indigo was used in the late 17th to early 19th century HUAM manuscripts 1984.501, 1984.511 and 1984.512 and in the Asian Art Museum manuscripts dated to 1725-75, 1993.11.2, and to 1791, 2006.27.85. In contrast, the Asian Art Museum manuscript dated to 1844, 1993.10, contained Prussian blue. Prussian blue also occurred as the only blue in several 19th century HUAM manuscripts, 1984.506, 1984.507, 1984.510 and 1984.516 and Asian Art Museum banner paintings 2006.27.122.7, F2002.27.69, F2002.27.80.1 and F2002.27.73. The remaining 19th and early 20th century examples contained ultramarine, HUAM manuscripts 1984.508, 1984.517, 1984.524 and 1984.433 and Asian Art Museum banner paintings 2006.27.76 and F2002.27.122.15. In general, the earlier examples contained Prussian blue whilst the later examples contained ultramarine. The 20th century HUAM manuscript 1984.433 also contained ultramarine. In all cases, the fine particle size and rounded shape indicate the use of synthetic rather than natural ultramarine.

Yellow

The only yellow identified in most manuscripts and banner paintings was gamboge mixed with huntite, kaolin and/or lead white. The main yellow in the 19th century HUAM manuscript 1984.521 also appears to be gamboge, again with lead white, but zinc yellow (zinc potassium chromate) was found in one sample and XRF analysis revealed low levels of zinc in some yellow areas. Visual examination indicates two different shades of yellow in some of these areas, suggesting gamboge was the main yellow with zinc yellow possibly used to highlight details. Another 19th century HUAM manuscript, 1984.524, was the only one to contain significant amounts of two different yellows, with gamboge and lead chromate, $PbCrO_4$, occurring in separate samples.

White

Lead white, basic lead carbonate $2PbCO_3.Pb(OH)_2$, was the most common white used in the manuscripts either alone or mixed with other colors. However, huntite, $Mg_3Ca(CO_3)_4$, was present in two HUAM manuscripts, as the only white in 1984.512 and together with lead white in 1984.506. Cerussite, lead carbonate $PbCO_3$, was associated with the lead white in some samples and calcite and kaolin were both present in many samples, both white and colored. Barite and gypsum occurred only in samples from the 19th century HUAM manuscripts containing chromates, 1984.521 and 1984.524.

Red, Orange and Others

Vermilion, sometimes combined with red lead, was the main pigment detected in all red areas, whilst all orange areas had red lead, sometimes combined with vermilion. Pink areas contained red ochre, red lead or vermilion mixed with lead white and/or calcite, whilst red-brown areas contained mainly red ochre.

Binders

The FTIR and Raman spectra generally lack any discernable organic compounds that could be present as a binder. In some instances, weak amide bands may be present in the FTIR spectrum indicating a proteinaceous material, but this cannot be clearly identified. Gas chromatography with mass spectrometry (GC-MS) may be employed in the final stage of the project to attempt to identify any organic binders. One manuscript dated to the 19th century has a translucent brownish organic material under some areas of gilding. FTIR suggested gutta percha as the best match but the origin remains obscure.

DISCUSSION

Previous work

Previous analytical studies of the materials used in Thai manuscripts or banner paintings are somewhat lacking. Most notably, in-situ Raman microscopy was used to study four Thai manuscripts from the British Library Oriental Department (5). In addition, three published analytical studies of Thai murals are relevant (6, 7, 8) as art historians' note parallel relationships between Thai murals, manuscripts, banners and sculptures. The results of these earlier studies are summarized in Table 1.

Study	17th century	18th century	Early 19th century	Late 19th century
5		red lead; vermilion; red ochre (1780)	indigo; orpiment; red lead; vermilion (1800)	ultramarine; red lead; vermilion (1880); ultramarine; lead chromate + barite + orpiment; anhydrite, gypsum, barite + lead white; red lead (1900)
6, 7	azurite; malachite; gamboge; clay; gypsum; calcite	indigo; cumengeite (lead copper chloride hydroxide hydrate); malachite; gamboge; lead white	Prussian blue; malachite; gamboge; lead white	ultramarine; emerald green; lead mixture (yellow); lead white + barite
8		lead white, calcium carbonate, silica, vermilion, green verditer		

Table 1: Summary of pigments identified in previous studies on Thai materials (6, 7, 8)

The green pigment used on three of the Thai manuscripts and the yellow on the three oldest manuscripts did not give a Raman spectrum, which was believed to rule out the common copper-based green pigments, inorganic yellow pigments, Indian yellow or berberine as these have well known spectra (5). These authors found that most of the pigments used in murals were not homogeneous in composition and various whites such as gypsum, kaolin, calcite or lead white were mixed into all of the colors. (6, 7, 8). The change from traditional to imported pigments (Prussian blue, emerald green and synthetic ultramarine) in the 19th century was compared to the decline in style of traditional Thai painting due to the influence and introduction of Western ideas after the third quarter of the 19th century (7).

Imported Pigments

The findings of these previous studies and of the current study of manuscripts and banner paintings show a transition from local to imported pigments during the 19th century. These imported pigments could come via a number of routes as Thailand traded actively with other countries in the area, China, Japan, India and Persia, and with the Western nations trading in the South China Seas, the Portuguese, Dutch and British. The most important sea routes were to China, Japan and India and the Europeans often traded between the key ports of the intra-Asian routes rather than direct to Europe. Trade with China was the most important, as shown by the detailed list of imports and exports to Thailand compiled by D. E. Malloch an employee of the British East India Company in 1852 (9). Malloch categorized his lists into four parts: *Production of Siam Annually, British imports, Imports of India Goods to Siam, and China goods imported into Siam,* and mentions artists' pigments specifically (see Table 2). Arthur Neale, a local resident in Siam, also stated that merchants imported boxes of watercolors, cakes of finest Indian ink, gum, and glue from Canton and Macao to Bangkok (10).

Table 2. Selected items from D.E. Malloch's record on Siam

Siam local resources	Dragon's blood, gamboges gold dust (for local temple), indigo (local use only), lead white and black, lac
Siam's import from India	Arsenic (white and yellow), red and yellow ochre, turpentine, gum tragacanth or gum dragon, from Turkey, gum Arabic, verdigris, safflower
Siam's import from China	Gold leaf, mock gold leaf, glue, varnish, paint (red, blue, black, white, and green), vermilion, coarse and fine, saffron
British Imports	Paints (green, blue, black, and red)

Artists could choose between foreign and local materials with the greatest volumes of imported paints apparently coming from China It is hence pertinent to consider the palette used in China and in Thailand's other main trading partners, particularly India, and when imported chemicals first appeared in these areas. For example, although some manufactured watercolors were probably available in Calcutta, India in the early 19th century, these became much more accessible in 1842 when N.C. Dutt and Aukhoy Coomar Laha opened the first specialist paint shop and eventually became the fist importers of Winsor and Newton brands, selling prepared paints and dry pigments (11). The date at which a number of synthetic pigments were discovered and commercially manufactured and at which they are know to have been used in China, Japan and India are given in Table 3.

Pigment	Discovered / Manufactured	China	India	Japan
Emerald green	Discovered early 19th century (12); Commercial sale from 1814 (12)	Introduced late 1840s to early 1850s (13, 14)	Found in 19th & 20th century (15, 16)	Found in 19th century paintings (17)
Prussian blue	Developed in 1704 (18)	Imported 1759 onwards (19, 20); Found in 19th century works (21)	Found in 19th century works (11)	Known import from 1782 but mentioned in 1763 (22); Found in late 18th and 19th century paintings (17, 23)
Synthetic ultramarine	Discovered 1826; Commercial sale from 1828 (24)		Found in 19th century works (11, 25)	
Lead chromate	Marketed in Europe from 1800's (4)		Found in 19th century works but may be natural mineral (11)	
Zinc chromate	Discovered 1809 (4); Developed as colorant 1847 (4)			

Table 3. Dates for discovery, manufacture and import to the East of some synthetic pigments

The available literature suggests some delay in the transfer of synthetic pigments from the Western to the Eastern art markets. These foreign pigments were popular as they were cheap, convenient and gave good results (14). Of course, the imported pigments did not completely replace the traditional pigments; in some instances, malachite and indigo were used at the same time as Emerald green and Prussian blue respectively (12, 14, 17) and synthetic ultramarine was used at the same time as indigo or Prussian blue (12, 25). Comparison with the other countries suggests that Prussian blue may have been imported into Thailand during the late 18th to early 19th century, and Emerald green, synthetic ultramarine and chrome yellow imported into Thailand during the later 19th century.

Chronological variations in materials

The results of our current study of Thai materials indicate a chronological variation in the green, blue and to a lesser extent yellow pigments. However, if the materials used by mural and manuscript artists are directly comparable and if the assumed dates of introduction for materials are correct, revision of the dates proposed for some manuscript may be necessary. For example, copper citrate appears to be used in the earliest manuscripts, late 17th to early 19th century, and then replaced by emerald green from the 1850s. This would suggest either that two HUAM manuscripts, 1984.506 and 1984.507, originally attributed to the mid to late 19th century and the late 19th century but containing copper citrate are either early 19th century or perhaps of more provincial manufacture and hence lack the more modern imported pigments used in the other manuscripts. An early 19th century date is also suggested for these by the presence of the imported Prussian blue rather than indigo (18th century) or synthetic ultramarine (late 19th century) in these manuscripts. A transition in the early 19th century appears to be confirmed but the presence of indigo in the Asian Art Museum manuscript dated to 1791, 1993.11.2, but Prussian blue on the Asian Art Museum manuscript dated to 1844, 2006.27.85. The presence of Prussian blue and emerald green in HUAM manuscripts 1984.510 and 1984.516 may also suggest a slightly earlier date than that originally suggested, mid 19th century rather than late 19th to early 20th century. In contrast, the presence of chromates in the HUAM manuscripts 1984.521 and 1984.524 specifically suggests a late 19th century date.

The mural paintings, banner paintings and manuscripts appear to have used a similar palette, with the exception of the early green pigments. All but one of the pre- and early-19th century manuscripts used copper citrate as the only green, with malachite found in only one late 18th century manuscript. In contrast, the early banner paintings analyzed employed malachite, which agrees with findings of studies on 18th to early 19th mural paintings. In the later 19th century, all painting types employed emerald green and/or mixtures of blue and yellow pigments. This greater uniformity of palette may be due to the dominance of the imported foreign pigments. The restriction of copper citrate to manuscripts is interesting and may be due to a number of factors, such as the instability of the copper citrate phases and the influence of humidity, the use of different binding media which could affect the stability of the copper citrate and/or distinct craft traditions prior to the wide-spread import of foreign pigments.

CONCLUSIONS

A new pigment, copper citrate, was identified for the first time on works of art. This was synthesized using laboratory chemicals and following a medieval recipe with verdigris and lemon juice to provide reference materials. Controlled dehydration of the synthesized phases showed that the color varies from dark green-blue to light blue depending on hydration. None of the forms produced was identical to the bright green material found on the Thai manuscripts, although the most hydrated form was the closest. Further work is required to understand the properties and stability of this phase and possible changes in the phase(s) used on the manuscripts.

A transition was observed in the green, blue and yellow pigments from the late 17th to 20th century: copper citrate to emerald green, indigo to Prussian blue to synthetic ultramarine, and chrome yellow joined, although it did not replace, gamboge. The results indicate a similarity between the materials used in murals, banner paintings and manuscripts, with the notable exception of the dominance of copper citrate for manuscripts but malachite for murals and banner paintings prior to the later 19th century. The reason for this difference is unknown and further investigation is required to understand how the various palettes were selected.

ACKNOWLEDGEMENTS

The authors would like to thank Robert Mowry, Anne Rose Kitigawa and Melissa Moy in the Asian department of the Harvard University Art Museums, Craigen Bowen and Anne Driesse in Straus Center for Conservation, Harvard University Art Museums, Richard Newman, Michelle Derrick, Tanya Uyeda and Joan Wright in the Museum of Fine Arts, Boston, Mark Fenn and Melissa Buschey at the Asian Art Museum, San Francisco, and Guoqiang Yang for providing the copper citrate FTIR data. J.S. thanks the Andrew W. Mellon Foundation for a post-doctoral fellowship.

REFERENCES

1. D.A. Scott, *Copper and Bronze in Art: Corrosion, Colorants, Conservation*, (Getty Publications, Los Angeles, CA.), 287-290 (2002).
2. G. Zhang, G. Yang and J. Shi Ma, *Crystal Growth and Design*, **6**, **2**, 375-381 (2006).
3. J. Giaccai, *Journal of the Walters Art Museum*, **46**, (in press).
4. H. Kuhn and M. Curran, in *Artists Pigments: A Handbook of their History and Characteristics*, **1**, ed. R.L. Feller, (National Gallery of Art, Washington), 187-218 (1986).
5. L. Burgio, R.J.H. Clark and P.J. Gibbs, *Journal of Raman Spectroscopy* **30**, 181-184 (1999).
6. C. Prasartset, *Journal of the National Research Council Thailand*, **22** (1), 73-86 (1990).
7. C. Prasartset, *Preprints ICOM-Committee for Conservation 11th triennial meeting, Edinburgh, Scotland, 1-6 September, 1996, ed. J.* Bridgland, (James & James Science Publishers Ltd), 430-434 (1996).
8. R. Suigisita, *Scientific Papers on Japanese Antiques and Art Crafts*, **28**, 20-25 (1983).
9. D.E. Malloch, *Siam: Some General Remarks on its Productions, and Particularly on its Imports and Exports and the Mode of Transecting Business with the People*, (J. Thomas at the Baptist Mission Press, Calcutta), 34-65 (1852).
10. F.A. Neale, *Residence in Siam*, (Office of the National Illustrated Library, London), (1850).
11. C. Mackay and A.N. Sarkar, in *Scientific Research on the Pictorial Arts of Asia, Proceedings of the Second Forbes Symposium at the Freer Gallery of Art*, ed. P. Jett, J. Winter and B. McCarthy, (Archetype Publications, London), 135-142 (2005).
12. I. Fiedler and M. Bayard, in *Artists Pigments: A Handbook of their History and Characteristics*, **3**, ed. E.W Fitzhugh, (Oxford University Press, Oxford), 219-272 (1997).
13. D. Wise and A. Wise, in *Proceedings of the Fourth International Conference of the Institute of Paper Conservation 6-9 April 1997*, (The Institute of Paper Conservation, London), 125-136 (1998).
14. Yu Feian, *Chinese Painting Colors: Studies of their preparation and application in traditional and modern times*, trans J. Silbergeld and A. McNair, (Washington: University of Washington Press) (1988).
15. C.W. Bowen with A. Snodgrass, in *Gods, Kings and Tigers: The Art of Kotah*, ed. S. C. Welch, (Asia Society Galleries and Harvard University Art Museums), 83-89 (1997).

16. E.W. Fitzhugh, in *An Annotated and Illustrated Checklist of the Vever Collection*, G.D. Lowry and M.C. Beach with R. Marefat and W.M. Thackston, (Arthur M. Sackler Gallery and University of Washington Press), 425-432 (1988).

17. E.W. Fitzhugh, in *Pigments in Later Japanese Paintings*, Freer Gallery of Art Occasional Papers, (Smithsonian, Washington, DC), 1-56 (2003).

18. B.H. Berrie, in *Artists Pigments: A Handbook of their History and Characteristics*, 3, ed. E.W Fitzhugh, (Oxford University Press, Oxford), 91-218 (1997).

19. C. Koninckx, *The First and Second Charters of the Swedish East India Company (1731-1766)*, (Van Ghemmert Publishing Company, Kortrijk, Belgium), (1980).

20. E.H. Pritchard, in *Journal of the Economic and Social History of the Orient*, 1, 1, 108-137 (1957).

21. J. Giaccai and J. Winter, in *Scientific Research on the Pictorial Arts of Asia, Proceedings of the Second Forbes Symposium at the Freer Gallery of Art*, ed. P. Jett, J. Winter and B. McCarthy, (Archetype Publications, London), 99-108 (2005).

22. H. Smith, *Ukiyoe Geijutsu*, 128, 3-26 (1998).

23. M. Leona and J. Winter, in *Pigments in Later Japanese Paintings*, Freer Gallery of Art Occasional Papers, (Smithsonian, Washington, DC), 57-81 (2003).

24. J. Plesters, in *Artists Pigments: A Handbook of their History and Characteristics*, 2, ed. A. Roy, (Oxford University Press, Oxford), 37-66 (1993).

25. R. Vasnatha, in *Scientific Research on the Pictorial Arts of Asia, Proceedings of the Second Forbes Symposium at the Freer Gallery of Art*, ed. P. Jett, J. Winter and B. McCarthy, (Archetype Publications, London), 143-148 (2005).

Mater. Res. Soc. Symp. Proc. Vol. 1047 © 2008 Materials Research Society   1047-Y06-05

Pigment Analysis of Two Thai Banner Paintings

Jennifer Giaccai

Conservation Division, Walters Art Museum, 600 North Charles Street, Baltimore, MD, 21201

ABSTRACT

The pigments used in two Thai banner paintings (*phra bot*) were examined using X-ray fluorescence (XRF), Fourier transform infrared microscopy (FTIR) and polarized light microscopy (PLM). The two paintings examined dated from the late 18th and the late 19th century. The paintings examined follow the trends observed on Thai wall paintings and manuscripts from the same time periods. Pigments identified include vermilion, iron oxide earths, red lead, lead white (hydrocerrusite), calcium carbonate, kaolin, Prussian blue, gamboge, artificial ultramarine, copper citrate and a copper-arsenic green.

INTRODUCTION

Despite interest in the use of pigments in Thai painting on the part of art historians, the few technical examinations of these works have focused on wall paintings and manuscripts. As a part of the recent conservation of Thai wall paintings, the pigments used in the wall paintings were identified.[1,2] One study of four Bangkok period Thai manuscripts held by the British Library has been published, and a study of twelve manuscripts from the 17th to the 20th centuries in the collection of the Harvard University Art Museums has recently been completed.[3,4] In the previous studies of Thai pigments, the most notable pigment differences found were the changes in green, white, and blue pigments. Prasartset observed a shift from malachite to the man-made emerald green pigment over the course of the 19th century. Eremin *et al* found that the most commonly used green pigment in manuscripts before the introduction of emerald green was a copper citrate pigment ($Cu_2(C_6H_4O_7)\cdot(H_2O)_x$). Emerald green (copper acetoarsenite, $3Cu(AsO_2)_2\cdot Cu(CH_3COO)_2$) quickly moved into the palettes of European artists by the mid-19th century and just as quickly was added to Asian palettes.[5,6]

The murals used a clay based white pigment in the early 18th century, and showed the addition of lead white (basic lead carbonate, $2PbCO_3\cdot Pb(OH)_2$) beginning in the 18th century. Although barite mineral is naturally occurring in Thailand, barium sulfate ($BaSO_4$) was only identified in late 19th-century mural paintings.[2] Similar changes in pigment use in manuscripts were described by Burgio *et al.*, who identified barite as well as anhydrite ($CaSO_4$) and gypsum ($CaSO_4\cdot2H_2O$) in an early 20th-century manuscript.[3] Eremin identified a variety of white pigments, including lead white, calcite ($CaCO_3$), kaolin ($Al_2Si_2O_5(OH)_4$), and huntite ($CaMg_3(CO_3)_4$).[4]

All previous studies observed a change in blue pigments, from azurite ($2CuCO_3\cdot Cu(OH)_2$) or indigo ($C_{16}H_{10}N_2O_2$) to Prussian blue ($Fe_7(CN)_{18}(H_2O)_x$) in the early 19th century and finally to ultramarine ($(Na,Ca)_8(AlSiO_4)_6(SO_4,S,Cl)_2$) in the late 19th century. Prussian blue very quickly made its entry into the artistic world; from its discovery around 1704 it was used by artists in Europe only twenty years later,[7] and was traded in Asia by 1775.[8] Prussian blue was found on paintings from the Late Historical Period 2 (1750-1900) in Sri Lanka.[9] It is not surprising that it was also used in Thailand as soon as it was available by trade with Europe; Prussian blue was identified on early 19th century wall paintings and mid-19th century manuscripts.[2,4] Its use continued throughout the 19th century and was ultimately substituted with artificial ultramarine.

The ultramarine found on late 19th century Thai wall paintings and manuscripts is undoubtedly artificial ultramarine, rather than ground lapis lazuli. The Société d'Encouragement pour l'Industrie Nationale awarded the prize for the discovery of a manufacturing method of artificial ultramarine at the beginning of 1828.[10] Artificial ultramarine was soon traded, as both pigment and an optical brightener in laundry bluing, worldwide.[11] It has been observed on Japanese paintings from the mid-19th century,[12] and was found on Thai wall paintings and manuscripts from the late 19th century.[2-4]

Pigments that were consistently used over the 18th and 19th centuries in Thai wall paintings and manuscripts, according to the previous discussed studies, are vermilion (HgS), red lead (Pb_3O_4), iron oxide reds and yellows (browns were either not present or unexamined), gamboge or orpiment (As_2S_3), and carbon-based blacks.

EXPERIMENT

Two Thai banner paintings, each approximately one meter by three meters, were selected for analysis from the Walters Art Museum collection. The paintings were chosen to straddle the 19th century and hopefully observe the shifts in pigment use identified in wall painting and manuscripts over the course of the 19th century. *Ten Birth Tales of the Buddha* (WAM 35.300) uses the traditional Thai painting hues of reds and browns, and is dated to the late 18th century. *Buddha's Descent from Tavatimsa Heaven* (WAM 35.187), Figure 2, is somewhat brighter in color, with vibrant blues and greens, and has an inscription that dates the painting to 1885.

X-ray fluorescence (XRF) was used to identify the elements present in the paint samples. XRF spectra were acquired with a Bruker AXS Artax system with a Rh X-ray tube and poly-capillary focusing lens to give a 70 µm spot size. XRF can be performed non-invasively or on removed samples. *Buddha's Descent* was examined non-invasively with 2-4 spectra obtained from each color, however because of space restrictions, samples were removed from *Ten Birth Tales* for XRF analysis. A sample of each visually distict color was removed from the paintings; a total of 16 samples were removed from *Ten Birth Tales*, 14 samples were removed from *Buddha's Descent*.

Samples from both paintings were examined by Fourier transform infrared microscopy (FTIR) for the presence of organic components. FTIR spectra were acquired with a Bruker Optics Hyperion microscope attached to a Tensor 27 FTIR spectrometer. Microscopic samples were crushed between two diamond windows in the low pressure Micro Compression Cell (manufactured by ThermoNicolet) and then analysed on a single diamond window. The analysis area of FTIR samples varied from 20-50 μm^2 depending on the amount available for each individual sample and if small heterogeneities were being examined. When spectra were dominated by carbonate peaks from calcite or lead white, the interference of the carbonate peak was minimized by treatment of a portion of the sample with HF vapor and reanalyzed in the FTIR microscope.

Select samples were further examined using polarized light microscopy (PLM) to determine the optical properties of the paint components. PLM samples were mounted using Cargille MeltMount (n=1.662) and examined with a Lietz Ortholux II microscope in transmitted, plane-polarized and cross-polarized light. Blue pigment samples were also examined using a Chelsea filter.

DISCUSSION

Ten Birth Tales of the Buddha (WAM 35.300)

Ten Birth Tales (Figure 1) is painted on a woven fabric cloth backing with a calcium carbonate ground. Colorants identified on the eighteenth century *Ten Birth Tales* generally agree with colorants used in mural paintings and manuscripts of the same time period. There are five distinctly different colors used, particularly observed in scene backgrounds: bright red, dark red, purple-brown, brown, and orange. The orange areas are colored by red lead and the bright red areas are colored by vermilion. The dark red, purple-brown and brown areas all contain iron oxides, as identified by XRF and PLM. Both the dark red and purple-brown areas also contain lead peaks in XRF and the characteristic green extinction of red lead observed by PLM. FTIR of the purple-brown paints show that Prussian blue was added to the pigment mixture to give a purple tone.

Figure 1. 35.300 *Ten Birth Tales of the Buddha.* Anonymous, Thai, 1790-1810. 248 x 93 cm. ©The Walters Art Museum, Baltimore.

FTIR of yellow paints clearly identify gamboge. All the blue samples were examined by FTIR and show the characteristic cyanide absorption band indicating Prussian blue. One of the green areas examined was composed of a mixture of Prussian blue and an organic yellow pigment which is most likely gamboge. In the dark green and dark blue areas examined, more Prussian blue is present than in the light areas. In the light blue areas, lead white and kaolinite were also identified by FTIR, in addition to Prussian blue.

A second green paint was also found on this painting. The green paint has a grayish cast and is lighter in color than the areas of green colored with the Prussian blue and gamboge mixture. XRF of the green paint has a strong copper peak, and FTIR shows strong absorption bands (Figure 2). Comparison of the FTIR spectrum obtained from the gray-green pigment with the copper citrate pigment newly discovered by Eremin *et al* on Thai manuscripts showed that the gray-green was copper citrate, in particular the peaks at 1645, 1570, 1440, 1398, 1373 1263, 1232, 1125, 1094/1083 (d) 920, and 858 cm^{-1}.[4,13]

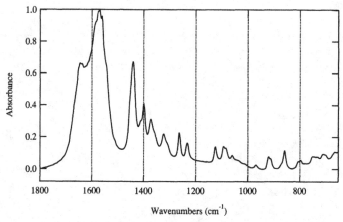

Figure 2. FTIR spectrum of gray-green pigment, copper citrate on *Ten Birth Tales* (WAM 35.300).

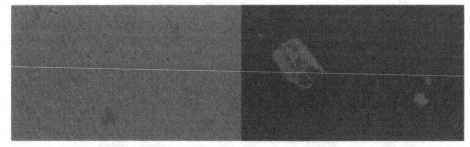

Figure 3. Polarized light microscopy of gray-green pigment, copper citrate, on *Ten Birth Tales* (WAM 35.300). Left—plane-polarized light, right—cross-polarized light.

Figure 4. 35.187 *Buddha's Descent from Tavatimsa Heaven.* Anonymous, Thai, 1885.
345 x 94 cm. ©The Walters Art Museum, Baltimore.

Buddha's Descent from Tavatimsa Heaven (WAM 35.187)

As in the previous painting, *Buddha's Descent* (Figure 4) is painted on a woven fabric cloth. The white ground layer was identified as calcium carbonate. In the white areas of the painting, for example the white arm of a bodhisattva, hydrocerrusite with small amounts of gypsum were identified by FTIR and applied over the white ground layer. There was no calcium carbonate identified in the paint layers.

PLM and XRF identify iron oxides in the brown samples. As in the earlier painting, the bright red pigment used is vermilion. Orange tints are composed of a mixture of red lead and barium sulfate. PLM of the orange-brown of the throne room indicate a mixture of an iron oxide with red lead, and FTIR also shows a small amount of barium sulfate. Barium sulfate was only identified in the orange and orange-brown paints.

In the two locations where yellow paint was sampled, XRF, FTIR and PLM identified the columns in the throne room to consist entirely of gamboge, while the other location, in the border of the painting, contained a mixture of gamboge and an arsenic-based pigment, most likely orpiment. It is possible that the gamboge/arsenic mixture was used more widely, but not observed in the first sampling location because of the small spot size of the XRF instrument. It is also possible that the border was painted separately from the rest of the painting, and using a slightly different pigment mixture.

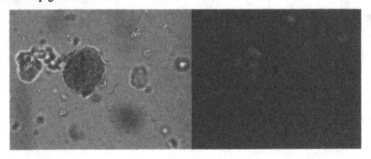

Figure 5. Polarized light microscopy of the bright green pigment showing a spherulite of the copper-arsenic green from *Buddha's Descent* (WAM 35.187). Left—plane-polarized light, right—cross-polarized light.

XRF of all the green areas examined contained both copper and arsenic peaks. PLM of the green pigment showed a mixture of pale blue-green spherulites and an organic blue dye. There was no presence of a yellow component (Figure 5). The FTIR spectrum from the copper-arsenic pigment has a peak due to Prussian blue and peaks from the gum binder, however there were no peaks at 1557 or 1454 cm^{-1} indicating the presence of emerald green, copper-aceto-arsenite $Cu(C_2H_3O_2)_2 \cdot 3Cu(AsO_2)_2)$. The bands at 875, 839 and 799 cm^{-1} may be indicative of Scheele's green or another copper arsenite (Figure 6).[5] The copper-arsenic green was always found in combination with Prussian blue, present as small blue particles mixed among the green pigment particles. In the dark green areas a thin layer of what appears to be a carbon-based black was painted over the bright green paint layer.

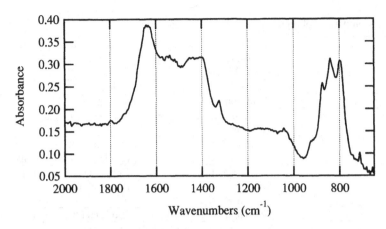

Figure 6. FTIR spectrum of copper-arsenic green on *Buddha's Descent* (WAM 35.187).

PLM of the gray areas also contained small amounts of Prussian blue mixed with a carbon-based black pigment as well as an iron-oxide earth and lead white. While both the green and gray areas had small amounts of Prussian blue in the paint, the samples taken from the light gray-blue areas and the blue sky did not contain any Prussian blue. FTIR of the blue sky and roof tiles show that the ultramarine that has been extended or mixed with gypsum. PLM of the blue sky showed the characteristic uniform small, rounded particles of artificial ultramarine.[10] PLM of the light gray-blue paint also confirmed the presence of artificial ultramarine.

CONCLUSIONS

The pigments identified on the two banner paintings investigated follow the general trends, discussed earlier, as observed by Prasartset on wall paintings and Burgio and Eremin in manuscripts. [1-4] A comparison table of pigments can be found in Table I. The red, brown, orange and yellow pigments remain generally unchanged over the two paintings, and agree with the pigments found during previous studies. Vermilion, red lead, gamboge and a variety of iron oxide earths were used alone and in combination on both paintings. Both paintings had a calcium carbonate ground on the cloth banner under the paint layer.

Although an artificial pigment, Prussian blue is observed on the late 18th century *Ten Birth Tales*, the use of brighter man-made pigments increased over the course of the 19th century, as is witnessed by the pigments identified on the late 19th century *Buddha's Descent*. The most notable changes in pigment use in the two banner paintings examined occur in the blue and green pigments.

Prussian blue is found on *Ten Birth Tales*, both as a blue pigment and in mixtures to create green paints. Prussian blue continued to be used in *Buddha's Descent* as a component of pigment mixtures found in the gray and green areas of the painting. However, the most prominent blue pigment on *Buddha's Descent* is artificial ultramarine. Ultramarine was used for the primary pigment in the sky and roof tiles, and as part of the mixture to create a gray-blue color.

Table I. Comparison of pigments found on the banner paintings.

	Ten Birth Tales of the Buddha (late 18th century) WAM 35.300	Buddha's Descent from Tavatimsa Heaven (1885) WAM 35.187
Brown	Iron oxide	Iron oxide
Bright red	Vermilion	Vermilion
Orange	Red lead (with some iron oxide earth)	Red lead (with barium sulfate)
Yellow	Gamboge	(1) Gamboge (2) Gamboge and orpiment
Green	(1) Prussian blue + gamboge (2) Copper citrate	Copper-arsenic green + Prussian blue
Blue	Prussian blue	Artificial ultramarine
White	Kaolin/clay (& some lead white)	Lead white
Ground	Calcium carbonate	Calcium carbonate

The white pigment on the paintings changed from a clay-based white, occasionally mixed with lead white, in the late 18th-century painting to a mostly or entirely lead white paint in the late 19th-century painting. Although barium sulfate was found in orange areas mixed with red lead on *Buddha's Descent,* it was not used as a white pigment.

Malachite has been found in Thai mural paintings in the past, but was not found on either of these paintings. *Ten Birth Tales* contained two green pigments, one was a mixture of Prussian blue and gamboge, the other copper citrate. Copper citrates are not a common paint pigment but have been identified on Thai manuscripts from this time period.[4] The previous studies on wall paintings and manuscripts mentioned above have identified the introduction of emerald green in the 19th century. Although the green pigment found on the late 19th-century painting is not true emerald green, copper aceto-arsenate, it is a related pigment, either Scheele's green or another undefined copper-arsenic green.

ACKNOWLEDGMENTS

The authors would like to thank the Andrew W. Mellon Foundation and the Stockman Family Foundation Trust.

REFERENCES

1. Chompunut Prasartset, Journal of the National Research Council of Thailand **22** (1), 73 (1990).
2. Chompunut Prasartset, "Materials and Techniques of Thai Wall Paintings: A Comparative Study of Late 19th Century Murals and Early-period Murals", in *11th Triennial Meeting, Edinburgh, Scotland, 1-6 September, 1996: Preprints (ICOM Committee for Conservation)*, edited by Janet Bridgland (James & James (Science Publishers) Ltd., London, 1996), pp. 430.

3. Lucia Burgio, Robin J. H. Clark, and Peter J. Gibbs, Journal of Raman Spectroscopy **30**, 181 (1999).

4. Katherine Eremin, Jens Stenger, Jo Fan Huang, Alan Aspuru-Guzik, Theodore Betley, Leslie Vogt, Ivan Kassal, Scott Speakman, and Narayan Khandekar, Journal of Raman Spectroscopy, *in press* (2007).

5. Inge Fiedler and Michael A. Bayard, "Emerald green and Scheele's green", in *Artists' Pigments: A Handbook of Their History and Characteristics. Volume 3*, edited by Elisabeth West FitzHugh (National Gallery of Art, Washington, DC, 1997), pp. 219.

6. David Wise and Andrea Wise, "Observations on Nineteenth-Century Chinese Pigments with Special Reference to Copper Greens", in *IPC Conference Papers London 1997*, edited by Jane Eagan (Institute of Paper Conservation, London, 1997), pp. 125.

7. Barbara H. Berrie, "Prussian Blue", in *Artists' Pigments: A Handbook of Their History and Characteristics*, edited by Elisabeth West FitzHugh (National Gallery of Art, Washington DC, 1997), Vol. III, pp. 191.

8. Earl H. Pritchard, Journal of the Economic and Social History of the Orient **1** (1), 108 (1957).

9. B. D. Nandadeva, "Recent Studies on Sri Lankan Mural Painting Technology", in *Scientific Research in the Field of Asian Art: Proceedings of the First Forbes Symposium at the Freer Gallery of Art*, edited by Paul Jett, Janet G. Douglas, Blythe McCarthy et al. (Archetype Publications Ltd., London, 2003), pp. 170.

10. Joyce Plesters, "Ultramarine Blue, Natural and Artificial", in *Artists' Pigments: A Handbook of Their History and Characteristics*, edited by Ashok Roy (National Gallery of Art, Washington DC, 1993), Vol. II, pp. 37.

11. Gerry Barton and Sabine Weik, Restauro: Zeitschrift für Kunsttechniken, Restaurierung und Museumsfragen **104** (5), 320 (1998); Nancy Odegaard and Matthew Crawford, "Laundry bluing as a colorant in ethnographic objects", in *11th triennial meeting, Edinburgh, Scotland, 1-6 September, 1996: preprints (ICOM Committee for Conservation)*, edited by Janet Bridgland (James & James (Science Publishers) Ltd., London, 1996), pp. 634.

12. Elisabeth West FitzHugh, Ars Orientalis **11**, 27 (1979).

13. Katherine Eremin (private communication).

Mater. Res. Soc. Symp. Proc. Vol. 1047 © 2008 Materials Research Society

Painted Decoration Studies in a Fourth Century BC Vergina Tomb

E. Pavlidou[1], A. Kyriakou[2], E. Mirtsou[3], L. Anastasiou[1], T. Zorba[1], E. Hatzikraniotis[1], and K. M. Paraskevopoulos[1]

[1]Physics, Aristotle University of Thessaloniki, Thessaloniki, 54124, Greece
[2]Faculty of History and Archaeology, Aristotle University of Thessaloniki, Thessaloniki, 54124, Greece
[3]Archeological Museum of Thessaloniki, Thessaloniki, Greece

ABSTRACT

Aegae, the first capital of the Macedonians, in Northern Greece, is being excavated since 1938. The most impressive findings come from the unlooted tombs of the Great Tumulus, where the grave of Philip II, father of Alexander the Great, was discovered. Not far from the Great Tumulus, in the "Tumuli cemetery", the most ancient part of the graveyard (1000-700 B.C.), recent excavations brought to light three looted graves dated in the mid-fourth century B.C., with very interesting finds such as weapons, gilded wreaths, pieces of jewelry, remains of decoration of wooden furniture, ceramic vases broken in small pieces and wall paintings. This paper describes studies carried out on the binding and the painting materials used for the decoration of the above wall paintings and ceramic vases. The characterization was performed through Optical Microscopy, Fourier Transform Infrared Spectroscopy (FTIR) and Scanning Microscopy (SEM-EDS). It was found that the fresco technique was used, while all the pigments were identified. The results are discussed and related with other findings in that period in the Greek area.

INTRODUCTION

In Northern Greece, not far from the city of Veroia, lies Vergina, a modern village, in the fields where the first capital of the Macedonians, Aegae is being excavated since 1938. Although the capital was transferred to Pella in the beginning of the fourth century B.C, Aegae was still the royal necropolis of the Macedonians. The most impressive finds come from the unlooted tombs of the Great Tumulus, where the grave of Philip II, father of Alexander the Great, was discovered.

Not far from the Great Tumulus, in the "Tumuli cemetery", the most ancient part of the graveyard (1000-700 B.C.) lies the second largest burial mound after the one with the royal tombs of the Great Tumulus. The excavation brought to light three looted graves with very interesting finds, since the looters took no interest in the iron weapons and the remains of the cremation pyres. These were carefully studied and revealed many elements concerning the social status of the deceased as well as aspects of the burial customs.

Two cist-graves enclosed with stone slabs bear painted decoration inside: a myrtle branch and white garland and red ribbons hanging from blue nails depicted with perspective in one grave, and a simple red zone in the other. In these two graves were deposed the bones of the dead after cremation in addition to weapons, gilded wreaths, pieces of jewelry, remains of decorated wooden furniture and ceramic vases broken in small pieces. The vases, a local, rather unique production, bear polychrome floral and geometric decoration with blue and red colors on a white fond. The grave in the center of the tumulus is a simple pit-grave, though large, that contained a rich burial as suggested by the quality of iron implements. The tombs are decorated with wall paintings with white, red and blue colors. This study presents the results from pigment and binder analysis and the technique used for the creation of the wall paintings and ceramic vases.

EXPERIMENT

A scientific investigation on a representative number of microsamples was carried out. Two types of the samples were studied: fragments from the wall paintings of the tombs with dimensions about 1-2 mm with white, red and blue color on the surface and broken pieces from ceramics vases with small painted areas colored, also, with white, red and blue (Figure 1). Their characterization was performed through the combined use of Fourier Transform Infrared Spectroscopy (FTIR) and Scanning Electron Microscopy (SEM-EDS).

(a)   (b)

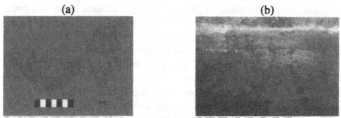

Figure 1. Representative photos from (a) parts of ceramic vase and (b) wall-painting

FTIR spectra were obtained using a Perkin-Elmer FTIR spectrometer, model Spectrum1000 connected with an i-series FTIR microscope, equipped with a nitrogen-cooled MCT detector. Grains from the painted surfaces of the samples and from the substrate of the wall-paintings were removed using a microscalpel under a microscope and placed on a freshly prepared KBr pellet. The IR transmittance spectra, were obtained from different areas of the specimens using a working aperture 40-100 μm. Compositional properties of the samples were obtained with energy dispersive x-ray analysis in a Jeol 840A scanning electron microscope (SEM-EDS).

RESULTS AND DISCUSSION

The wall-paintings

FTIR analysis revealed characteristic $CaCO_3$ peaks at 1407, 872 and 712 cm^{-1} for all colored surface samples from the wall-paintings, attributed to asymmetric stretch, out-of-plane bend and in-plane bend modes respectively of the CO_3 group of $CaCO_3$. The presence of Ca was confirmed by EDS analysis, and was ubiquitous throughout the samples. These analyses indicate that the wall paintings are made with fresco technique where figures are painted onto a layer of probably wet, lime mortar or plaster.

The pigments
Blue pigments:

The EDS and FTIR spectra collected from the blue colored surface of the wall-paintings are presented in Figures 2 and 3, respectively. Peak frequencies at 1280 and 1000 cm^{-1} are attributed to Si-O-Si stretching vibrations while subsequent peaks can be attributed to calcite vibrational modes, already, mentioned. The special technical feature of the mixture of pigment with the plaster of the wall, is a strong indication of the use of wet fresco as the painting technique. Comparison of FTIR spectra with spectral library and literature [3, 4] suggest the presence of Egyptian blue (CaCuSi$_4$O$_{10}$). The elemental analysis obtained by EDS, confirm the presence of Egyptian blue by the detection of Ca, Cu, and Si. This pigment is a particularly suitable pigment for use with all kinds of supports and techniques [5].

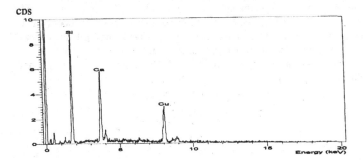

Figure 2. EDS spectra obtained from the blue specimen of wall painting

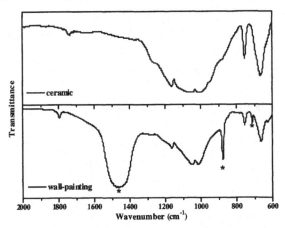

Figure 3. FTIR spectra obtained from the blue specimens of wall-painting and ceramic vase (* indicates the peaks of calcite)

In the Figure 3, a representative FTIR spectrum from the blue grains obtained from the painted area of the ceramic vases, is also presented. As it can be seen the spectrum is almost identical with that from wall-paintings and only the peaks of calcite are absent. In accordance with the EDS results the blue pigment is identified as Egyptian blue, as well.

Egyptian blue is the first synthetic pigment manufactured by fusing sand and other materials together with copper compounds used as coloring agents. The invention of the manufacturing process goes back to the first dynasties of ancient Egypt about 3100 B.C. Egyptian blue –brought in Greece by trade- was an important blue pigment in Greek area, used from the first half of the third millennium B.C. until the Roman period [6, 7].

White pigments:

The main elements that are detected from EDS measurements for the both white colored surfaces, in wall-painting and ceramic vases, are Ca, Si, Mg and Al, with Ca in quite much lower amounts in ceramics. In the EDS spectra the Ca peak is dominant (Figure 4). In the FTIR spectra (Figure 5) beyond the calcite peaks, the presence of a broad band at 1200-900cm^{-1} and the weak peaks at 3700-3600 cm^{-1} are indicative for the existence of an aluminosilicate material probably kaolinite [8]. Additionally the presence of Mg in great amounts suggests the use of a magnesium silicate material [9] -such as talc- as well, but this is difficult to be confirmed from FTIR measurements due to overlapping peaks.

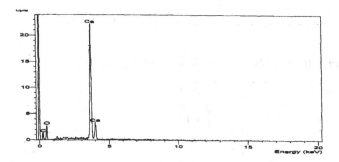

Figure 4. EDS spectra obtained from the white specimen of wall painting

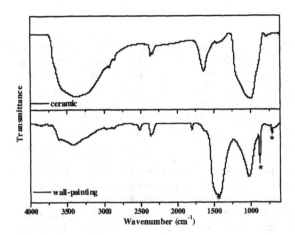

Figure 5. FTIR spectra obtained from the white specimens of ceramic vase and wall-painting

Red pigments:
Elemental analysis on the red coloured surfaces identified Fe along with Al and Si, indicating the existence of iron oxide as the possible material producing the red colour. The presence of alumino-silicates is supported with the FTIR spectra(Figure 6), where the peaks at 1150-950 cm^{-1} are attributed to Si-O-Si and Si-O-Al stretching vibrations. Also, in this case, the calcite is present only in the spectra of wall-paintings. The coexistence of the alumino-silicate materials with the iron oxide, leads to the conclusion that the pigment is red ochre.

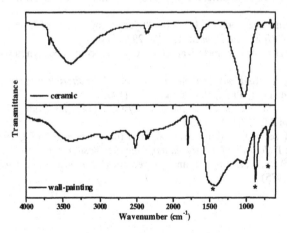

Figure 6. FTIR spectra obtained from the red specimens of wall-painting and ceramic vase

CONCLUSIONS

The tombs that are dated in the mid-fourth century B.C., a period during which Macedon under the reign of Philip II takes up a leading role in the history of Greece, were paint decorated. All the finds and details of the funeral procedure indicate that the occupants of the tombs were high-ranking army officials belonging to the upper class of the Macedonian society in this thrilling period. From our first results, it can be concluded for the tombs decorated with wall paintings, that the technique used was the fresco technique, probably wet. The pigments that were used for the creation of the wall paintings are common in that period in the Greek area, and it must be noticed the use of the imported Egyptian blue pigment that continued to be the source of blue coloring material during that period. The ceramic vases were also decorated with the same type of pigments, with a small difference for the white colored surfaces in which probably talc is used, as well. It is remarkable the limited number of the pigments used, expressing a simplicity, despite the high social status of the deceased. The work, that is under continuation, is a part of a bigger project studying the wall paintings in Balkan peninsula from the prehistoric period till post Byzantine, in order to find similarities in the materials and the techniques used.

REFERENCES

1. E. Pavlidou, M. Arapi, T. Zorba, M. Anastasiou, N. Civici, F. Stamati, K.M. Paraskevopoulos, "Onoufrios, the famous XVI's century iconographer, creator of the "Berati School": Studying the technique and materials used in wall paintings of inscribed churches" *Appl. Phys. A*, **83**, 709 (2006).
2. W.S. Taft, "Fresco to Oils" *MRS Bulletin*, **21**, 18 (1996).
3. S. Bruni, F. Cariati, F. Casadio, L. Toniolo, "Spectrochemical characterization by micro-FTIR spectroscopy of blue pigments in different polychrome works of art" *Vibrational Spectroscopy* **20**, 15 (1999).
4. M.C. Edreira, M.J. Feliu, C. Fernández-Lorenzo, J. Martín, "Spectroscopic study of Egyptian blue mixed with other pigments" *Helvetica Chimica Acta* **86**, 29 (2003).
5. D. Ulrich, "Egyptian Blue and green frit characterization, history and occurrence, synthesis" *PACT* **17**, 323 (1987).
6. Riederer, J., *Artists' Pigments, A Handbook of their History and Characteristics* E. West Fitzhugh (Ed), vol.3, Oxford University Press, New York, 23-45 (1993).
7. Calamiotou, M., Siganidou, M., Filippakis, S.E., "X-ray analysis of pigments from Pella, Greece" *Studies in Conservation* **28**, 117-121 (1983).
8. R. Newman, "Analysis of red paint and filling material from the sarcophagus of Queen Hatshepsut and King Thutmose I" *Journal of the Museum of Fine Arts Boston*, **5**, 62 (1993).
9. W. Noll, R. Holm, and L. Born, "Painting of ancient ceramics" *Angewandte Chemie International Edition* **14**, 602 (1975).

Mater. Res. Soc. Symp. Proc. Vol. 1047 © 2008 Materials Research Society   1047-Y03-05

Study of Painting Materials and Techniques in the 18th Century St. Athanasius Church in Moschopolis, Albania

E. Pavlidou[1], N. Civici[2], E. Caushi[3], L. Anastasiou[1], T. Zorba[1], E. Hatzikraniotis[1], and K. M. Paraskevopoulos[1]

[1]Physics, Aristotle University of Thessaloniki, Thessaloniki, 54124, Greece
[2]Institute of Nuclear Physics, Tirana, Albania
[3]Institute of Cultural Monuments, Tirana, Albania

ABSTRACT

In this paper is presented the study of the painting materials and techniques used in 18th century wall paintings that originated in the orthodox church of St Athanasius, in the city of Moschopolis, Albania. The church was painted in 1745 by Konstantinos and Athanasios Zografi, and during recent years, conservation activities has been performed at the church. A total number of 8 samples that include plasters and pigments of different colors were collected from important points of the wall paintings. Additionally, as some parts of the wall-paintings were over-painted, the analysis was extended to the compositional characterization of these areas. The identification of the used materials was done by using complementary analytical methods such as Optical Microscopy, Fourier Transform Infrared spectroscopy (FTIR), Scanning Electron Microscopy (SEM-EDS) and Total Reflective X-Ray Fluorescence (TXRF).

Common pigments used in Balkan peninsula at 15-16th centuries, such as cinnabar, green earth, manganese oxide, carbon black and calcite were identified. The presence of calcite in almost all the pigments is indicative for the use of the fresco technique however, some areas were identified as containing egg tempera. The detection of gypsum and calcium oxalate indicates environmental degradation and biodeterioration.

INTRODUCTION

Historical Background

Albania is a country with long history in Byzantine art, history that began at the foundation of small churches decorated with plants, animals and went on until the end of the ninetieth century, many centuries after the fall of the Byzantine Empire and the conquest of the country by the Ottomans. During this long period, many churches were built mainly at the cities and villages of central and southern Albania. The interior of many of these churches has been decorated with wall paintings of exceptional quality that were the work of famous – not only in the albanian area – artists like Onoufrios, Nikolaos, Onoufrios Cypriot, the brothers Konstantinos and Athanasios Zografi and David Selenicasi. Besides the orthodox tradition, all these painters followed their artistic inclination in order to improve both their art and the technology of painting.

The eighteenth century was for Albania a period of strong economical activity at the cities. The merchants, who were the main sponsors in the domain of art, could no longer be satisfied with the former typical Byzantine reverence. The traditional content of wall-paintings underwent a new formation, through the creative activity of Konstantin Shpataraku, the brothers Konstantinos and Athanasios Zografi –that have also painted several churches in Mount Athos– in the period 1740-1764 and David Selenicasi. Their creative work was not accidentally developed at Moschopolis. Although, as a village, existed from the tenth or eleventh century, it is a fact that at the eighteenth century Moschopolis was a flourishing economical and cultural center that maintained commercial relations with Leibzig, Vienne, Budapest and other important European cities. Moschopolis had its own academy, a very rich library and more that 20 luxurious churches (basilicas). This city of 12000 houses and 60000 inhabitants was burnt by the Ottomans at the end of eighteenth cent, and was also subjected to extended damage during the First and Second World Wars. Nowadays, from this culture, remain 7 churches.

The church of St Athanasius was painted in 1745 by Konstantinos and Athanasios Zografi, as it is testified by an inscription, to the side porch of the church and most scholars attribute, also, to them the wall-paintings of the whole interior. During the last years restoration activities supported by an international project are being performed in the church.

As part of the restoration process, the identification of the original pigments used for the preparation of the wall paintings of St Athanasius church was requested and moreover, as some parts of the wall-paintings were over-painted, the analysis was extended to the compositional characterization of these areas.

EXPERIMENTAL

At first, a general overview of the used pigments was obtained by in-situ measurements performed at several areas with different colours using a field portable X-ray fluorescence (EDXRF) system. This allowed us to select carefully some important points at which representative samples were collected for further and more detailed examinations using Optical Microscopy, Fourier Transform Infrared Spectroscopy (FTIR), Scanning Electron Microscopy - Energy Dispersive Spectroscopy (SEM-EDS) and Total Reflective X-Ray Fluorescence TXRF. A total number of 8 characteristic samples that include plasters and pigments of different colors, with dimensions from about 4 to 10mm^2, were collected from different areas of the wall paintings (Figure 1a, b).

The TXRF analysis system is composed of a total reflection module (Vienna Atominstitute) attached to a tube excitation system (Philips PW 1729 X-ray generator and PW 2215/20 Mo anode X-ray tube) and an X-ray spectrometer. FTIR spectra were obtained using a Perkin-Elmer FTIR spectrometer (Spectrum1000), connected with an FTIR microscope (i-series), equipped with a nitrogen-cooled MCT detector. Tiny species from the painted layers or from the substrate of the samples were removed with a sharp tip of a micro-scalpel using a microscope and placed on a freshly prepared KBr pellet. IR spectra, in transmittance mode, were received from different areas of the specimens with an aperture 50-100 μm in the MIR spectral region. A database of FTIR spectra of reference inorganic and organic materials created in our laboratory was used to characterize the unknown materials (pigments and binders) of the samples. The same sample specimens were coated with carbon, using a Jeol JEE-4X vacuum evaporator and were studied by EDS analysis, using a Jeol 840A Scanning Microscope with an Energy Dispersive Spectrometer attached by Oxford, model ISIS 300.

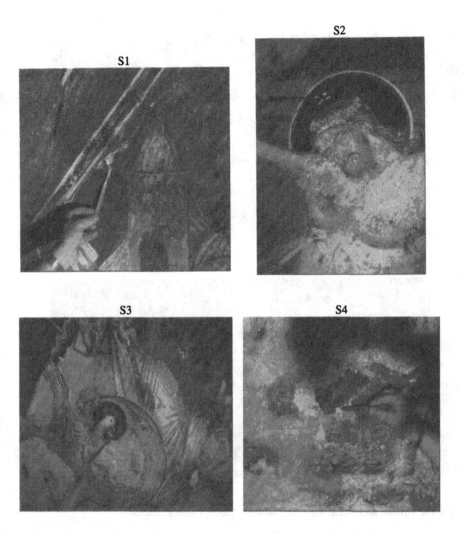

S1

S2

S3

S4

Figure 1a. Painted areas from where the samples (S1-S4) were extracted

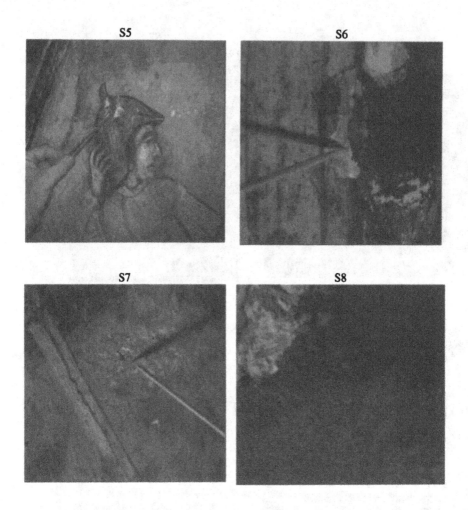

Figure 1b. Painted areas from where the samples (S5-S8) were extracted

RESULTS AND DISCUSSION

Plasters: From the combination of FTIR measurements (Figure 2) and EDS analysis (Figure3) of the plasters it is confirmed the presence of calcite [1] with the additional presence of traces of silicate material (probably sand). In some places gypsum and calcium oxalate were detected by FTIR spectra, indication of an environmental degradation along with a biodegradation [2, 3].

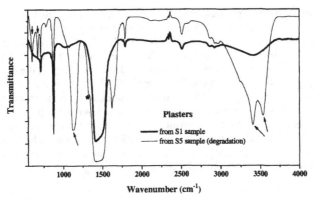

Figure 2. Representative spectra from the plaster of the samples (↓ gypsum, * calcium oxalate)

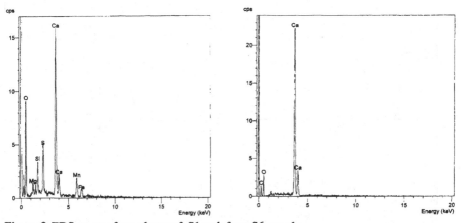

Figure 3. EDS spectra from plaster of S1 and from S6 sample

Sample S1: The analysis with TXRF (Figure 4) and EDS gives Mn and Ca as essential elements along with traces of S, Al and Si, while the presence of Hg, As, Co in TXRF measurements must be attributed to the small quantities of blue and red pigment (Figure 1a). From the FTIR spectrum calcite and gypsum [1] with small participation of quartz (1000-1100cm^{-1}), could be identified. Therefore, from the above results could be extracted the use of MnO$_2$ as black pigment. The presence of MnO$_2$ could not be identified by FTIR because it presents its characteristic broad peak below 550cm^{-1}, which is the lower limit for the MCT detector.

Samples S2 and S3: For the samples S2 and S3, which are extracted from the halo of Jesus Christ and Holy Spirit (Figure 1), it can be seen that the gold painted area is covered inartistically, with black. A detailed observation, with an optical microscope, reveals the presence of three layers; a white-yellow layer above the plaster, a gold layer and finally a black layer on the surface. From the EDS analysis of the white-yellow layer are detected C as major element and traces of Ca, S and Si, while in the FTIR spectra (Figure 5) except of $CaCO_3$ and few $CaSO_4$ is detected –from the peaks at ~1730 and at 3100-2800cm^{-1}- the presence of an organic material. The gold layer consists of pure Au as it is confirmed from EDS and TXRF measurements. Finally, for the black layer, the elemental analysis suggests the existence of C mainly, with small quantities of S, Ca and Pb. From the FTIR spectra it was possible to have only indications of calcium oxalate and Pb white. Consequently, from the above results we can estimate that $CaCO_3$ was used as white pigment at the lower layer and organic medium as adhesive for the gold leafs. As for the overpainted layer, we can consider that carbon black is used, along with Pb-white part of which is turned to PbS due to the atmospheric conditions [4]. The absence of any organic material in the FTIR spectra, along with the bad adhesion of this layer, impose questions about the composition and the reasoning of this layer.

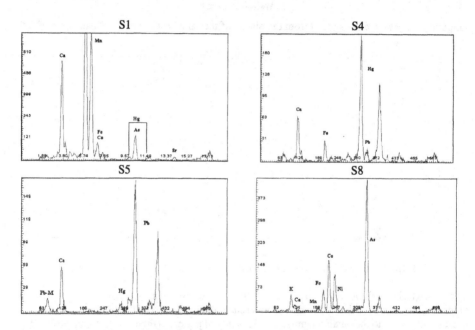

Figure 4. TXRF spectra from S1, S4, S5 and S8 samples

Sample S4: This sample presented a light red surface and a very decayed plaster beneath. From the TXRF (Figure 4) and EDS measurements it is evident that the red pigment is cinnabar (HgS) due to the detection of Hg in great amount, while the presence of Ca and S is identified by FTIR as calcite and gypsum. Moreover, in the FTIR spectra the peak at ~1320cm^{-1} [3] is attributed to calcium oxalate.

Sample S5: The FTIR spectra obtained from the red, grey and white grains of this sample present mainly the characteristic peaks of calcite along with gypsum and calcium oxalate without any other characteristic peaks attributed to red or black color. On the other hand the EDS and TXRF (Figure 4) analysis detected Hg, Ca, S and Pb. The combination of the two methods leads us to consider that the used red pigment is cinnabar, while the black is carbon black. Moreover, the presence of Pb is an indication of the use of lead white although that, in the FTIR spectrum, it isn't so clearly evident -due to the very strong peaks of the calcite.

Sample S6: The EDS analysis of the yellow layer showed mainly, the presence of Pb, Sb, and Ca (Figure 3). This suggests the presence of lead antimonite as yellow pigment. The FTIR spectra (Figure 6) confirm the use of lead antimonite from the strong peak at about 700cm^{-1} [5]. In the same spectrum we identify also, calcite and an organic material, probably egg, from the peaks at 3010-2900, 1734cm^{-1} [6]. That suggests the use of tempera technique for this painted area. Finally, for the black layer only carbon black is extracted from EDS measurements.

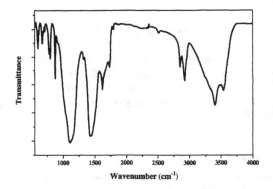

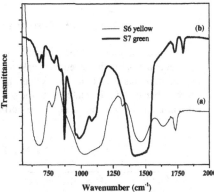

Figure 5. FTIR spectrum obtained from the white-yellow layer of the sample S2.

Figure 6. FTIR spectra obtained from (a) the yellow layer of the sample S6 and (b) the green layer of the sample S7

Sample S7: The detection of Si, Mg and Al by EDS and TXRF (Figure 4) analysis suggests the use of green earth as green pigment. Additionally, the FTIR spectra (Figure 6) confirm the presence of green earth and in particular form of celadonite [7] along with an organic medium probably egg [6].

Sample S8: The area from where the sample S8 is extracted is blue partially covered with black. Its observation, by the optical microscope, showed aggregates of small blue crystals with a black cover at the external surface. The elemental analysis detected Si (42-30%), Mg (1-0.4%), Co (38-7%), As (0.5-2%), Ca (3.5-0.4%) and the FTIR spectrum (Figure 7) presents a broad band at 1200-1000cm^{-1} attributed to glass silicate matrix [8].

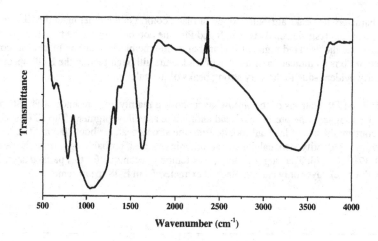

Figure 7. FTIR spectra obtained from the blue of the sample S8

With the combination of these results we may safely conclude that the blue pigment is smalt, but it is unexplainable the absence of calcite or an organic binder, although many FTIR spectra from different points were obtained. Regarding the identification of blackish area we suppose that is carbon black and it is related with the soot of the cantles, oil lamps or any other kind of illumination by organic combustible burning [9].

Table 1. List of analyzed pigments

Sample code	Description	Identified Pigments
S1	Black	MnO_2, calcite
S2,S3	White/Gold /Black	Calcite, Au, carbon black
S4	Red	Cinnabar, calcite, gypsum, oxalate
S5	Grey, Pink	Carbon black, calcite, lead white, cinnabar
S6	Yellow/Black	Lead antimonite, carbon black
S7	Green	Green earth, calcite
S8	Blue	Smalt

CONCLUSIONS

According to our previous works [10, 11] in this area of Balkan peninsula, the majority of pigments are common, as green earth, cinnabar and carbon black, while the use of manganese oxide as black pigment and lead antimonite are observed for the first time in the present study. The fresco technique is used for the most of the samples we studied, but there are also painted areas, yellow and green, where the tempera technique is used. The deterioration, evidenced by gypsum and calcium oxalate, is limited in comparison to other older churches. The over-painting attributed to early restoration attempts or human-caused destruction or deterioration is obvious, in certain areas, but historical questions of why and how cannot be answered clearly by differences in painting materials or techniques.

REFERENCES

1. Brian Smith, *Infrared Spectra Interpretation, A Systematic Approach* (CRC, 1999) pp.165-174.
2. R. Van Grieken, F. Delalieux, and K. Gysels, *Pure & Appl. Chem.* **70**, 2327 (1998)
3. L. Dei, M. Mauro, and G. Bittossi, Thermechimica Acta **317**, 133 (1998)
4. Roy Ashok, *Artists Pigments: A Handbook of Their History and Characteristics,* vol.2, (Oxford University, 1993) pp.67-77
5. R. Feller, *Artists Pigments: A Handbook of Their History and Characteristics,* vol.1, (Oxford University, 1993) pp.242-243
6. M. Fabbri, M. Picollo, S. Porcinai, and M. Bacci, *Applied Spectroscopy* **55**, 428 (2001)
7. T. Zorba, K.M. Paraskevopoulos, D.I. Siapkas, E. Pavlidou, S. Angelova, D.B. Kushev, in *Materials Issues in Art and Archaeology VI* edited by P. Vandiver, M. Goodway, J. Mass (Mat. Res. Symp. Proc. **712**, 2002) pp. II10.10.1-8.
8. V. Ganitis, E. Pavlidou, F. Zorba, K.M. Paraskevopoulos and D. Bikiaris, *J. Cul. Heritage* **5**, 349 (2004)
9. M. Pérez-Alonso, K. Castro, I. Martinez-Arkarazo, M. Angulo, M.A. Olazabal, J.M. Madariaga *Anal. Bional. Chem.*, **379**, 42 (2004)
10. E. Pavlidou, M. Arapi, T. Zorba, M. Anastasiou, N. Civici, F. Stamati, K.M. Paraskevopoulos, *Appl. Phys. A*, **83**, 709 (2006).
11. N. Civici, K.M. Paraskevopoulos, T. Zorba, T. Dilo, F. Stamati, M. Anastasiou, M. Arapi in *Non-Destructive Testing and Microanalysis for the Diagnostics and Conservation of the Cultural and Environmental Heritage,* edited by C. Parisi, G. Burrancca, A. Paradisi (ART05, Lecce, Italy, 2005), pp. B-165.

Mater. Res. Soc. Symp. Proc. Vol. 1047 © 2008 Materials Research Society

A Ceramic Plaque Representing a Part of the Moses Panel by Lorenzo Ghiberti in the East Baptistery Doors (Florence, Italy)

Pamela B. Vandiver
Dept. of Materials Science and Engineering, University of Arizona, Program in Heritage Conservation Science, Tucson, AZ, 85716

ABSTRACT

A ceramic plaque was studied that depicts the figurative part of the lower half of the Moses Panel from the gilt bronze doors that Lorenzo Ghiberti and his workshop installed on the east side of the San Giovanni Baptistery in Florence, Italy. The doors were completed in 1452, and thermoluminescence dating of two areas of the ceramic relief panel gave a broad, but consistent fifteenth century date. No differences were found in the composition, microstructure or phase assemblage of the two stylistically distinct parts of the ceramic panel. Microscopy and radiography were used to reconstruct the forming methods and sequence of steps in manufacture and restoration.

INTRODUCTION

Lorenzo Ghiberti (ca 1379-1455 CE), the Renaissance Florentine sculptor, is best known for his two sets of bronze doors for the Baptistery of San Giovanni in Florence. The first set on the north side of the Baptistery was sculpted and cast between 1404 and 1424, and the east doors facing the Duomo were contracted in 1425 and cast, chased, gilded, mounted and installed in the period from 1437 to 1452. Since Vasari's time the latter doors have been called the "Gates of Paradise," a title ascribed by legend to Michelangelo, and it is this appellation by which they have been known to the present day [1].

We have only the most vague idea of how Ghiberti worked, and this is based solely on what he himself wrote [2]. In his autobiography written during the winter of 1447-1448, he mentioned that models were made of wax and clay as preparations for his castings of sculpture. Because the models were studies made of base materials and not meant to be permanent or art objects, they presumably were destroyed. In the opinion of some scholars, no wax or terracotta models made by Ghiberti or members of his workshop have survived.

One work that has challenged this opinion is a little known terracotta sketch of a part of the panel depicting Moses on Mount Sinai. The bronze panel shows the laws given by the Lord to Moses on the burning mountain, as trumpets sound and as onlookers view the sight above them (Fig. 1). The ceramic plaque was purchased by Allan Marquand in Siena during 1892 and is now housed in the Princeton Art Museum. Marquand justified the authenticity of the sketch in a short article published in 1894 [3], citing stylistic similarities as the basis of his argument. For instance, he supported his view by stating that the construction of the different, individual characters, built in relief by adding layers that increase in number and thickness from the base and feet to the heads cross section near the base and the presence of a basal flange, is similar. Marquand also described the "individualizing" of the figures as a Ghiberti trait. However, the ceramic plaque depicts only the lower middle group of figures in the Moses panel, including a soldier with his back turned toward the audience, a bearded man to his and our proper left and a woman holding a small child to the right of the soldier. At right a standing child clutches the folds of his mother's robe in the sketch, but the composition is different from the bronze panel in which the child stands to the left of the mother and next to the soldier. Other differences in the arrangement, gesture and presence or absence of figures are visible in the bronze and ceramic reliefs (Figs. 2a and b). Marquand's opinion was that the ceramic plaque cannot be a copy or replicate of part of the Moses panel in the Baptistery door.

Fig. 1. Terracotta relief plaque depicting a group of figures facing in various directions that is similar in design to those in the Moses panel of the "Gates of Paradise;" (a) before conservation, (b) after conservation (The Princeton Art Museum, no. 50-13).

(a)

(b)

Figures 1a and 2a clearly show that the upper part has broken into many pieces that are infilled with plaster. The fracture separates the two women on the left from the rest of the composition. Furthermore, it crosses the knees of the more forward of the two women. The two women are modeled with less relief and without the anatomical sense of the other figures in the large fragment.

Fig. 2. Drawing (a) comparing the design of the ceramic plaque to (b) the larger composition of figures in the Moses panel composition. The fracture between the two areas of the ceramic plaque is drawn in black showing the approximate separation of the fragments. The labels are placed below the soldier with his back toward the viewer. Scales are in cm. Dotted lines (c) show the outlines of the 12 fragments and extent of the plaster repair of the soldier's shoulder. Drawn onto a photograph (d) of the cleaned reverse view are outlines of the 3 oval-shaped repairs near the base (marked R) and one joining the two parts of the two women that were made from coils of about the same diameter and length. The pattern of iron and iron oxide firing setter marks is shown and indicated by a circle drawn around each one. The double circle indicates a ring shaped mark. Clay has been added prior to firing to the lower right corner and along the left and top edges to make the plaque rectangular.

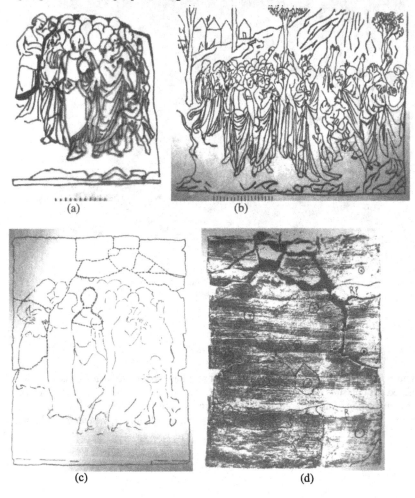

(a)

(b)

(c)

(d)

Many years after Marquand's article appeared, Richard Krautheimer referred to the terracotta plaque in a footnote in his well-known monograph on Ghiberti [4]. He concluded that, because of certain compositional and anatomical weaknesses, it could not possibly have been made by Ghiberti but probably did date from the period. In particular, Krautheimer observed differences between the two female figures to the left and the main group at center (Figs. 1 and 2a). He suggested that the terracotta plaque was neither a copy, nor a preliminary sketch, but rather a free variant of a slightly later date, perhaps the second half of the fifteenth century. In light of recent conservation and exhibition of "Gates of Paradise," the ceramic sketch and its attribution to Lorenzo Ghiberti or artisans in his workshop warrant further examination.

The methods and sequence of manufacture and restoration were studied and documented using the techniques of compositional, microstructural and radiographic analysis, and samples from the two stylistically different areas were shown to have a contemporaneous date using thermoluminescence. The study was conducted from 1983 to 1985 but its publication was delayed due to a lack of confidence in the conclusions. The paper was presented at a MRS conference in 1993 and at an American Ceramic Society conference in 1986.

ANALYTICAL METHODS AND THE RESULTS OF DATING, CHEMICAL AND MICROSTRUCTURE CHARACTERIZATION

The terracotta relief plaque measures 29.8 cm x 37.5 cm and varies in thickness from 1 cm at the edges to 8.5 cm near the center at the soldier's head and shoulder. In 1985 it was conserved by removal of a slate slab backing held in place with gypsum plaster. Gypsum was also used to adhere the 12 fragments that can be seen in Figs. 1 and 2d. After several days of immersion in water, the softened plaster and slate were removed. The plaque had been broken into several fragments (Figs. 2c and d). Several 2-5 mm samples were removed from the two female figurines in the background to the left, from the smaller fragments and from the largest fragment.

In 1984 two samples were submitted for thermoluminescence dating to Doreen Stoneham at the Research Laboratory for Archaeometry and the History of Art, Oxford. They proved to be contaminated with gypsum plaster and resulted in a wide range of overlapping dates. For the sample from the smaller fragment containing the two women, analysis yielded dates of 605 to 405 years ago (1393-1603 CE) and the large fragment yielded dates between 675 and 445 years ago (1333-1563 CE). Due to contamination, calculation of average dates (1498 and 1448, respectively) is not appropriate. However, both fragments yielded a Renaissance period date, and neither is of modern manufacture.

Semi-quantitative emission spectroscopy (Oxford Instruments Flame Emission Spectrograph 20) showed that samples of the same two fragments contain the same elements in similar proportions (Table 1). Energy dispersive x-ray analysis (KEVEX 5500 EDX) mounted on a Cambridge Instruments Stereoscan II scanning electron microscope and having a detection limit of about 1 wt% showed the following elements in order of decreasing peak height: Si,Ca,Al,Fe,K,S,Ti,Mg. The x-ray diffraction patterns (taken with a Phillips Diffractometer 30, run at 50KV, 40 mA at slowest speed from 12-80° 2-theta) of the same two samples are the same, with d-spacings at 4.2496 (intensity of 10), 4.1295 (12), 3.7354 (8), 3.2639 (90), 2.9664 (100), 2.4531(10), 2.2521 (25), 2.065

(15), 1.8900 (20), 1.8579 (20), 1.7990 (12), indicative of quartz, hematite and a calcium-containing phase.

Table 1. Elemental analysis by emission spectroscopy shows similarity of trace element composition between clay bodies from two parts of the terracotta sketch.

Concentration Range	Large Fragment with Soldier	Fragment with 2 Women
>10 weight%	Si	Si
>1%-10%	Ca,Al	Ca,Al
>0.1-1.0%	Fe,K,Mg,S	Fe,K,Mg,S
~0.01-0.1%	Na,Ti	Na,Ti
~0.001-0.01%	Cu,Cr,Mn,Sr	Cu,Cr,Mn,Sr,Zn
~0.0001-0.001%	Ag,Ni,V,Zn,Rb	Ag,Ni,V,Rb
<0.0001%	Be	Be

The color of the fired clay body in cross section is light red, and this is consistent with an iron impurity fired in oxidation. If calcium had been present in the fired clay body, the color would be yellow or yellow-orange. It is reasonable to assume that the Ca and S are due to a secondary gypsum plaster contamination that would have been drawn into the pores during mounting, drying and removal of the gypsum backing. Given this assumption, the elemental identification indicates clay that is low in alumina and, therefore, one that is not kaolinite.

Examination of the microstructures of freshly fractured surfaces of both samples reveals similar fine-grained platy clays with a range of particle size of 0.05 to 4 microns (Fig. 3a). The size and shape of the particles is characteristic of an illite, and this clay type is consistent with the presence of potassium as a flux. Interspersed in the clay matrix are quartz and other gritty inclusions ranging in size from 5 to about 70 or 80 μm, with most below 50 μm (Fig. 3b). These particles are smaller than those in most ceramics termed terracottas. The presence of coarse particles decreases drying shrinkage and prevents cracking during drying, and the absence of coarser particles allows modeling in finer detail with ease of clay flow during sculpting.

Based on refiring of small fragments of the two samples at 700°, 800° and 900°C, and observation in the scanning electron microscope of freshly fractured cross sections of these 6 samples, the pores are angular and the surfaces of clay particles are rough, indicating a low firing temperature characteristic of an earthenware clay body. Very little sintering has occurred because very little glass can be seen holding the particles together. In the refired samples, the microstructure starts to soften and the particles and pores begin to round and smooth at 900°C. Therefore, the probable firing temperature is about 800°C, and similar for samples from both areas of the plaque.

These results indicate a range of date that is appropriate for the Renaissance, a similarity in composition and microstructure of two stylistically different parts of the plaque. The ceramic body is made of a ferruginous, illitic clay with fine quartz inclusions that was fired in oxidation to about 800°C, or in the earthenware range of temperatures. During firing parts were probably together because of the similar rough, horizontal wooden plank pattern on the reverse and because of the continuous pattern of setter marks. However, the variations in a very active modeling style (Fig. 4a) and the anatomical aberrations made further investigation essential.

Fig. 3. (a, upper) Microstructure of the freshly fractured surface of a platy illitic clay from the large fragment (6000x), and (b, lower) fine, semi-rounded 20 μm quartz particle in the center and upper right in the clay matrix (1400x).

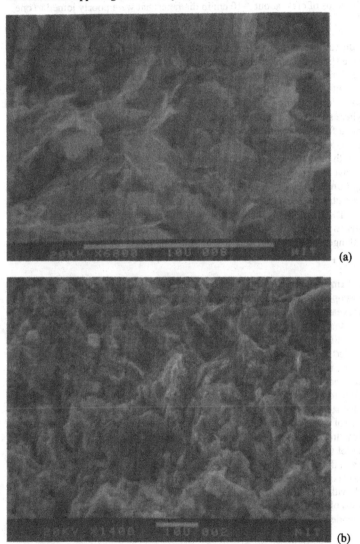

(a)

(b)

FORMING METHODS AND SEQUENCE OF MANUFACTURE AND RESTORATION

Based on detailed examination of the macrostructure and microstructure of the earthenware plaque, the following sequence of manufacture and subsequent history of restoration can be reconstructed.

Basal Clay Slabs Formed on a Partially Planed Wood Substrate: The plaque was built with small lumps of clay, about 3-10 cm in diameter that were poorly joined to one another (Fig. 4b). These bits of clay were pressed against a wooden plank backing or easel that had a bottom shelf-like support (Fig. 2d). The panel's wood grain was impressed into the reverse clay surface and became a permanent feature after firing (Fig. 2d and 4c). Some shallow, horizontally aligned grooves on the back surface, seen in raking light, are similar to jack plane marks found on Italian panel paintings. Partially rusted iron pieces of about 2-3 mm are attached in a regularly spaced array and used to support the plaque during firing (Figs. 4c and 2d). By examining the plaque on edge in cross section where separation between clay layers occurred, one observes that two layers of clay were added for the background where the plaque is thin (Fig. 5d) and up to four layers where it is thick (Fig. 4c). The imprint of the wood is less pronounced on the small fragment with the two women and on the upper part of the panel, implying that dryer clay may have been used for these features.

Supporting Stand or Easel: The thickness of the plaque gradually increases near the base and is undercut, with a flange protruding along the underside and on the reverse side (Figs. 2c and d). The flange was probably meant to facilitate display in an upright position, perhaps on a stand. The impression of a wooden clamping device is seen at the far right of the flange. The flange continues intermittently along the sides and may have served to retain the plaque in a framing device. These impressions occurred prior to firing.

Various Tool Impressions: A study of the surface features and textures with a binocular microscope with continuous magnification from 7 to 140x (Bausch and Lomb Stereozoom 7) revealed impressions in the large fragment made by at least three different tools: one wedge-shaped (Fig. 5a), one blunt-ended and irregular or worn (Fig. 5b), and a pointed tool of small diameter, 0.2-2 mm (Fig. 5c). Six instances were found where the narrow, blunt-ended tool, measuring 6-8 mm across, was impressed into plastic wet, slightly sticky clay; the edges of the impressions are somewhat rounded as the clay flowed a little around the tool. In ten instances, a wedge-shaped tool that was angled at the tip, and measured 8-10 mm in width and 1.5 mm in thickness, left clear, sharp-edged impressions of the stepped and rough tip as it impressed and scraped the dryer surface. Several examples of finely drawn, incised lines commonly of about 0.2 mm in width also are present. Most areas preserve the fresh quality of modeling (Fig. 4a), but in some areas a smooth tool was rubbed across a leather-hard surface, clay was moved into an incised groove and the clay surface was burnished or smoothed until somewhat glossy. No evidence was seen of large coarse particles having been dragged through the clay leaving a trough with a particle at on end. In the fragment with the two women were found impressions of the 0.2 mm pointed tool and a 6-8 mm wide flat-ended tool but not the same rough-ended one used in the large fragment.

Fig. 4. Illumination with extreme raking light from two sides reveals surface undulations (light to dark patches) and actively modeled surfaces with varied overlying tool marks. Reverse view (b) showing lumps of clay joined together and impressed onto a striated wooded backing, and (c) showing the iron setter or support for the plaque during firing, (d) cross section of large fragment showing method of building up the figures in relief to give the impression of perspective and depth (cm scale).

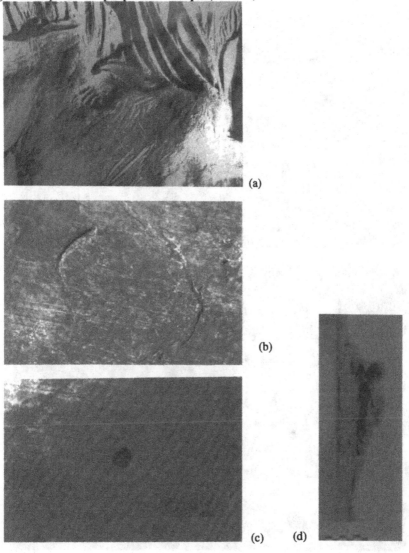

(a)

(b)

(c)   (d)

Fig. 5. (a, upper left) Arrow points to wedge-shaped tool mark in wet clay. (b, upper right) Impression in drier clay of large blunt-ended tool with stepped tip. (c, lower left) At center top and following drapery is pointed linear tool mark. More are present in the separate drapery overlay to the right; also in the upper left is a fold of drapery with an added coil joined to build up the ridge. (d, lower right) two-layer basal plaque with added layers of clay in the upper left, and haphazard joint of basal lumps near top right.

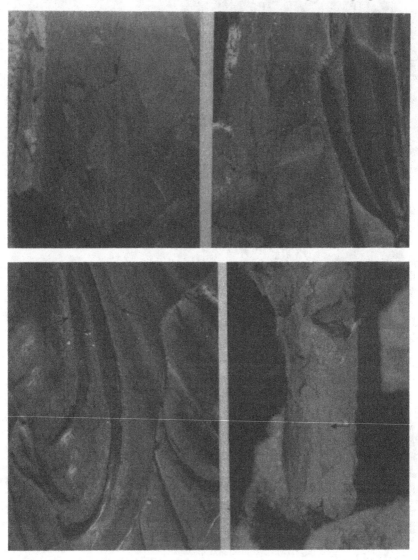

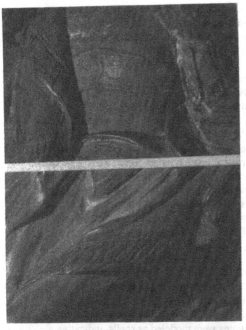

Fig. 6. The high quality of modeling and presence of modeled features, such as (a, upper) embroidery and fringe, and (b, lower) fur and leather is critical to the artistic quality and overall effect of the plaque. Sufficient carved and incised details are present to indicate that the artisan consciously imitated chasing or hammering of the bronze surface as a way of showing the finish of the final composition. The rendering of such features by incising and carving requires a clay body that is dryer than that required for shaping and forming drapery.

Fig. 7. (a, upper) Three small tool marks with blunt ends (elongated in the direction of the arrows) indicate smoothing of the vertical surface from which the figures rise. The meeting of diagonals gives the impression that the foot could actually stand on a flat vertical surface. (b, lower) Increasing the density of details at joints of the figures with the picture plane, and exaggerating the relief of the figures and drapery at joints gives the illusion of right angles, reinforcing the viewer's the tacit assumption of a picture box and the appearance of depth.

In the small fragment with the two women, the modeling has tool marks that are less pronounced, and the surface is smoother and does not undulate. The modeling is less direct and spontaneous. The tool marks are fewer and not as deep, and the lines are more tentative. The background is recessed from the relief figures by the technique of many strokes carefully placed perpendicular to the figure's outlines. In the large fragment the background and foreground are recessed from the figures in relief by strokes that are at all directions to the outline of the figures. For instance, in Fig. 4a, one wonders if a misplaced foot was not removed and placed to the figure's proper right side.

Tool Impressions Made at Intervals of Differential Drying: This same microscope showed variation in surface textures of the large fragment that can only be made at time intervals from the beginning to the end of the drying process (Figs. 6,7). The clay was worked several different times during the creative process: when wet and easily deformed plastically, then reworked when the clay was dryer, and finally in a leather-hard or semi-dry state, when enough water films have evaporated between particles that the particles touch one another and the strength increases such that plastic deformation is no longer possible. Anna Shepard presented images of these various textures in her classic reference on archaeological ceramics [6]. The small fragment of the two women seems to have been made from a stiffer, dryer clay body and only has impressions of tools that were made in a somewhat wet state and a near leather-hard state. In addition, many of the tool marks have been smoothed and are less distinct than in the large fragment.

Variation in Forming Heads: Most of the heads, such as that of the soldier, were freely modeled as a solid lump of clay and then stuck on the plaque by moistening the neck and working the clay surfaces together in a joint. Those of the two women in the fragment to the left of the larger group of figures were modeled as shells, indicating a different technique altogether. The two different methods of modeling heads imply two different artisans at work, and that considerable effort was expended to obtain the final form of the plaque.

'In-Process' Changes to the Composition: Krautheimer objected to such stylistic anomalies as the unconvincing gesture of the small boy to the right clutching his mother's skirt (Fig. 1a,b, lower left). This may be explained by considering changes made during the process of modeling the plaque as evidence for a sketch. Presently, the shoulder of the boy is anatomically incorrect. If the boy's head were rotated so that he looked upward at his mother, then the shoulder would be correct. The fracture line at the neck was made when the clay was wet and suggests that the head position was changed during the process of modeling without remodeling of the shoulder. Such changes support the plaque's use as a working sketch.

Cracks Developed During a Long Period of Forming: Working and reworking of the semi-dry clay interrupted by periodic storage under damp conditions permitted the artisan to make changes in design of the composition. A large horizontal crack is visible on the reverse side of the larger fragment (Fig. 2d). An unsuccessful effort was made to patch this crack with short coils of somewhat dry clay while the object was still in a partially dry state but after the plaque had been removed from the panel. There are four such patches; the other two are at the feet of the soldier (the wettest of the patches) and another between the neck and shoulders of the small fragment with the two women.

Using clay with a finer texture than terracotta enables fine surface detail, but the trade-off is a larger drying shrinkage and the danger of cracks. Using a long working period and dampening or rewetting the surface exacerbated the cracking. Lastly, restraining the clay slab on a highly textured wooden panel prevents shrinkage and release of moisture evenly from both sides, thus allowing internal stresses to develop, again making cracking likely.

Patching of the cracks in the lower part of the plaque was successful. The current fracture at the shoulders of the two women supports the conjecture that this repair was unsuccessful. However, the reworking of the surface during later restoration does not allow us to determine how closely the pieces fit. The crack at knees of the forward woman was not patched, and differential drying occurred because the folds in the upper skirt are closer together than the folds in the lower part. This unpatched crack occurred prior to firing and indicates that the added fragment with the two women was modeled at a later time than the large fragment. The differences in modeling style described previously reinforce the conclusion that sculpting took place in two separate periods of work, with the second session executed by a less skilled hand.

A radiograph (Fig. 8) revealed three kinds of cracks, each with a separate cause: the large horizontally banded sets of drying cracks from variable rates of drying, the fine networks of small elongated pores and drying cracks within the added lumps that result from poor clay body preparation common among sculptors but not potters, and large fractures from a postfire event in which the relief may have fallen forward.

Repair, Drying, Firing and Fracture: Once the plaque was nearly dry, it had to be removed from the wooden panel. If the plaque had been removed before it was dry, we should, but do not, find spatula marks on the reverse side that interrupt the wood texture. The major break probably occurred during removal from the panel and followed along the line of heads of the main group, leaving the main group of figures intact. This is exactly where the largest thickness gradient occurs and concentrates stress. This fracture extended just below the knees of the female figure to the proper far left, continuing at a place where a drying crack was initiated due to the two different modeling sessions that had occurred. Differential drying shrinkage on the order of 1 to 2 mm disregistration occurred on opposite sides of this crack with the upper part shrinking more.

When the plaque was turned over, repairs were made by inserting horizontal coils across the cracks that formed the band at the feet, a vertical coil below the soldier and a horizontal across the upper torsos of the two women. For some reason, no coil was added to the knees of the two women, probably because the upper and lower parts separated at this time.

After drying, a decision was made to preserve the plaque, and it was fired on metal supports, one of which was an iron ring. If one more iron support were used in the upper left of the backside, then the two pieces could have been independently supported. Due to the low firing temperature no distortion or firing shrinkage would have been measurable.

Fracture Into Many Pieces and Possible Damage to the Soldier's Shoulder: The upper part may have broken into several pieces during removal from the panel, but this scenario is unlikely. The few fracture surfaces that are still intact display brittle fractures caused by impact, and they appear to have been made postfire because they still fit together. If the plaque had suffered an impact or rotated forward prior to firing, the fracture surfaces would have been complex fractures splintering into many small fragments with much crushed, powdered clay forming at the joints.

Fig. 8. Radiograph, taken prior to the 1985 conservation treatment and with the proper image reversed, shows a plethora of coarse drying cracks that occur in horizontal bands at the feet, waist and heads of the figures, at joints where coils of clay were added at the top and sides, and at the joints between lumps of clay that were poorly worked together. Also present are webs of fine cracks and elongated pores caused by air pockets within the added lumps of clay. The large fractures in the background above the heads are post-fire cracks from a later postfire event.

If the plaque rotated forward after firing, presumably accidentally at a later time, then the warrior's shoulder, as the feature in highest relief, also would have been damaged as well as a simultaneous severing of his head. The upper surface would have been thrust forward after the soldier, and the impact above an already weakened area could have resulted in the pattern of brittle fracture we see today in the upper part of the plaque.

Replacement of the Damaged Plaque: If the plaque had broken into many pieces, it probably would not have been fired. If only the fracture of the upper background section and the two women occurred prefire, it is likely that a new upper section would have been fabricated, once the decision to preserve the sketch was made. Additionally, the plaque could have been discarded and another formed anew, but this did not occur.

Summary of and Scenario for the Modeling of the Two Female Figures: The apprentice who was entrusted with the initial repair or replacement of the two female figures and the area above their heads was presumably from the same workshop because examination reveals that the same wooden support was used. There is a continuous pattern of horizontal grooves in the wood from the large fragment to the two females, and the same spots of hematite from rusting iron supports used in firing continue across the two sections, implying contemporaneity and a single firing of both parts. In addition the chemical composition, microstructure, firing temperature are the same for both sections.

The clay for the two women was added after shrinkage of the main fragment, and then the added clay shrank, about 1-2 mm across the skirt of the woman. Thus, the folds of the upper and lower portions of the dress of the woman to the left do not align. The upper garment has smaller spacing, indicating drying shrinkage. The dryer working consistency of the clay body can be seen at the break adjacent to the woman's girdle where tool marks indicate scraping occurred. The handling of this section is not of the same high quality of modeling and finish as that found in the other figures of the main section of the relief. For instance, the figure is outlined with strokes perpendicular to the outline that are short and monotonous. Such a treatment is nowhere else found on the sketch. Had the two female figures been left to the first artisan, probably a more satisfactory handling of the spatial problems and anatomy would have resulted.

Krautheimer objected to the drapery of the girl and the elderly woman to the left. We propose that this section was remodeled or added at a slightly later time of days to a week or two, most likely by an apprentice in the workshop because the microstructure, composition and clay texture are the same. However, reforming this section at this late stage in drying meant that the two women did not fit or articulate with the large fragment and perhaps were not even attached during the subsequent firing.

Restoration and Application of Brown Wash and Paint: Once the postfire fracture of the upper section occurred, the soldier's head was reattached and his shoulder was rebuilt with plaster. Its surface was redecorated to continue the pattern of chasing, but the line width and depth changed. The damaged edges of the upper section were carved both with a knife and a file, and beveled; then all the pieces were reassembled on a bed of moist plaster of Paris spread on top of a gray slate backing. A thin mortar of quite liquid gypsum plaster was spread on top of the backing and worked into the joints (Fig.1). Next the entire modeled surface of the sketch was covered with a very thin brown iron-containing organic wash in order to unify the color of the sketch (Fig. 5d, left surface). Brown-colored plaster was added to some broken edges to make them appear even. Finally, or more likely at a later time, a "tempera" paint was added to the drapery of the various figures. Much of this color has now been washed or worn away, but traces remain of a red organic lake, a natural ultramarine blue and a yellow.

In summary, the evidence of spontaneous, in-process compositional changes where the same or similar materials were used when design changes were made, the variation in craftsmanship, tools, modeling practices and the two different craft methods of forming heads (hollow for the two women and solid for the soldier and his near compatriots), displays little of the control and uniformity expected in the production of a replica. In addition, an unusually long working period is implied by variation in surface textures that occur with different stages in drying and the presence of pre-firing cracks and repairs made prior to firing. Such defects are not desirable in copies.

DISCUSSION

When the Baptistery doors were made towards the end of his life, Lorenzo Ghiberti's workshop has a reputation as a studio that was busy bidding for and fulfilling many commissions and usually a bit behind in meeting its commitments, and recent scholarship has enriched our understanding of his workshop practices [6]. Many interactions involved collaboration with other artists on commissions, provision of designs (e.g., cartoons for stained glass windows) for execution in other workshops and the management and transfer of funds in a timely manner to maintain an honorable reputation and a reasonable standard of living [6]. In the workshop, several teams of artisans were carrying out a variety of tasks simultaneously, as in the chasing and finishing of the Baptistery doorframes. In this case, three teams of workmen were employed, one probably lead by Ghiberti's son, Vittorio, Michelozzo, another, and Ghiberti himself, the third [7]. There is some evidence from tax records to suggest that Donatello and Luca della Robbia also worked on the doors [4, p. 39]. Records indicate that some tasks, like furnace building, were let to subcontractors and other tasks requiring less skill were contracted to day laborers. In addition, deliverymen brought metal, wood and other materials as required. However, we have no insight into the apprentice system. One young man of twenty, Benozzo Gozzoli, was hired from Fra Angelico's workshop where he was learning painting. He was hired as an assistant for three years from 1444 to 1447, and presumably he would have learned tasks related to casting, but at that time the workshop was mostly occupied by cleaning, chasing and finishing (7, p. 93).

CONCLUSIONS

The complex history of manufacture, damage and repair that the ceramic plaque of a part of the Moses panel has undergone supports a view that it was held in high esteem and warranted preservation, in spite of its being incomplete and in less than perfect condition. The artisan had knowledge of the design of the gilt bronze Moses panel, for although certain figures were moved about and altered, he maintained the same proportions, degree of relief and quality of modeling as that found on the bronze original. Workshops studied one another's distinguished productions, and the ceramic plaque certainly could be a free or experimental copy. The Moses panel was placed at eye-level in the Baptistery doors, thus facilitating the making of such a study copy.

Evidence for time of production includes two thermoluminescence dates of the proper century. The sculptural practices found in the ceramic plaque are in accord with what is known for this period, such as the similar methods and sequence of manufacture described by Rees-Jones [8], Bewer [9] and others. For instance, panels are built in layers of several small lumps of clay, rather than from slabs that are thrown or rolled and trimmed.

The stylistic anomalies that the Ghiberti workshop would have considered substandard can be explained. The two women to the left were reformed by a different, less proficient hand in a separate fragment joined to the larger group of figures. The soldier's shoulder was restored in plaster, and his head has been glued in place—both acts serving to distort the proportions. The anatomical anomaly of the young boy to the left is due to the repositioning of his head during the process of modeling but without adjusting his body. Such changes are common in working sketches. This in-process change of the figure's focus is difficult to explain if the plaque is a copy, but such evolution in design is consistent with our expectations of a sketch. The plaque being only a partial representation of the Moses panel, and focusing not on the major subject and action of the panel but only on the observers, is an unlikely subject for a copyist. Was some special problem being elucidated through modeling a sketch? Our attention is focused on the identity, relationship and significance of the central figures, but we have no text or correspondence as a guide.

The plaque was worked over a long period of time, probably several days to weeks, and it was sculpted with several tools. Various operations were carried out at different stages in the drying process. Joints between bits of clay were made haphazardly, and a network of cracks, some of which have been patched, is visible on the reverse side. The clay body formulation was not optimized to prevent cracking and reduce shrinkage. Longevity of the object was not of paramount importance, although considerable time and effort were expended to obtain the final form of the plaque. Such a damaged plaque probably was not meant to be exhibited as an object but may have served as a working model or at some intermediate step. It has a complex history of repair and restoration, so it also must have been significant enough to preserve.

The one suggestion consistent with this tortured object history is that the plaque is the result of the collaboration in the fifteenth century of a master modeler, who made the large section and demonstrated the process of modeling in clay, making changes in design and producing a finished surface, and a beginning apprentice, who modeled the two women, and then learned through craft practice the effects of a not quite sandy enough clay body, poor clay preparation, defects caused by drying shrinkage, detachment from a support, the attempted repair of cracks, problems with clamping the clay during modeling sessions and the repair of edges and corners, as well as the process of firing and handling the object at various stages of fabrication. Students usually do not succeed until they make a sufficient number of mistakes, and this plaque could have been a just such a training exercise.

Hopefully, this article will promote consideration of this fascinating, enigmatic ceramic plaque, so that it will be studied and discussed further in light of new scholarship on Lorenzo Ghiberti and the "The Gates of Paradise."

ACKNOWLEDGMENTS

This study would not have been possible without the collaboration of Norman Muller, Doreen Stoneham (formerly of the Laboratory for Archaeology and the History of Art, Oxford) and Stuart Fleming (MASCA, University of Pennsylvania) and insightful discussions with Frederick Hartt and John Larson. I thank Norman Muller for use of the photographs in Figures 1b, 2d, and Fig. 9, the radiograph, Doreen Stoneham for performing the thermoluminescence dating tests, and Walter Correia, Center for Materials Science, Massachusetts Institute of Technology, for helping me with the emission spectroscopy. All other analyses, drawings, photographs and errors of fact and interpretation are my own.

REFERENCES

1. Vasari, G., Lives of the Painters, Sculptors and Architects, trans. G. du C. de Vere, A.A. Knopf, New York, 1996, 294ff.
2. Ghiberti, Lorenzo, I Commentarii, ed. L. Bartoli, Giunti, Florence, 1998, 97. Marquand, A., "A Terracotta Sketch by Lorenzo Ghiberti," Am. Jour. of Archaeology, 9 (1984) 206-211.
3. Krautheimer, R., Lorenzo Ghiberti, Princeton Univ. Press, 1956, p. 191.
4. Radke, G.M., ed., The Gates of Paradise: Lorenzo Ghiberti's Renaissance Masterpiece, Yale University Press, New Haven, 2007.
5. Shepard, Anna O., Ceramics for the Archaeologist, Carnegie Institution, Washington, D.C., No. 609, 1954, pp. 183-193.
6. Radke, G.M., "Lorenzo Ghiberti: Master Collaborator," in G.M. Radke, The Gates of Paradise, op. cit., pp. 51-71.
7. Haines, M. and F. Caglioti, "Documenting the Gates of Paradise," Ibid., p. 83-84, 93-94.
8. Rees-Jones, S., "A Fifteenth Century Florentine Terracotta Relief: Technology, Conservation, Interpretation," Studies in Conservation, 23/3 (1978) 95-113.
9. Bewer, F.G., "Studying the Technology of Renaissance Bronzes," Materials Issues in Art and Archaeology IV, eds., P.Vandiver, J. Druzik, J-L. Galvan, I.C. Freestone and G.S Wheeler, MRS Symp. Proc. 352, 1995, p. 701-709.

Conservation Science

Mater. Res. Soc. Symp. Proc. Vol. 1047 © 2008 Materials Research Society   1047-Y05-03

Controlling Swelling of Portland Brownstone

Timothy Wangler[1], and George W. Scherer[2]
[1]Chemical Engineering, Princeton University, Eng. Quad. E-211, Princeton, NJ, 08544
[2]Civil & Env. Eng., Princeton University, Eng. Quad. E-319, Princeton, NJ, 08544

ABSTRACT

Many clay-bearing sedimentary stones such as Portland Brownstone will swell when exposed to water, and this can generate damaging stresses as differential strains evolve during a wetting cycle. Current swelling inhibitors, consisting of α,ω-diaminoalkanes, can reduce swelling in Portland Brownstone up to 50%. In this study, through X-ray diffraction and swelling strain experiments, we demonstrate that the α,ω-diaminoalkanes inhibit swelling by substituting for interlayer cations and partially hydrophobicizing the interlayer, then rehydrating on subsequent wetting cycles. We also introduce the copper (II) ethylenediamine complex as a potential treatment for swelling inhibition.

INTRODUCTION

Portland Brownstone, a sandstone widely used throughout the northeastern United States in historic buildings and monuments, shows damage that is related to the swelling of clay in the stone. These clays cause differential stresses and strains that lead to buckling and cracking, and the clays may also create small pores that make the stone susceptible to damage by salt crystallization or frost [1]. As the exterior surface of the stone absorbs water, the wet layer expands with a strain ε_s relative to the dry interior, which creates a compressive stress, σ_x, in the wet layer. When that layer is thin compared to the dry interior, the stress is [2]

$$\sigma_x = \frac{E_w \varepsilon_s}{1 - v_w}$$

where E_w is the elastic modulus of the wet stone (which may be much less than that of the dry stone [3], and v_s is Poisson's ratio. If a treatment is applied that reduces ε_s, without increasing E_w, then the stress is reduced.

Most of the damage observed is of the buckling type (as seen in Figure 1), where large surfaces buckle away from the surface during wetting, although exposed elements can crack upon drying. It has been demonstrated that treatment with surfactants can reduce swelling, and in this paper, we show the mechanism of strain suppression through tests on pure clays and whole stone.

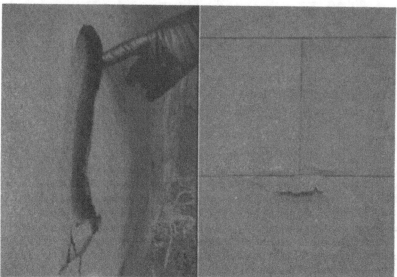

Figure 1. Examples of buckling damage as seen on the Victoria Mansion (Portland, ME), a 19th century building constructed with Portland Brownstone.

It is quite common to find swelling clays in sedimentary stones, and these clays can lead to many engineering problems in the fields of tunneling, oil well borehole stability, and foundation stability [4-6]. It was recognized as a problem in conservation by Delgado [7], among others. Wendler and Snethlage devised and tested an α,ω-diaminoalkane treatment that was found to be effective on some swelling German sandstones and on swelling stones on Easter Island [8-9]. This treatment consisted of a linear hydrophobic alkane chain with protonated amine groups at each end; the amines are believed to substitute for the alkali in the interlayer of the clay and act to bind opposing sheets together, while the hydrophobic chain discourages water entry. Gonzalez and Scherer tested and found this treatment to be effective on Portland Brownstone and also found that there is sensitivity to the order of addition of these surfactants, and that mixtures of surfactants could be a more effective treatment [10]; these observations were confirmed by Wangler et al. [11]. Gonzalez and Scherer also found that because the clay is in the cementing phase of the stone, it has a very large effect on the elastic and viscoelastic properties [10]. They found that the diaminoalkane (DAA) treatment raised the modulus of the stone while increasing the viscoelastic relaxation rate, so any immediate reduction in swelling from the treatment may be offset by an increase in the modulus, although the stress may relax away faster. The DAA treatment has also been found to be quite stable with respect to washout [12], but all attempts at optimizing it have been unsuccessful at fully eliminating swelling [11]. The goal of the present study is to understand the mechanism by which the DAA reduces swelling, with the idea that this knowledge can be used to further optimize the treatment or potentially reveal other options. Most evidence [8-12] seems to indicate that swelling is reduced by intercalation of the DAA in the interlayer via ion exchange. The effect on the mechanical properties indicates that intercalation is taking place, and the resistance to washout indicates that ion exchange is taking place rather than complexation of the amine to the cation or the negatively charged clay surface of the clay interlayer. However, the source of the residual swelling has been a mystery. A recent study proves that Portland Brownstone swells almost entirely by intracrystalline swelling [13], so it is possible that the DAA in the interlayer are still hydrating upon wetting.

There is too little clay in Portland Brownstone for direct study of the stone by X-ray diffraction (XRD). Therefore, we examine clays separated from the stone, as well as a highly swelling clay, sodium montmorillonite, during wetting and drying, with and without exposure to DAA. Of course, neat clays exhibit much larger strains than clay confined between the grains in a stone, where the network of rigid grain contacts exerts a confining stress on those grain boundaries containing clay. It is clear that some grain contacts contain clay, because the elastic modulus of Portland Brownstone decreases strongly (~50%) when the stone is wet; however, the modulus does not drop to zero, so many of the grain contacts must be relatively free of clays (as indicated schematically in Figure 2).

Figure 2. The cementing material at some grain boundaries is rich in clay (dark brown) while other boundaries are relatively free of clay (light brown); the latter remain rigid during wetting, and form a network that constrains the expansion of the clays.

EXPERIMENT

Materials

The materials studied in these experiments were laboratory grade bentonite powder (primarily Na-montmorillonite, obtained from Fisher Scientific) as the neat clay and Portland Brownstone (from Pasvalco Corp., Closter, NJ) for the stone studies. Chemicals were obtained from Acros Organics and included α,ω-diaminoalkanes (DAA) with carbon chain lengths of 2, 3, 4, 6, and 8. Also obtained were hexamethonium bromide, an α,ω-diquaternaryammonium alkane with a carbon chain length of 6, and a copper (II) ethylenediamine complex solution.

Clay separation

For separation of the clay-sized fraction from the stone, a procedure adapted from USGS Open File Report 01-041 was used [14]. It consisted of manually grinding a few hundred grams of stone with a mortar and pestle and then performing a series of sonication, sedimentation (settling time determined by Stokes settling), and decantation cycles until the decanted portion was visually clear. A dispersant (Calgon) was added to the mixtures, and because there is little to no calcite content, no acid pretreatments were necessary.

X-Ray diffraction studies

X-ray diffraction studies were performed on a Rigaku Miniflex (Cu-Kα) diffractometer. Samples were prepared for XRD studies by drying several mL of clay suspension (either bentonite or separated clay fraction from Portland Brownstone) on a glass slide in order to obtain preferred orientation along the basal planes. Scans were performed of the expected 001 basal spacing region, from about 4 to 12 degrees 2θ. Treatments on clays were performed by direct application of the 0.31 M diaminoalkane solution to the oriented clay on the slide followed by drying. Scans were performed under dry and wet condition; for wet conditions, samples had water applied and were observed before and after the experiment to ensure some water was still present in the sample.

Treatment procedure

The treatment procedure for stone samples followed that of previous studies [10-11]. Samples (roughly 5 x 5 x 45 mm) were soaked in a solution of a particular concentration of treatment for 1-3 hours to ensure full saturation, dried at ambient for approximately one hour, and then placed into a 60 C oven overnight. Most treatments were performed at 0.31 M in total diaminoalkane concentration, corresponding to the diaminoalkane concentration of 5 wt% diaminobutane used by Wendler in the initial studies of this treatment.

Swelling experiments

Swelling experiments were performed with a linear variable differential transformer (resolution ~0.2 μm) obtained from Macrosensors (Pennsauken, NJ) monitoring the displacement of the sample in a flat bottomed cup after swelling fluid was added. The experiment was allowed to proceed until saturation was complete, which was about 15-30 minutes. Samples were always measured after coming out of a 60 C oven and equilibrating with ambient temperature while sealed in a closed container (to avoid atmospheric moisture). For drying performed in some experiments, a nitrogen flow was added above the cup to speed the process.

RESULTS

XRD

The XRD experiments performed on neat bentonite demonstrate a shift in the d-spacing between untreated wet and dry samples, indicating intercalation by water. Most diaminoalkanes increased the basal spacing to about 13 Å. Ethylenediamine had a smaller basal spacing, probably owing to the size of the smaller molecule. Under wet conditions, peak shifts were observed for the untreated, ethylenediamine-treated, and diaminopropane-treated clays. The diaminobutane and diaminohexane treated samples did not show any shift under the wet conditions of this experiment. Figure 3 shows the XRD scans. An experiment was run in which a lower concentration (~0.08 M) of diaminobutane treatment was applied to the clay. Figure 4 is the diffractogram of that experiment, showing a sharp peak at about 16 Å under wet conditions.

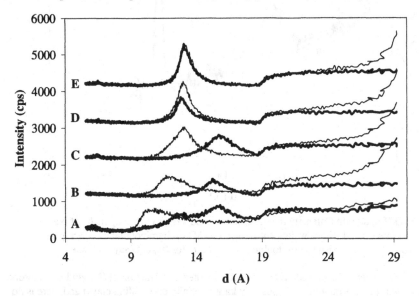

Figure 3. Diffractograms (Intensity in counts per second versus d-spacing in Å, curve A is baseline and all others progressively offset by 1000) of bentonite clay peaks under wet and dry conditions and with various diaminoalkane treatments. Darker lines are wet scans. Curves are labeled: A) untreated, B) ethylenediamine (C2), C) diaminopropane (C3), D) diaminobutane (C4), E) diaminohexane (C6).

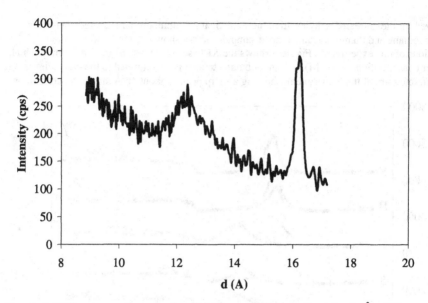

Figure 4. Diffractogram (Intensity in counts per second versus d-spacing in Å) of Na-montmorillonite treated with low concentration (0.08 M) of diaminobutane. Sharp peak at about 16 Å confirms existence of double layer hydrate for larger diaminoalkane.

Figure 5 shows the diffractogram for the separated clay fraction of Portland Brownstone. Peaks are seen for chlorite, illite, and possibly kaolinite (all non-swelling clays) and there is no peak shift when wetted or glycolated. The 14 Å peak does not collapse after heating to 550 C, indicating that it is not a vermiculite peak [14]. The absence of a peak shift after glycolation is disconcerting at first; however the previously cited study [13] confirms the existence of intracrystalline swelling in clay layers via experiment with organic solvents and cation substitution. This implies the existence of randomly interstratified swelling layers throughout the chlorite fraction.

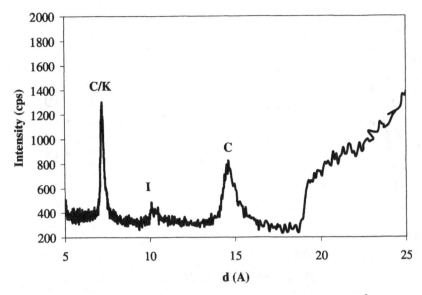

Figure 5. Diffractogram (Intensity in counts per second versus d-spacing in Å) of Portland Brownstone oriented clay fraction. Chlorite, illite and chlorite/kaolinite peaks are marked. No peak shift was observed upon glycolation.

<u>Swelling experiments</u>

Standard swelling experiments were performed to characterize the untreated swelling of Portland Brownstone, and it was found to be 1.0±0.05 mm/m. (Examination of a wide variety of Portland Brownstone samples in our lab yields swelling strains ranging from about 4-10 x 10^{-4}, so the present sample is at the upper end of the range for this stone.) A series of experiments were then performed to see the effect of concentration of DAA treatment on the swelling reduction. The results shown in Figure 6 indicate a ceiling of about 40-50% reduction of the swelling. Since it is believed that the DAA substitute for random charge sites in the interlayer, it was hypothesized that a mixture of varying chain lengths would be more effective. A "soup" of equimolar concentration of DAA (total DAA concentration = 0.31 M) utilizing carbon chain lengths of 2, 3, 4, 6, and 8 was tested and the swelling reduction was 50%, making it the most effective single DAA treatment performed, but still having a residual swelling strain. Finally, because it was shown in ref. [13] that almost all the swelling in Portland Brownstone is intracrystalline, and since organometallic complexes are known to bind strongly in the interlayer space [15], the copper (II) ethylenediamine complex was tested for its effect on swelling. It showed a 70% reduction in swelling for the duration of this experiment, making it the most effective swelling inhibitor in Portland Brownstone to date.

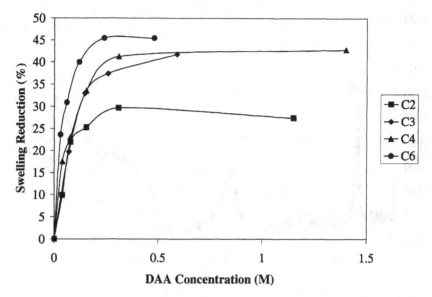

Figure 6. Swelling reduction vs. diaminoalkane treatment concentration. Each curve corresponds to an α,ω-diaminoalkane with the indicated carbon chain length.

<u>Swelling/drying/rewetting treatment experiments</u>

These experiments consisted of monitoring the displacement of a stone sample throughout an entire treatment cycle. The treatment was directly applied, then excess liquid was removed with a syringe; the sample was dried with nitrogen, and then rewetted with water. Three treatments were tested with this experiment: the 0.31 M "soup" of DAA mentioned previously, 0.31 M diaminohexane, and (to differentiate the amine chemistry's effects on the treatment mechanism) 0.31 M hexamethonium bromide. The results of these three experiments are shown in Figures 7a-c. The important features are: 1) all curves swell to approximately 70% of the full water swelling strain, 2) additional swelling does not occur (in fact, sometimes slight reduction occurs) during extended exposure to the treatment, 3) drying reduces the strain to near the baseline level, but not completely, and 4) subsequent rewetting shows the "treated" swelling strain rising to a level lower than the initial "treatment" swelling strain. It was not possible to heat the sample during the experiment, but the final swelling strain was remeasured after oven-drying the treated sample at 60°C and the swelling was unchanged.

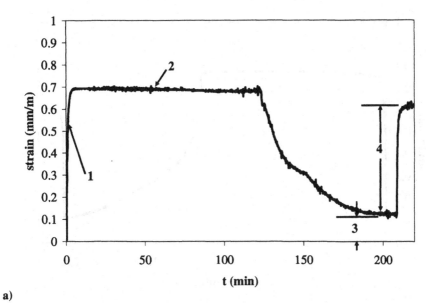

a)

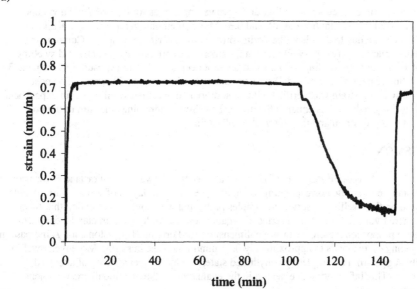

b)

99

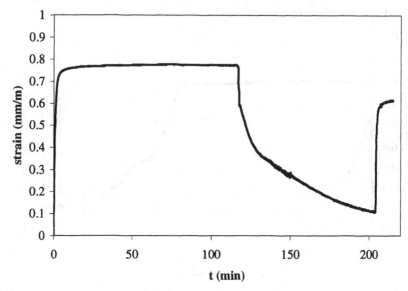

c)

Figure 7a-c. Direct treatment swelling/drying/rewetting curves for 0.31 M diaminoalkane "soup" (a), 0.31 M diaminohexane (b), and 0.31 M hexamethonium bromide (c). Kinks in drying curves are due to details in the drying process for individual samples. Common curve features are marked in (a): 1) swelling with treatment does not proceed to untreated swelling level of 1.0 mm/m, 2) swelling remains restricted or decreases during exposure to treatment, 3) residual "intercalation" strain, and 4) post-treatment swelling strain with water. The post treatment swelling strain was verified with an additional oven-dried swelling strain experiment. Actual swelling reduction is equal to the original swelling strain minus the sum of the post-treatment swelling strain and the "intercalation" strain.

DISCUSSION

A separate study [13] found that almost all the swelling in Portland Brownstone is intracrystalline swelling, corresponding to about 4 pseudo-monolayers of water. This conclusion was based on the swelling observed in samples pretreated with solutions of cation that form inner-sphere hydrates, as well as a series of organic liquids. An increase in clay layer spacing produces a proportional increase in stone dimension: for Portland Brownstone, a 1 Å increase in layer spacing corresponds to approximately 0.11 mm/m increase in strain. Moreover, swelling typically begins from a completely dehydrated state, or a basal layer spacing of about 10 Angstroms. This information is essential in the evaluation of the diaminoalkane treatment mechanism.

The results of XRD studies on the separated clay fraction of Portland Brownstone highlight the potential difficulties encountered in studies of this sort on the clays from stones. All the clay peaks in Portland Brownstone are non-swelling clay peaks, as evidenced by the lack of peak movement upon glycolation. It has already been demonstrated that intracrystalline swelling is responsible for almost all swelling observed in Portland Brownstone [13], so it can reasonably be assumed that the swelling clay layers are randomly distributed throughout the chlorite clay fraction, as it is known that chlorites can weather into swelling clays [16]. XRD studies on Na-montmorillonite served as a model system, and they demonstrated that intercalation takes place and increases the basal spacing by 2-3 Å. However, the most interesting result from the XRD experiments is the fact that the DAA demonstrate that they can hydrate upon wetting, although only the shorter chain length DAA did so at the normal treatment concentration. The fact that both short chain DAA hydrated to a basal spacing of approximately two pseudo-monolayers of water (~15-16 Å) is significant. All studies of DAA swelling reduction in stone indicate a ceiling of 50% reduction [10-11], which would correspond to two pseudo-monolayers of water. A major inconsistency, however, is that the larger chain length DAA did not show the double layer hydrate in the neat clay, although from all indications in the stone swelling experiments, it exists. We suspected that the neat clay could take up enough DAA to become hydrophobic, whereas the confined clay in the stone could not. Therefore, we tested the expansion of neat clay exposed to a lower concentration of DAA When the diaminobutane treatment solution was diluted by about 4X, a sharp peak corresponding to the double layer hydrate did appear under wet conditions. This supports the idea that the diaminoalkane molecules are hydrating in the stone clay layers (however, we cannot exclude the possibility that the negatively charged clay layer itself is hydrating).

The results of the concentration vs. swelling reduction studies for various diaminoalkanes verify the ceiling to the swelling reduction achievable with this treatment, as most treatments reduced swelling up to about 40-50%. The fact that the "soup" mixture of varying chain lengths was the most effective single treatment indicates that a mixture of chain lengths is more efficient at fully exchanging the randomly situated charge sites in the interlayer. The experiments that are the most useful in understanding the treatment mechanism are the direct treatment/drying/rewetting experiments. The initial rise of each curve is lower (~25-30% lower) than the full, untreated, water swelling strain. This is an indication that the DAA are entering the interlayer immediately, although the full exchange does not necessarily have to occur instantaneously. Indeed, the fact that the swelling curve remains constant, and in some cases begins to decrease, throughout the exposure to the treatment indicates that the exchange/intercalation is occurring. It may be that the limit to treatment time is not necessarily the saturation time of the stone's porosity, but is controlled by the diffusion of DAA into the interlayer of clay particles and subsequent displacement of the cations currently residing there. Drying the sample reduces the strain to a level that is above the original dimension of the sample, indicating that intercalation has taken place and the molecules are propping open the layers. However, using the results of the neat clay XRD studies (showing an intercalation basal spacing increase of about 3 Å) along with the scaling factor of 0.11 mm/m-Å, one would expect to see a residual "intercalation" strain of 0.33 mm/m, and the observed strain is much lower than that in these experiments. This is an indication that, although intercalation is occurring, the clay layers are collapsing around the diaminoalkanes. In fact, the neat clay XRD studies may be deceptive because they are not performed under pressure, and in the case of the stone the surrounding stone matrix is applying pressure to the clay layer and pushing the clay layers down around the intercalating DAA molecules. The final swelling strains upon rewetting agree with the swelling reductions that we observe, so the actual swelling reduction is equal to the difference between the full swelling strain (1.0 mm/m) and the top of the rewetting curve (or, the total swelling strain minus the sum of the rewetting strain and the residual "intercalation" strain) –

approximately 30-40%. The degree of collapse, or the magnitude of the "intercalation" strain, is probably the reason that ethylenediamine's ceiling to swelling reduction was much lower than all other DAA – the shorter chain length DAA probably allows the layers to collapse more readily around it. The "treated" swelling strain is restricted (in contrast to swelling behavior with monofunctional amines and quaternary ammonium ions) very likely because in some cases, opposite ends of the difunctional molecule are balancing charges on opposite layers, just as Wendler hypothesized [8]. The DAA molecules are lying flat in the interlayer and some of them must have opposing ends "keyed" into opposite sheets. Lastly, the similar behavior of hexamethonium bromide to diaminohexane is a final verification of the mechanism of ion exchange for the DAA, as hexamethonium bromide is unable to form amine complexes with the interlayer cations or negatively charged clay surfaces.

It should be noted that for all of these experiments, the degree of exchange is unknown. It is possible that if incomplete exchange is occurring, the residual "intercalation" strain could increase without increasing the magnitude of the final rewetting swelling curve. This would act to increase the apparent swelling reduction. All indications are, however, that complete exchange is occurring, as very high concentrations of DAA (> 1M) produce the same results. (These high concentrations of DAA are unrealistic as treatments for monuments, because high concentrations can alter the appearance of the stone.) In light of the neat clay XRD results, it may still be possible to achieve complete hydrophobicization of the interlayer, although this may be dependent on some factors that differ between the neat clay and the Portland Brownstone swelling clay, such as interlayer charge density and location.

Given this ceiling to the effectiveness of the DAA treatment, it was supposed that an organometallic complex, such as the copper (II) ethylenediamine complex, could suppress swelling more effectively. The complex is very strongly bound, and in fact has been suggested as an alternative for use in cation exchange capacity experiments [17]. The complex consists of a copper cation surrounded by two bidentate ethylenediamine ligands. The ligands could act as a hydrophobic "shell" around the cation, lowering the hydration energy. The complex is the best performing Portland Brownstone swelling suppressor to date, at least for the duration of the swelling experiment (~0.5-1 h). It still allows swelling, however, so its effect on the mechanical properties must still be tested. The same reasoning applies to the potential use of hexamethonium bromide as a treatment – if there is no increase in the modulus, it may be a viable candidate for treatment of monuments.

CONCLUSIONS

Tying it all together, it seems clear that the diaminoalkane molecules are either partially or fully exchanging and intercalating into the interlayer via an ion exchange reaction; the clay layers are collapsing to a certain degree around the intercalant, and then rewetting produces subsequent hydration to a double layer hydrate which is restricted from further increase by the difunctionality of the diaminoalkane molecule. This places a ceiling of about 50% reduction on the effectiveness of the treatment in Portland Brownstone. There still exists the possibility that full exchange has not occurred, or possibly that a higher chain length diaminoalkane or polyamine will be able to completely hydrophobicize the interlayer. Organometallic complexes may be more effective at making the interlayer hydrophobic, as the copper (II) ethylenediamine complex seems to be. Another organometallic complex may be more effective, or possibly a mixture of any of these treatments may ultimately prove the best. Whatever the case, unless a treatment completely eliminates swelling, the effect on the modulus must always be tested.

ACKNOWLEDGMENTS

This work was supported in part by grant MT-2210-07-NC-05 from the National Center for Preservation Technology and Training.

REFERENCES

1. "Internal stress and cracking in stone and masonry", G.W. Scherer, pp. 633-641 in *Measuring, Monitoring and Modeling Concrete Properties*, ed. M.S. Konsta-Gdoutos (Springer, Dordrecht, The Netherlands, 2006).
2. "Hygric Swelling of Portland Brownstone", I. Jimenez Gonzalez, M. Higgins, and G.W. Scherer, pp. 21-27 in *Materials Issues in Art & Archaeology VI*, MRS Symposium Proc. Vol. 712, eds. P.B. Vandiver, M. Goodway, and J.L. Mass (Materials Res. Soc., Warrendale, PA, 2002).
3. A.N. Tutuncu, A.L. Podio, A.R. Gregory, and M.M. Sharma, "Nonlinear viscoelastic behavior of sedimentary rocks, Part I: Effect of frequency and strain amplitude", *Geophysics* 63 (1) 184-194 (1998).
4. N. Barton, R. Lien, and J. Lunde, "Engineering classification of rock masses for the design of tunnel support", *Rock Mechanics and Rock Engineering*, 6 (4) 189-236 (1974).
5. E.S. Boek, P.V. Coveney, and N.T. Skipper, "Monte Carlo Modeling Studies of Hydrated Li-, Na-, and K-Smectites: Understanding the Role of Potassium as a Clay Swelling Inhibitor", *J. Am. Chem. Soc.*, 117 12608-12617 (1995).
6. A.M.O. Mohamed, "The role of clay minerals in marly soils on its stability", *Engineering Geology*, 57 (3-4), 193-203 (2000).
7. J. Delgado Rodrigues, "Swelling behaviour of stones and its interest in conservation. An appraisal", *Materiales de Construcción*, 51 (263-264) 183-195 (2001).
8. "Consolidation and hydrophobic treatment of natural stone", E. Wendler, D.D. Klemm, R. Snethlage, in *Proc. 5th Int. Conf. on Durability of Building Materials and Components*, eds. J.M. Baker, P.J. Nixon, A.J. Majumdar, and H. Davies (Chapman & Hall, London), 203-212.
9. "Easter Island tuff: laboratory studies for its consolidation", E. Wendler, A.E. Charola, B. Fitzner, *Proc. of the 8th Int. Congress on Deterioration and Conservation of Stone*, ed J. Riederer (Berlin, Germany) (2), 1159-1170.
10. I. Jiminez Gonzalez, G.W. Scherer, "Effect of swelling inhibitors on the swelling and stress relaxation of clay bearing stones", *Env. Geo.* 46 364-377 (2004).
11. T. Wangler, A. Wylykanowitz, and G.W. Scherer. "Controlling stress from swelling clay", pp. 703-708 in *Measuring, Monitoring and Modeling Concrete Properties*, ed. M.S. Konsta-Gdoutos (Springer, Dordrecht, The Netherlands, 2006).
12. "Evaluating the potential damage to stones from wetting and drying cycles", I. Jiménez González and G.W. Scherer, pp. 685-693 in *Measuring, Monitoring and Modeling Concrete Properties*, ed. M.S. Konsta-Gdoutos (Springer, Dordrecht, The Netherlands, 2006).

13. T. Wangler and G.W. Scherer, "Clay swelling mechanism in clay-bearing sandstones", *Env. Geo.,* submitted.
14. L.J. Poppe, V.F. Paskevich, J.C. Hathaway, and D.S. Blackwood. "U.S. Geological Survey Open-File Report 01-041: A Laboratory Manual for X-Ray Powder Diffraction" (2001). http://pubs.usgs.gov/of/2001/of01-041/index.htm
15. M. Stadler and P.W. Schindler, "The effect of dissolved ligands on the sorption of Cu(II) by Ca-montmorillonite", *Clays Clay Miner.* **42** 148-160 (1994).
16. A.L. Senkayi, J.B. Dixon, and L.R. Hossner, "Transformation of Chlorite to Smectite Through Regularly Interstratified Intermediates", *Soil Sci. Soc. Am. J.,* **45** 650-656 (1981).
17. F. Bergaya and M. Vayer, "CEC of clays: Measurement by adsorption of a copper ethylenediamine complex", *Applied Clay Science,* **12** 275-280 (1997).

Mater. Res. Soc. Symp. Proc. Vol. 1047 © 2008 Materials Research Society   1047-Y04-03

Predicting Efflorescence and Subflorescences of Salts

Rosa Maria Espinosa Marzal[1], Lutz Franke[2], and Gernod Deckelmann[2]
[1]Civil and Environmental Engineering, Princeton University, Eng. Quad E-320, Princeton, NJ, 08544
[2]Institute of Building Materials, Physics and Chemistry of Buildings, Hamburg University of Technology, Eissendorfer Strasse 42, Hamburg, 21071, Germany

ABSTRACT

Crystallization of salts is a common cause of damage in porous building materials. Understanding of the crystallization mechanism of salts is important in order to prevent or avoid the problem. Subflorescence of salts (i.e., crystallization within the pores of the body) can induce scaling and cracking, while efflorescence (i.e., crystallization in a film of solution on the exterior surface of the body) does not generally affect the coherence and endurance of the building materials.

In this paper, we deal with the crystallization behavior of two salts, sodium sulfate and sodium chloride, in two bricks with different capillary porosity. The results reveal quite different crystallization behavior depending on salt and substrate.

The supersaturation of the solution is induced in our experiments by evaporation. Indeed, the main reason for the different behavior of these salts is their different ability to supersaturate. Thus, the sodium sulfate solution is prone to be much more supersaturated than the sodium chloride solution. Furthermore, the solution transport, which depends on salt properties, material porosity, pore-clogging and environmental conditions, affects the position of the drying front and, with it, the crystallization front, leading to the formation either of efflorescence or of subflorescence. Simulation of the experiments is used to understand the effect of the influencing factors on the crystallization pattern. Therefore, considering both factors, supersaturation ratio and solution transport, it is possible to predict the different crystallization behaviors observed in the experiments.

INTRODUCTION

Efflorescence occurs in masonry construction, when water moving through a wall or other structure brings salts to the surface and evaporates there, causing salt to crystallize on the material surface. In some cases, salts crystallize beneath the material surface and build subflorescences (also called cryptoflorescence).

Although efflorescence does not generally affect the coherence and endurance of building materials, it impairs the surface appearance, which can be critical in the case of historical buildings. Indeed, there are some efforts to prevent the formation of efflorescence. For example, a penetrating sealer can help prevent or lessen the occurrence of efflorescence by soaking in and blocking the pores below the surface, thus preventing the water from moving to the surface bringing the salts with it. However, this process might lead to the crystallization of salts some millimeters beneath the surface (as subflorescence) or even deeper, causing possibly more severe damage.

In previous works [11], different crystallization behavior for sodium chloride and sodium sulfate was observed. Indeed, to predict crystallization and, finally, to prevent material damage due to salt crystallization it is necessary to understand the crystallization mechanism as well as the interaction between the different involved processes. It is necessary to determine which properties both of the salts and of the materials influence the position of the crystallization front as well as the potential resulting stress, kind of damage and its intensity.

Numerical simulation of experiments carried out in the laboratory is used to understand the interaction between the influencing factors on the crystallization process. Moreover it allows making a quantitative analysis.

EXPERIMENTS

Two different bricks (A and B) were selected for the experimental investigation. A significant difference is the presence of large pores (~10 μm) in brick A, while most pores in brick B are smaller than 2 μm. Thus, the measured sorptivity of brick A is approximately 3 times larger than that of brick B: $S_A=1.71$ g^2/s and $S_B=0.57$ g^2/s.

Two kind of experiments were carried out:

- Drying experiments. Samples of brick A (prisms, 14 cm x 6 cm x 6 cm) were impregnated with sodium sulfate solution (1.61 mol/kg) or with sodium chloride solution (5.83 mol/kg). The evaporation occurred either through two opposite sides (14 cm x 6 cm) or through only one side under constant temperature (23 °C) and relative humidity (50 %).

- Capillary rise tests were performed with samples of brick A and B in contact with sodium sulfate solution (1.61 mol/kg). The bottom side was immersed to a depth of 1 cm in the salt solution at known and constant concentration.

1. Drying Experiments

Figure 1 shows the measured weight change of the sample with respect to the dry and salt free sample, i.e. the total mass of solution and salt crystals (L+C) in the material pores. The drying curve for the case of pure water is depicted in the same diagram. The significant difference of the starting point is given by the mass of salt.

The results reveal a strongly different drying behavior for brick A depending on the salt (NaCl or Na$_2$SO$_4$). Moreover, they show clearly that the drying of the sample impregnated with sodium chloride is the most strongly hindered.

Efflorescence of halite and mirabilite is visible after just 10 hours. This can be expected since the capillary transport in brick A proceeds so fast, that the crystallization front is placed on the surface

Figure 1 shows efflorescence of halite (right, top) and of mirabilite (bottom). While NaCl-crystals on the surface have a compact structure with high density, which impedes evaporation, the mirabilite crystals on the surface are thin prisms (acicular) and form a (vapor) permeable layer. Thus, the kind of efflorescence is quite dependent on salt (and substrate) and determines further evaporation rates, obviously affecting further crystallization.

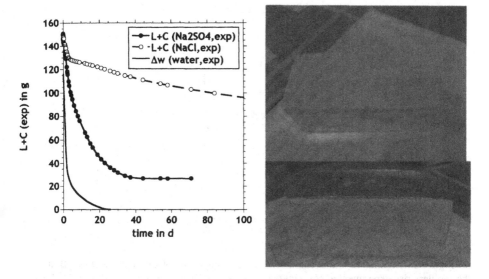

Figure 1. (left) Weight change due to evaporation of water in brick A impregnated with pure water, sodium sulfate or with sodium chloride solution. Evaporation takes place through one surface (14cm x 6cm). Efflorescence of halite (right, top) and of mirabilite (right bottom) after drying 24 hours at 50 % RH and 23 °C.

Furthermore, the results of the simulation of this experiment show a more effective pore clogging with halite than with mirabilite, which is also confirmed experimentally in [11].

2. Capillary rise test of sodium sulfate solution in brick A and B.

The measured weight change (figure 2) indicates that the evaporation in brick B is significantly slower, which can be expected because of its lower permeability.

The change of the slope of the evaporation rate in brick B (approximately after 120 h) is due to the increase of the airflow velocity during the experiment.
In addition, the results show a considerably different crystallization behavior for mirabilite in those bricks. While severe efflorescence forms in brick A, it is negligible in brick B even after 2 weeks. Indeed, the transport properties of the porous material influence the position of the crystallization front, as will be described in the next sections.

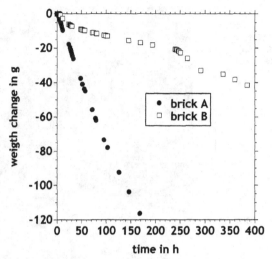

Figure 2. Evaporation rate of sodium sulfate solution (1.6 mol/kg) in bricks A and B during the capillary rise experiment at constant temperature (23°C) and relative humidity (50 % RH).

MATHEMATICAL MODEL

The system of coupled partial differential equations that describe heat and water transport in porous materials subjected to climatic conditions, was derived from the thermodynamical point of view in [9]. For the prediction of salt crystallization in porous materials an additional model for the ion transport based on the Nernst-Plank-model [1,2] coupled with the kinetics of crystallization of salts in porous materials was developed.

In a previous work [2], we investigated the phase change of salts (crystallization, dissolution, hydration, dehydration and deliquescence) in capillary porous materials and developed a model for the average rate. This model says that kinetics of crystallization in porous materials can be described using eq. (1). This equation is based on the fact that the driving force for crystallization is the supersaturation ratio of the solution with respect to a salt $U_{a,s}$:

$$\sigma_{c,s} = K_{c,s} \cdot \left(U_{a,s} - 1\right)^{g_{c,s}} \quad \text{with } U_{a,s} > U_{s,start} \quad \text{and } U_{a,s} = \frac{Q}{K} \qquad (1)$$

with $\sigma_{c,s}$ the average rate of crystallization $K_{c,s}$ and $g_{c,s}$, kinetic parameters, $U_{s,start}$, the necessary supersaturation for crystallization to start, Q the ion activity product and K the solubility constant. For the dissolution, hydration, dehydration and deliquescence, similar equations were deduced.

Three kinetic parameters are necessary to calculate the crystallization rate ($K_{c,s}$, $g_{c,s}$, $U_{s,start}$), which generally depend on both the pore structure and salt and must be obtained experimentally. The starting supersaturation ratio gives the threshold condition for the phase change. Actually, it must be remarked that eq. (1) does not give the growth rate of a single crystal but the average crystallization rate per unit volume of liquid.

The coupled transport-crystallization model was implemented in a simulation program (called ASTra). Computed results are in good agreement with experiments [5,6].

DISCUSSION

In this section the influence of the nucleation, solution and material properties on the resulting crystallization pattern during the drying and capillary rise experiments is discussed. Further, the numerical simulation of those experiments is used to understand the effect of each of them.

Nucleation

The application of the theory of heterogeneous nucleation [7] to halite and mirabilite reveals the quite different behavior of these salts. Thus, the nucleation rate per unit surface and time ($I_{a,het}$) results from:

$$I_{a,het} = \frac{k_B T}{3\pi\eta\Omega^{5/2}}\exp\left(-\frac{f(\theta)\Delta F_i^{*}}{k_B T}\right)$$

with $f(\theta) = \frac{(2 - 3\cdot\cos(\theta) + (\cos(\theta))^3)}{4}$

and $\Delta F_i^{*} = i^{*}\cdot k_B T \ln U_a + 4 i^{*2/3}\cdot\gamma_d\cdot\Omega^{2/3}$

θ=contact angle
Ω = volume of a molecule
k_B=Boltzmann-constant
T=temperature
η=viscosity
ΔF_i^{*} = free energy barrier for nucleation
γ_d=interface energy
i^{*}=critical size of an embryo

(2)

The influence of the substrate, in this case the pore wall, on nucleation results from the surface area of the nucleating sites and the contact angle. Since we do not have any information about the contact angle, we assumed the same value for both bricks.

The computed nucleation rate per unit surface of the substrate as a function of the supersaturation ratio S is depicted in figure 3. While the nucleation of NaCl requires a low theoretical supersaturation (about 1.21 [-] for 1 nuclei/m^{2}s), the results show the mirabilite to first nucleate at a higher supersaturation (S=4.4 [-] for 1 nuclei/m^{2}s). Further, the nucleation rate of NaCl changes strongly at small variations of the supersaturation.

The ability of mirabilite to achieve a high supersaturation was confirmed experimentally in [2]. Unfortunately it is not possible to measure the threshold supersaturation of NaCl according to the method explained in [2] since the solubility of NaCl is not dependent on temperature. However the experimental results for other salts (like KNO$_3$, KCl, CaCl$_2$.6H$_2$O)[1], are also in agreement with the theoretical results from this model.

---

[1] Crystallization commences for CaCl$_2$.6H$_2$O at high supersaturation and for KCl and KNO$_3$ at low supersaturation [2]

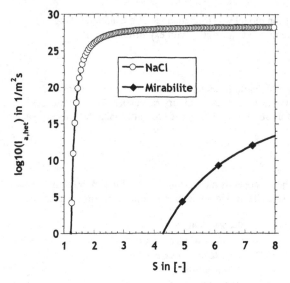

Figure 3. Heterogeneous nucleation rate per unit surface and time assuming a contact angle equal to 1.75 rad.

Since the necessary supersaturation of the solution for nucleation of both salts is so different, the crystal growth will take place in a different way, leading to equilibrium NaCl crystal morphology and non-equilibrium mirabilite crystal morphology (acicular) according to [10]. In fact, equilibrium NaCl-crystals in large pores are observed very often, while non-equilibrium mirabilite crystals are often reported in the pores [11].

Transport properties

The higher viscosity and surface energy of the solution compared to pure water leads to a slowdown of the drying process of the samples impregnated with the solution. Thus, the ratio of solution to pure water flow velocity v_{sol}/v_{H2O} is given by:

$$\frac{v_{sol}}{v_{H2O}} = \frac{\eta_{H2O}}{\eta_{sol}} \frac{\gamma_{sol}}{\gamma_{H2O}}$$

γ=surface energy in N/m
η =viscosity in Pa•s
v=flow velocity in m/s

(3)

Experimental values for the viscosity were found in [8] and for the surface energy of the solution in [12]. Figure 4 shows v_{sol}/v_{H2O} for sodium sulfate and for sodium chloride solutions for different values of the ratio between concentration and solubility. This figure shows clearly that the flux of both solutions is slower than the flow of pure water.

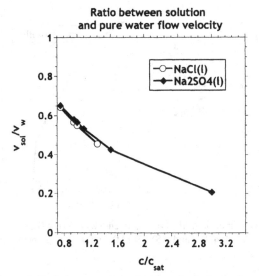

Figure 4. Ratio between the velocities of solution and pure water flow for sodium sulfate and sodium chloride solution at different values of the ratio between concentration and solubility c/c_{sat}, considering the influence of viscosity [8] and surface energy [11].

Further, within the usual range of values of the supersaturation ratio of halite and mirabilite, the flow velocity of sodium sulfate is slower. That means, the solution properties (viscosity and surface tension) do not explain the faster drying behavior of mirabilite found in figure 1. Therefore, *pore clogging* as well as the change of the surface properties due to efflorescence may be used to account for the decrease of the drying rate.

In this work we assume that the crystals are distributed in all pore sizes. This leads to a reduction of the permeability k even at a low pore filling.

Figure 5 shows the assumed permeability as a function of the pore filling pf according to:

$$k(pf) = k_0 \cdot g_p \text{ with } g_p = \left(1 - (pf)^a\right)^b \text{ and } pf = \frac{\theta_s}{\psi} \qquad (4)$$

with k_0, the permeability of the salt-free material in m^2/s, θ_s, the volume of precipitated salt per unit volume (m^3/m^3) and ψ, the pore volume (m^3/m^3). The parameters for this model (a=2.77 and b=1.77 for mirabilite) were obtained empirically.

Further, at low water contents, the pores will depercolate and the liquid flow will be hindered. That happens in brick A at a saturation ratio of about 10 % and in brick B at about 18 %. Therefore, it is assumed that the permeability tends to zero if salt crystals fill more than 90% of the total porosity of brick A and more than 80 % of the total porosity of brick B.

This empirical model leads to a good agreement between computed and experimental results [6].

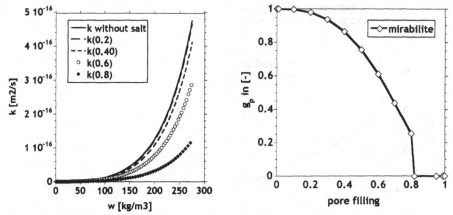

Figure 5. (Left) Liquid permeability k as a function of the water content (w) in brick B at different values of pore filling with salts (0.2, 0.4, 0.6 and 0.8). (Right) Resistance against liquid flow depending on pore filling given by g_p.

Numerical Simulation of capillary rise experiment

During capillary rise in brick A, the evaporation front is located at the surface because of the high capillarity of this material [4]. The results of the simulation show the severe formation of mirabilite efflorescence. Further, they show that there is no dehydration of mirabilite to thenardite in the simulated period of time (200 h) since the water content on the surface is still very high (70 % of saturation ratio) due to the steady water uptake. Thus, the rest of the pore volume is filled with salt, which grows from the solution into the surrounding air.

Apart from the differing permeability of bricks A and B, the distribution of nucleation sites may be also quite distinct in the bricks and with it the crystal distribution and the resulting pore clogging. Unfortunately, we do not have any information about the distribution of nucleation sites now. Therefore, the same resistance against the liquid flow as a function of the pore filling (see figure 5, right) is assumed for both bricks.

The influence of the kinetic parameters on the computed results for brick B was examined. Assuming $K_{c,s}=0.0064$ mol/m^{3}s and $g_{c,s}=1$ in eq. (1), crystallization takes place 1 cm beneath the surface after 196 hours and no appreciable efflorescence is expected.

Figures 6 shows, how the average crystallization rate affects the results of the simulation. If crystallization takes place 3 times slower ($K_{c,s}=0.002$ mol/m^{3}s), both, the resulting total amount of salt and the evaporation rate are larger. The reason for that is simply the reduction of the amount of crystals on the surface at the first drying stage due to the lower crystallization rate, which leads to a smaller reduction of the evaporation rate since pore clogging is less effective. Thus, the higher evaporation rate compared to the capillary transport moves the crystallization front into the interior of the sample of brick B. Due to this, the width of the crystallization zone with $K_{c,s}=0.002$ mol/m^{3}s reaches 1.8 cm beneath the surface after 196 hours.

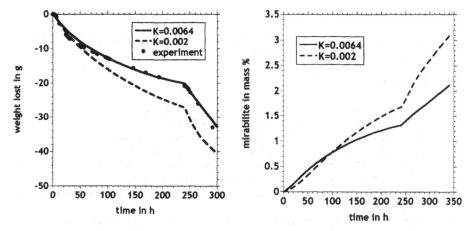

Figure 6. Evaporation rate (left) and total amount of mirabilite (right) in brick B at two different average crystallization rates ($K_{c,s}$=0.002 and $K_{c,s}$=0.0064).

It must be remarked that the situation may be quite different in the same material. Thus, a reduction of the evaporation rate due e.g. to a change of the climatic conditions may lead to the formation of severe efflorescence.

CONCLUSIONS

The model presented in this work reveals that the combination of salt crystal properties, solution transport and environmental conditions determines whether crystallization will result in damaging subflorescence or in efflorescence. The interaction between those factors can be examined and/or predicted with the help of the numerical simulation.

In particular, the ability of each salt to achieve high or low supersaturation determines the heterogeneous nucleation rate in the pores. Further, the rate of nucleation affects not only the crystal distribution in the pore network but also the resulting pore clogging.

Moreover, it was shown that liquid transport and crystallization processes will influence each other. Thus, liquid transport depends on the kind and intensity of pore clogging and determines the concentration gradient and supersaturation and with it the crystallization process at the same time.

Finally, it must be remarked that the average crystallization kinetics [2] does not say anything about the amount of crystals and the distribution in the pore structure. A high crystallization rate can be given by a large number of crystals, even if they have a low growth rate but also by a small number of crystals with a large growth rate caused by a high supersaturation. However the consequences on the pore clogging and on the damage of the materials are quite distinct depending on the crystal distribution. Therefore, the goal of future investigations is the study of the nucleation process and of the crystal distribution in the pores.

ACKNOWLEDGMENTS

The authors thank DFG for financial support and Prof. Scherer, Princeton University for his comments, suggestions and helpful discussions.

REFERENCES

1. Samson, E.; Marchand, J. : Numeric Solution of the extended Nernst-Planck Model, In: Journal of colloid and interface science 215 (1999), pp. 1 – 8.
2. Gunstmann, C.: Rechnerische Simulation von Säurekorrosionsprozessen zementgebundener Materialien. PhD, Hamburg University of Technology, 2007
3. Espinosa R. M., Franke L., Deckelmann G.: Phase changes of salts in porous materials, "Construction and Building materials, Elsevier", 2007 in press.
4. Scherer, G. W.: Stress from crystallization of salt, Cement and Concrete Research 34 (2004) 1613-1624
5. Espinosa R. M., Franke L., Deckelmann G.: Damage due to phase changes of salts in porous materials. Proceedings of 5th International Essen Workshop: Transport in Concrete: Nano-to-Macrostructure, 2007.
6. Franke, L., Kiekbusch, J., Espinosa, R., Gunstmann, C.: Cesa und Astra-two program systems for cement and salt chemistry and the prediction of corrosion processes in concrete. Proceedings of 5th International Essen Workshop: Transport in Concrete: Nano-to-Macrostructure, 2007.
7. Christian, J.W., The theory of transformation in metals and alloys, Part I: Equilibrium and General Kinetic Theory, 2nd edition, Pergamon Press, Oxford, 1975.
8. CRC, Handbook of Chemistry and Physics, CRC Press, 83rd Edition, 2002-2003
9. Bear, J. Dynamics of fluids in Porous Media. Dover Pubn Inc, 1988.
10. Sunagawa I: Crystals: growth, morphology and perfection. Cambridge University Press, 2005.
11. Rodriguez-Navarro C., Doehne E.: Salt weathering: influence of evaporation rate, supersaturation and crystallization pattern. Earth Surface Processes and Landforms, 24 (1999) 191-209.
12. Abramzon, A., Gaukhberg, R.: Surface Tension of Salt solutions. Russian Journal of Applied Chemistry, Vol. 66 (6, 7 and 8), part 2, 1993.

Mater. Res. Soc. Symp. Proc. Vol. 1047 © 2008 Materials Research Society

Focused-Ion Beam and Electron Microscopy Analysis of Corrosion of Lead-Tin Alloys: Applications to Conservation of Organ Pipes

Catherine M. Oertel[1], Shefford P. Baker[2], Annika Niklasson[3], Lars-Gunnar Johansson[3], and Jan-Erik Svensson[3]

[1]Department of Chemistry and Biochemistry, Oberlin College, Oberlin, OH, 44074
[2]Department of Materials Science and Engineering, Cornell University, Ithaca, NY, 14853
[3]Department of Chemical and Biological Engineering, Chalmers University of Technology, SE-412 96 Göteborg, Sweden

ABSTRACT

Across Europe, lead-tin alloy organ pipes are suffering from atmospheric corrosion. This deterioration can eventually lead to cracks and holes, preventing the pipes from producing sound. Organ pipes are found in compositions ranging from >99% Pb to >99% Sn. For very lead-rich (>99% Pb) pipes, organic acids emitted from the wood of organ cases have previously been identified as significant corrosive agents. In order to study the role of alloy composition in the susceptibility of pipes to organic acid attack, lead-tin alloys containing 1.2-15 at.% Sn were exposed to acetic acid vapors in laboratory exposure studies. Corrosion rates were monitored gravimetrically, and corrosion product phases were identified using grazing incidence angle X-ray diffraction. In a new method, focused-ion beam (FIB) cross sections were cut through corrosion sites, and SEM and WDX were used to obtain detailed information about the morphology and chemical composition of the corrosion layers. The combination of FIB and SEM has made it possible to obtain depth information about these micron-scale layers, providing insight into the influence of acetic acid on alloys in the 1.2-15 at.% Sn range.

INTRODUCTION

As the earliest western keyboard instrument, the pipe organ has its roots in rudimentary devices with wind-blown pipes invented in Alexandria during the third century B.C.E. and found throughout Greco-Roman culture. Use of the organ faded with the fall of the Roman Empire and was not resumed in western Europe until the eighth century A.D. Over the following centuries, the role of the organ in sacred settings increased, and the instrument grew in both musical and technological sophistication, reaching a high point during the late Renaissance and early Baroque period [1]. While interest in the organ has continued since that time, the industrial revolution led to changes in organ design and construction, and many historic organs have been lost to changing aesthetics and to wartime damage. Baroque pipe organs that remain today are valued for their ties to noted composers and performers of the past and for the record that they provide of pre-industrial organ building techniques. In organs across Europe, atmospheric corrosion has resulted in the buildup of heavy corrosion layers on valuable historic pipes. As shown in Figure 1, this has in some cases led to formation of cracks and holes that prevent the pipes from producing sound.

Organ pipes are typically made from lead-tin alloys, with compositions ranging from >99% Pb to >99% Sn. Work carried out as part of the European Commission-funded

COLLAPSE (Corrosion of Lead and Lead-Tin Alloys of Organ Pipes in Europe) project showed that acetic acid and formic acid emitted from the wood of organ cases are the primary corrosion agents for very lead-rich (>99% Pb) pipes [2-6]. Corrosion has been observed in organ pipes across the lead-tin phase diagram [7], but the role of organic acids in corrosion of tin-containing pipes has not been studied in detail. Several previous reports have focused on corrosion of lead-tin alloys by inorganic pollutants including Cl_2, NO_2, and SO_2 [8-10], but the effects of organic acids have not been examined.

Figure 1. Photograph of a corroded pipe from an organ in Lübeck, Germany. The pipe was part of the original organ constructed in 1467 and was retained when the instrument was rebuilt by Friedrich Stellwagen in 1636-37. (Photo: Ibo Ortgies, Göteborg Organ Art Center).

We report here on microscopic characterization of corrosion films resulting from acetic acid attack on lead-tin alloys containing 1.2-15 at.% Sn. Samples were exposed to controlled atmospheric compositions, temperatures, and humidity levels in the range of those measured in organ cases. A new focused-ion beam (FIB) cutting method was used to prepare cross sections through corrosion crusts. The combination of FIB milling and analysis by scanning electron microscopy (SEM) with wavelength-dispersive X-ray detection (WDX) has made it possible to obtain chemical maps of the micron-scale corrosion layers grown under realistic conditions.

EXPERIMENT

Atmospheric exposure of samples. Laboratory exposure experiments were carried out using 1.2, 3.4, and 15 at.% Sn (0.70, 2.0, and 9.7 wt.% Sn, respectively) lead-tin alloy coupons with dimensions of 30 x 30 x 2 mm (total surface area 20 cm^3). Coupons were produced and supplied by organ builder Marco Fratti (Campogalliano, Modena, Italy). Prior to exposures, samples were polished with 220 grit SiC paper using water as a lubricant and then with 1000 grit SiC paper using absolute ethanol as a lubricant. Samples were cleaned ultrasonically in two washes of 30 seconds each in absolute ethanol. Coupons were then dried and stored over silica gel until just before the start of the exposure.

The exposure apparatus, made entirely of glass and Teflon, consists of eight glass chambers in which samples are suspended on nylon threads without contact with the chamber wall (Figure 2). The gas is sequentially distributed through the parallel chambers so that the whole gas flow passes through each chamber for 15 s at a time. The flow rate was 1000 mL/min for all exposures. The chambers were held at a constant temperature of 22 °C by immersion in a water bath. The relative humidity (RH) was 95% for all exposures, achieved through mixing of dry air and air saturated with water vapor. Carbon dioxide was introduced to the mixture from a tank, and a CO_2 analyzer (BINOS 100) was used to adjust the concentration to 350 ppm (v/v). For exposures in acetic acid environments, acetic acid vapor was introduced by passage of the gas through a temperature-controlled chamber containing a permeation tube.

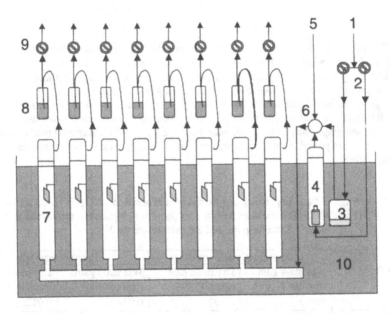

Figure 2. Schematic of chamber used for exposure experiments. (1) inlet of dried, purified air, (2) flow rate control, (3) humidifying chamber, (4) vessels for permeation tubes, (5) inlet of CO_2, (6) mixing point, (7) sample chambers, (8) outlet bubblers, (9) eight-channel solenoid valve system, and (10) constant-temperature water tank.

After the start of each exposure, mass gain data were obtained gravimetrically after 0.5, 1, 2, 3, and 4 weeks of exposure time. These "wet" mass gains were obtained without drying of samples in order not to disrupt the corrosion process. At the end of the four-week experiment, samples were dried over silica gel for at least 24 hours before obtaining final "dry" mass gain measurements. All samples were exposed in duplicate.

X-ray analysis. Phase identification of corrosion product layers was done using grazing incidence angle X-ray analysis with an incidence angle of 0.3°. A Siemens D5000 powder diffractometer (CuK$_\alpha$ radiation) equipped with a Gobel mirror and a grazing incidence attachment was used.

Scanning electron microscopy. Scanning electron microscopy (SEM) imaging was carried out using a Zeiss SUPRA 55VP SEM. Elemental mapping was performed on a JEOL 8900 EPMA microscope with wavelength dispersive X-ray spectrometers. Plan views were obtained by mounting the sample horizontally on the stage. For cross-sectional views, samples were mounted on a 45° pre-tilt sample holder before being introduced into the microscope chamber.

Focused-ion beam sectioning. Focused-ion beam (FIB) cross-sectioning of corrosion sites was done using an FEI 611 FIB with a Ga$^+$ ion beam. Sites of interest were selected by using the instrument in low-current (60-100 pA) imaging mode. Initial rough cutting was done with a 2000 pA beam, and finer cutting and polishing steps were done with beams of 1000 and 500 pA. Each cross section was cut at 45° with respect to the sample surface normal.

RESULTS

Table I shows mass gains for the alloys examined in this study as well as data for pure lead previously collected in our laboratory [2]. Both wet and dry mass gains for all compositions were enhanced by more than an order of magnitude when exposed to 1100 ppb acetic acid as opposed to pure air. In an atmosphere of pure air, mass gains were quite low, and both wet and dry mass gains decreased with increasing tin content. Under an atmosphere of 1100 ppb acetic acid and 95% relative humidity, an increase in tin content led to an increase in wet and dry mass gains. As shown graphically in Figure 3, mass gains were generally linear with time during the four-week exposure period, with a slight acceleration in mass gain rates for the samples containing 15% Sn.

Table I. Average wet and dry mass gains for samples exposed to 1100 ppb acetic acid or pure air at 95% RH and 22°C. The standard deviation of data from duplicate samples was about 5%. Data for pure lead are from reference 2.

Tin content of alloy (at.%)	1100 ppb acetic acid exposure		Pure air exposure	
	Wet mass gain (mg/cm^2)	Dry mass gain (mg/cm^2)	Wet mass gain (mg/cm^2)	Dry mass gain (mg/cm^2)
0.0%	0.910	0.890	0.0240	0.0120
1.2%	0.992	0.945	0.0195	0.0100
3.4%	1.189	1.127	0.0170	0.0080
15%	1.386	1.264	0.0065	0.0020

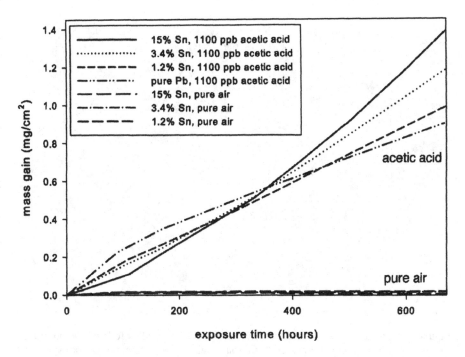

Figure 3. Wet mass gain as a function of exposure time for samples exposed to 1100 ppb acetic acid or pure air at 95% RH and 22°C for four weeks (672 hours). Data for pure lead are from reference 2.

Phases comprising the thick oxide layer formed on samples exposed to 1100 ppb acetic acid at 95% RH were examined using grazing incidence angle X-ray diffraction. The layer was characterized as a mixture of lead acetate oxide hydrate ($Pb(CH_3COO)_2 \cdot 2PbO \cdot H_2O$, ICDD no. 18-1740), plumbonacrite ($Pb_{10}O(OH)_6(CO_3)_6$), and a small amount of massicot (PbO). Neither metallic tin nor tin-containing corrosion products were observed by X-ray diffraction. It was not possible to identify phases making up the very thin corrosion layer on samples exposed to pure air.

Corrosion product morphologies as examined by SEM plan views are the same as those reported for pure lead exposed to acetic acid [2]. As shown in Figure 4, thick, cauliflower-like crusts are interspersed among regions of thin oxide coverage. Scratches from grinding prior to exposure are visible in the thin oxide regions. At high magnification, individual crystallites can be observed within the crusts.

Figure 4. Plan view secondary electron SEM image of 15% Sn alloy following exposure to 1100 ppb acetic acid at 95% RH and 22 °C for four weeks. The combination of corrosion crusts and thin oxide regions is typical of alloys exposed within this study.

Composites of SEM secondary electron images and elemental maps of cross sections in alloys containing 1.2, 3.4, and 15% Sn are shown in Figures 5-7. The elemental maps for Pb, Sn, and O were obtained using WDX from the cross section produced in each sample using FIB. Note that each micrograph is wider than the FIB-milled region so that the unmilled sample surface can be compared with the FIB cross section. In all three alloys studied, the crust penetrates beneath the surface of the bulk metal. In samples containing 3.4% and 15% Sn, tin-rich inclusions are visible in the bulk metal, consistent with the low solubility of tin in lead at room temperature [11]. For each alloy, Pb is most concentrated in the bulk metal, and O is most abundant in the oxide layer. Sn content is enriched in the region at the oxide-metal interface in each case. Mapping of a cross section cut in a thin oxide region of the sample containing 3.4% Sn (Figure 8) does not show significant patterns in oxygen content or tin segregation beyond several tin-rich inclusions.

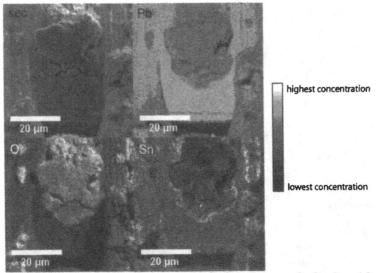

Figure 5. Secondary electron SEM image and WDX elemental maps for Pb, O, and Sn for an FIB cross section through a corrosion crust in an alloy containing 1.2% Sn and exposed to 1100 ppb acetic acid for four weeks.

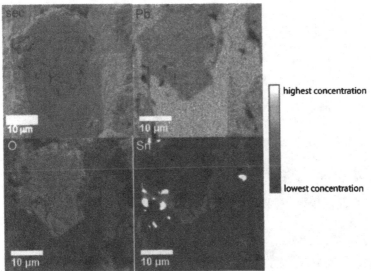

Figure 6. Secondary electron SEM image and WDX elemental maps for Pb, O, and Sn for an FIB cross section of a corrosion crust in an alloy containing 3.4% Sn and exposed to 1100 ppb acetic acid for four weeks.

Figure 7. Secondary electron SEM image and WDX elemental maps for Pb, O, and Sn for an FIB cross section of a corrosion crust in an alloy containing 15% Sn and exposed to 1100 ppb acetic acid for four weeks.

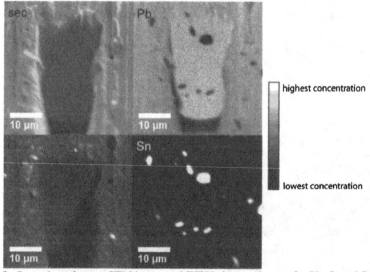

Figure 8. Secondary electron SEM image and WDX elemental maps for Pb, O, and Sn for an FIB cross section in a thin oxide region in an alloy containing 3.4% Sn and exposed to 1100 ppb acetic acid for four weeks.

DISCUSSION

Trends in extent of corrosion with tin content

In this study, mass gains for lead-tin alloy samples exposed to acetic acid increased as the tin content increased from 1.2 at.% to 15 at.%. The mass gain for each alloy was larger than that for pure lead under the same laboratory conditions [2]. This result is in contrast with empirical evidence from conservation studies that suggests that tin has a protective effect in lead-tin alloys. Lead seals and coins in the collection of the British Museum suffered from acetic acid corrosion after having been in contact with wood or stored in wooden cases. X-ray fluorescence analysis of the artifacts revealed that objects containing 2-5% Sn had suffered markedly less corrosion than those containing >99% lead [12]. A similar observation has been made concerning pipes in a sixteenth-century organ in Ponte in Valtellina, Italy. A portion of a pipe containing 1.3% tin developed a thin corrosion layer while a portion of the same pipe containing no detectable tin became more heavily corroded [4, 5].

The difference between the laboratory and field results may be explained by differences between the laboratory exposure conditions and those found in organ cases. Field studies carried out through the COLLAPSE project measured temperatures, humidities, and corrosive agent concentrations in several organ cases throughout Europe [4-6]. Acetic acid concentrations ranged from 100 ppb to nearly 1500 ppb; thus the concentration of 1100 ppb used in laboratory exposures falls within a realistic range. Mean relative humidites over a year of monitoring fell between 53 and 62%, with summertime maxima mostly in the range of 70-80%. The RH of 95% used in the laboratory exposure experiments is outside the observed range and may explain the difference in corrosion trends. Indeed, as will be described in detail in a future report, laboratory exposure experiments carried out at 60% RH show the expected protective role for tin.

Microscopic analysis of corroded lead-tin alloy surfaces

The SEM plan views and grazing incidence angle X-ray diffraction results indicate that the morphology and composition of the corrosion crusts formed on the lead-tin alloys examined are the same as those for pure lead. Although the presence of tin enhances the extent of corrosion in samples exposed at 95% RH, tin-containing corrosion products are not identifiable by X-ray diffraction. Tin-containing products may be of insufficient thickness or crystallinity to produce diffraction of observable intensity. Tin-based products deposited beneath lead-containing corrosion crusts may also be undetectable by X-ray diffraction due to the strongly absorbing character of the lead-based salts.

The FIB cross-sectioning method employed offers several advantages in obtaining chemical information about corrosion crusts. Previously, surface and near-surface methods had been used to obtain depth information for nanometer-scale native oxides on lead-tin alloys [13-15], but micron-scale crusts had not been accessible through these experiments. As compared with mechanical cross-sectioning and polishing, FIB is a gentle method that keeps the oxide layer and metal-oxide interface intact. This is of particular concern for soft lead-tin alloys and soluble corrosion products, both of which could be damaged by solvent-cooled mechanical polishing. Further, because the sample is imaged in the low-current mode of the instrument prior to milling, it is possible to carefully select the cross-sectioning site so that it coincides with either a corrosion crust or an area of thin oxide. Impregnation of the exposed cross section by the gallium beam was initially a concern, but a gallium map obtained using WDX shows that gallium is not present in detectable amounts on the cross-sectional surface. Preparation of a cross-sectional surface at 45° to the surface normal is essential in allowing analysis through SEM/WDX. This geometry makes the analysis surface horizontal when the sample is tilted by 45°. When the sample is tilted toward the X-ray detector, the cross section is accessible to the electron beam, and generated X-rays are not blocked from reaching the detector. The horizontal orientation of the cross-section also enables semi-quantitative X-ray analysis.

In all alloys studied, corrosion crusts penetrate beneath the surface of the bulk metal, confirming that the corrosion is taking place through an electrochemical corrosion mechanism. Enrichment of tin is consistently observed at the oxide-metal interface in corroded samples. A cross section cut through a thin oxide region on a sample exposed to acetic acid does not display tin enrichment, indicating that the segregation occurs at the site of corrosion attack and is not a uniform feature of the metal surface. The oxidation state of the tin in this region has not been ascertained, but the area of tin enrichment appears to overlap with the area of oxide enrichment, particularly in the alloys containing 3.4 and 15% Sn. Because of the passivating nature of tin oxides, the enriched area might be expected to act as a diffusion barrier to further corrosion. This type of protective behavior for tin has been observed in other alloys. In an example from conservation science, formation of regions of SnO_2 in Chinese bronze mirrors has been shown to produce a corrosion-resistant surface [16]. Under the high humidity employed in our study, presence of an area of tin enrichment did not afford corrosion inhibition, but formation of the layer may have a role in the protective effect of tin that has been observed under other conditions.

CONCLUSIONS

As has been observed previously for pure lead, lead-tin alloys in the range of 1.2-15 at.% Sn are susceptible to attack by acetic acid, leading to formation of lead-based corrosion products. Under an atmosphere of 95% relative humidity and 1100 ppb acetic acid, the mass gain due to corrosion is enhanced with increasing tin content within the compositional range studied.

FIB milling can be used to prepare cross sections through corrosion sites on lead-tin alloys. Examination by SEM and chemical mapping by WDX reveal corrosion crusts that penetrate beneath the surface of the bulk metal and segregation of tin at the metal-oxide interface. Characterization of this enriched layer and its role in the susceptibility of lead-tin alloys to organic acid attack will be the objects of future work.

ACKNOWLEDGMENTS

C.M.O. thanks the National Science Foundation Discovery Corps Program (grants CHE-0412181 and CHE-0631552) for support. This research made use of shared facilities of the Cornell Center for Materials Research, part of the National Science Foundation MRSEC program (grant DMR-0520404). This work was also performed in part at the Cornell NanoScale Facility, a member of the National Nanotechnology Infrastructure Network, which is supported by the National Science Foundation (grant ECS-0335765).

Part of this work was supported within the European Commission Fifth Framework Program: Energy, Environment, and Sustainable Development (EVK4-CT-2002-00088, COLLAPSE).

REFERENCES

1. B. Owen, P. Williams, and S. Bicknell in *The New Grove Dictionary of Music and Musicians*, 2nd ed., edited by Stanley Sadie (Grove, New York, 2001), Vol. 18, pp. 565-650.
2. A. Niklasson, L.-G. Johansson, and J.-E. Svensson, J. Electrochem. Soc. **152**(12), B519-B525 (2005).
3. A. Niklasson, L.-G. Johansson, and J.-E. Svensson, J. Electrochem. Soc. **154**(11), C618-C625 (2007).
4. A. Niklasson, Ph.D. Dissertation, Chalmers University of Technology, Göteborg, Sweden, 2007.
5. C. Chiavari, C. Martini, D. Prandstraller, A. Niklasson, J.-E. Svensson, A. Åslund, and C.J. Bergsten, Corros. Sci., accepted for publication.
6. A. Niklasson, S. Langer, K. Arrhenius, L. Rosell, C.J. Bergsten, L.-G. Johansson, and J.-E. Svensson, Studies in Conservation, accepted for publication.
7. C. Chiavari, C. Martini, G. Poli, and D. Prandstraller, J. Mater. Sci. **41**, 1819-1826 (2006).
8. H.G. Tompkins, Surf. Sci. **32**, 269-277 (1972).
9. H.G. Tompkins, Solid State Ionics **120**(5), 651-654 (1973).
10. V. Brusic, D.D. DiMilia, and R. MacInnes, Corrosion **57**(7), 509-518 (1991).
11. W.F. Gale and T.C. Totemeier, editors, *Smithells Metals Reference Book*, 8th ed. (Elsevier Butterworth-Heinemann, Burlington, MA, 2004) p. 457.
12. L. Green in *Conservation of Metals: Problems in the Treatment of Metal-Organic and Metal-Inorganic Composite Objects*, Proceedings of the Seventh International Restorer Seminar, Veszprém, Hungary, 1989, edited by Márta Járó, (István Éri, Veszprem, Hungary, 1990) pp. 121-130.
13. R.J. Bird, Met. Sci. J. **7**, 109-113 (1973).
14. R.P. Frankenthal and D.J. Siconolfi, J. Vac. Sci. Technol. **17**(6), 1315-1319 (1980).
15. R.P. Frankenthal and D.J. Siconolfi, Corros. Sci. **21**(7), 479-486 (1981).
16. M. Taube, A.J. Davenport, A.H. King, and W.T. Chase in *Aqueous Chemistry and Geochemistry of Oxides, Oxyhydroxides, and Related Materials*, edited by J.A. Voigt, T.E. Wood, B.C. Bunker, W.H. Casey, and L.J. Crossey, (Mat. Res. Soc. Symp. Proc. **432**, Pittsburgh, PA, 1997) pp. 283-288.

Mater. Res. Soc. Symp. Proc. Vol. 1047 © 2008 Materials Research Society   1047-Y04-01

Delamination of Oil Paints on Acrylic Grounds

Yonah Maor, and Alison Murray

Art Conservation Program, Queen's University, Kingston, K7L 3N6, Canada

ABSTRACT

In a set of composite samples of oil or alkyd paints, over acrylic grounds, naturally aged for eight years, some of the samples delaminated. Samples were analyzed with X-ray fluorescence (XRF), inductively coupled plasma (ICP), Fourier transform infrared - attenuated total reflectance (FTIR-ATR) and atomic force microscopy (AFM), as well as other techniques not detailed in this paper. Results indicate the main cause of delamination is metal soaps in the oil paint and particularly zinc soaps. The ground is a minor consideration as well, rougher grounds providing better adhesion than smooth ones.

INTRODUCTION

Since the 1950s the traditional lead white in oil ground has mostly been abandoned in favor of titanium white in acrylic or oil-modified alkyd grounds. The acrylic grounds are used for both oil and acrylic paints. The bond between an acrylic ground and an oil paint is of a mechanical nature, rather than there being a chemical bond: the oil paint enters crevices in the rough acrylic ground and interlocks as it dries. Painting in oil over acrylic is usually considered a safe practice. In some cases, however, there has been severe delamination a few months or years after painting (see figure 1).

This paper examines a set of composite samples, some of which are delaminating, in order to find what the delaminating samples have in common. Once theories were established, severely delaminated paintings were examined as well.

 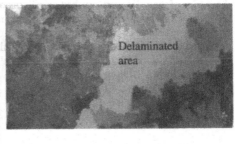

Figure 1. Michael Skalka, Colarado Lagoon, Long Beach, CA, Oil and alkyd on canvas, 12"x16", 2006. Photo taken immediately after painting. Detail of delamination with ground showing, taken Nov. 2007

EXPERIMENT

Samples

This research made use of samples prepared in 1999 by Marion Mecklenburg at the Smithsonian Institution. The samples were combinations of eight different acrylic grounds (table I) and 20 paints, on Mylar® (polyethylene terephthalate). The paints were either oils or alkyds, from several manufacturers, and included the following pigments: verdigris, titanium white, lead white (also called flake white), raw sienna, cobalt blue, ultramarine, light red oxide, yellow ochre, terre verte, Indian red and burnt sienna. This paper focuses on ten paints where there were samples on all the different grounds. None of the samples with the other ten paints delaminated. The samples were prepared identically and stored together so that variables related to technique or environment may be ruled out.

Table I. Ground and paints in composite samples

Ground		Paint	
A	Aaron Brothers White Gesso	1	Custom manufactured Verdigris in oil
B	W&N* Clear Gesso Base for Acrylics	2	W&N, Titanium White in oil
C	W&N Acrylic Gesso Primer	3	Grumbacher, Flake White in oil
D	Utrecht Artists' Acrylic Gesso	4	W&N, Flake White in oil
E	Utrecht Professional Acrylic Gesso	5	W&N, Raw Sienna in oil
F	Liquitex Acrylic Gesso	6	Custom manufactured Cobalt blue in oil
G	Dick Blick Artists' Acrylic Gesso	7	W&N, Titanium White in alkyd
H	Grumbacher Artists' Acrylic Gesso	8	W&N, Flake White in alkyd
		9	Gamblin, Ultramarine blue in oil
		10	W&N, Raw Sienna in alkyd

* W&N = Winsor Newton

The samples were prepared by drawing down parallel strips of acrylic ground on Mylar® and then strips of the paint were brushed over in the other direction, in a cross-hatch pattern (figure 2). There are therefore areas of paint alone and areas of uncovered ground, as well as the composite structures. In addition, two samples from severely delaminating paintings were examined by non-destructive methods. The paintings were donated by artists, and are designated by their main colors. Skalka's 'Colarado Lagoon' is 'the blue painting' and the second painting is 'the grey/brown painting'.

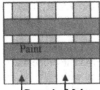

Ground Mylar   **Figure 2**. Schematic of composite samples.

The samples had been kept in storage, so it is not known when they began to delaminate. Currently only samples with four oil paints are delaminating: verdigris, titanium white and two lead-white paints from different manufacturers (figure 3). None of the alkyd samples are delaminating. The distribution of samples by type of ground shows at least one sample delaminating of every type of ground and four at the most. There are approximately ten samples with each ground and eight with each paint.

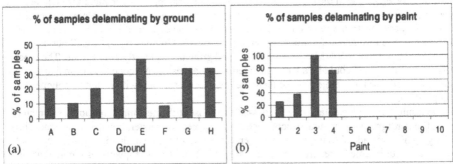

Figure 3. Distribution of delaminating samples: (a) by ground, (b) by paint.

Methods

Two methods of elemental analysis were used, X-ray fluorescence (XRF) and inductively coupled plasma - optical emission spectroscopy (ICP-OES). XRF data was collected with a handheld, Innov-X systems XT-440L and accompanying Innov-X PC 1.53 software. No sample preparation was needed with this method. For ICP-OES, samples of approximately 100 mg were dissolved in 2 ml nitric acid and 6 ml hydrochloric acid. The samples were heated overnight (18 hours) until almost dry. Double de-ionized water (25 ml) was mixed in and the solutions were filtered. Analysis was done with Vista AX CCD simultaneous ICP-AES.

The non-destructive, XRF technique is much faster and can analyze for more elements simultaneously; however, there are no standards for artists' materials, so it is not quantitative. The 1999 samples were analyzed by both methods and linear calibration curves were created for Zn and Pb. There are too few samples to claim these curves allow exact quantification of XRF results, and therefore the quantities reported for paintings examined only by XRF are approximations.

The paint and ground medium were examined by Fourier transform infrared (FTIR) spectroscopy. Analysis was performed with a Nicolet Avatar 320 FTIR instrument equipped with a Nicolet SMART Golden Gate diamond accessory for attenuated total reflectance (ATR). Data was collected with EZ OMNIC 5.2a software and converted into GRAMS/32 AI (6.00) software format for comparison.

The surface morphology was examined with atomic force microscopy (AFM) in tapping mode, with a Di Digital Instruments / Veeco Metrology Group Multimode AFM. The probe was VistaProbe T300-10 with tip radius <10nm. Images were collected and analyzed with NanoScope software (version 5). Roughness values were taken as the mathematical mean of deviation from the median line (Ra). Samples were cut into squares of approximately 5x5 mm. Scan size was 20x20 µm. Readings were taken at a minimum of three random locations on each sample and the roughness results were averaged.

Results

The methodology was simply to analyze the samples to find what the delaminating ones have in common. Many additional techniques were used besides those detailed above, and did not show any correlation with the distribution of delaminating samples. Those methods included gloss measurements, pyrolysis - gas chromatography - mass spectrometry and laser desorption/ ionization mass spectrometry of surfactants leached from the surface of grounds. One result of note was the identification of binding media in the grounds. Grounds B to H are all butylacrylate – methylmethacrylite (nBA-MMA) copolymers and ground A is mostly polystyrene. The type of polymer in the ground could be a factor in determining adhesion. No conclusions on that aspect can be derived form this project, since there are not enough samples with differing polymer compositions.

The results of the elemental analysis showed that most of the delaminating samples contained zinc. The verdigris paint obviously contained a high percent of copper, but also some lead, since litharge, PbO, was added to the paint as a drier. Since two of the delaminating paints were lead-white paints, the elements reported in table II are Pb, Zn and Cu. The quantities reported are the results of ICP-AES and are accurate within approximately 50 ppm, unless noted otherwise. None of the grounds in the 1999 samples contained any of those metals.

Table II. Quantities of zinc, lead, and Copper

Sample	% delaminating	Zn (ppm)	Pb (ppm)	Cu (ppm)
Paint 1	25	13.23	2446.52	272786
Paint 2	37.5	63488.7	312.21	25.65
Paint 3	100	34785.5	578680	9.45
Paint 4	75	124263	485923	17.37
Paint 5	0	556.42	61.92	231.84
Paint 6	0	35.14	78.24	21.56
Paint 7	0	593.28	13.52	19.94
Paint 8	0	172.05	636407	19
Paint 9	0	58.25	6.69	18.4
Paint 10	0	121.74	21.48	134.71
Blue paint from 'blue painting'*	Completely	~190000	~43000	none
Green paint from 'blue painting'*	Slightly	~70000	~40000	none
Ground from 'blue painting'*	N/A	~110000	~58000	none
Paint from 'grey/brown painting'*	Severely	~160000	~15000	A little
Ground from 'grey/brown painting'*	N/A	~210000	~26000	none

* Measured with XRF and converted to approximate quantities.

Zinc, lead and copper all form metal soaps with fatty acids in oil [1]. The metal soaps have major absorbance peaks in the IR in the 1530-1600 cm^{-1} range. All the delaminating samples examined contain metal soaps (figure 4). For the sake of clarity, spectra of only copper stearate and zinc stearate are provided as references. It should be noted that overlap between the various metal soap peaks is expected, creating broad absorption around 1530-1650 cm^{-1}. In some cases, for instance in paint 3, there are peaks of specific metal soaps super-imposed on the broad peaks, indicating a greater amount of that material. If more than one metal is present in paint, the interpretation is difficult. For instance, lead stearate absorbs at 1541cm^{-1}, practically identical to zinc stearate. Zinc oleate, on the other hand, has a medium peak at 1589cm^{-1}, quite close to the copper stearate peak at 1586cm^{-1} [2].

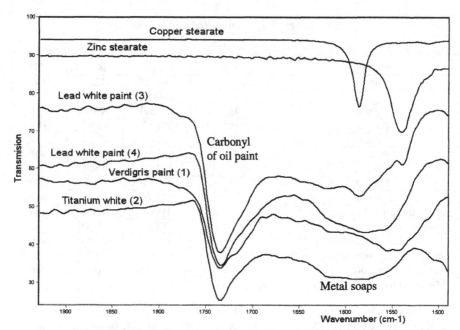

Figure 4. FTIR-ATR spectra of oil paints from delaminating samples.

The non-delaminating samples were examined by FTIR-ATR as well and none of them showed signs of metal-soaps (results are not displayed). Flakes of paint from the delaminated paintings were examined and so was the exposed ground. All showed absorption in the 1530-1600cm^{-1} range (figure 5). In the spectra of the grey/brown paint, there is a succession of three small peaks around 1600cm^{-1}, characteristic of alkyd paints and not indicative of additional metal soaps; however, the peak at 1539 cm^{-1} matches the zinc stearate reference perfectly.

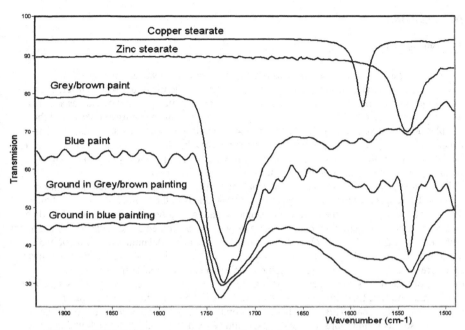

Figure 5. FTIR-ATR spectra of paints and grounds from delaminating paintings.

The results of the roughness analysis from the AFM data indicated that the roughness of the ground can be significant. The mean roughness results are displayed in figure 6. Ground B was too rough to measure with the equipment used in this study (it was rougher than the others). It is included in the graph for the purpose of illustrating the results as fully as possible, with a value slightly higher than any of the other grounds.

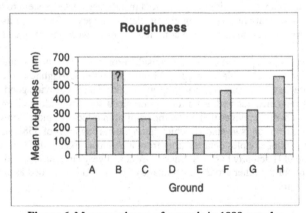

Figure 6. Mean roughness of grounds in 1999 samples

DISCUSSION

The 1999 samples

Additives can migrate to both surfaces of an acrylic ground; however, on Mylar®, a nonabsorbent substrate, they are likely to migrate mostly to the top of the acrylic layer. This means the build-up of a surfactant layer could be greater than in a painting on canvas and this may increase the tendency to delaminate. This could explain why a significant percent of the samples delaminated, even though this is a rare problem in actual paintings. The distribution of delaminating samples shows that the problem stems from the choice of oil paint. These samples included only acrylic grounds by the major manufacturers. The grounds appear to be only a minor consideration in determining adhesion, as shown by figure 3.

The elemental analysis shows the most problematic paints contain significant quantities of zinc and FTIR spectra show that at least some of that zinc is in the form of zinc-soaps. Among the zinc-containing paints, the number of samples delaminating seems to be greater when there is lead present in addition to zinc. Paint #3 has a very high lead content, so despite containing about half the zinc of the TiO_2 paint #2, 100% of the samples with are delaminating, while only 37.5% of those with paint #2 delaminated. The FTIR spectra show a wide absorption area where metal soaps are expected, so identifying a specific metal soap is only possible if there is a larger quantity of that metal soap. When comparing the spectra of the different paints, it is seen that the zinc stearate peak at 1539 cm^{-1} is more prominent in paint #3. This indicates that this paint contains more zinc-soap. This is another indication that zinc soaps are a cause of delamination.

The reason why zinc soaps, or lead soaps, could cause delamination may be found in a new theory that explains the drying of oil paints. According to the traditional theory the paint dries by cross-linking [3], and there is hydrolysis breaking up the cross-linked paint. Research on the rate of hydrolysis showed this should bring about complete failure of the paint layer. Since that doesn't happen, a 'repair mechanism' was suggested by the MOLART group [4]. Lead or other metals bind the loose ends in a metal coordinated network. This could explain why researchers found paints with lead soaps were stronger [5]. Another study found that zinc white pigment catalyses the hydrolysis of glycerol-esters [6]. Perhaps the faster hydrolysis in zinc-rich paints causes the transition from a cross-linked system to a metal-coordinated system to happen quickly, forming paints that are very strong and brittle. The cohesive forces in this metal-coordinated layer are then stronger than the adhesion forces between the paint and ground and the adhesion fails.

As mentioned above, the ground is a minor consideration as well. If the distribution of delaminating samples by ground layer (figure 3) is compared to the roughness, seen in figure 6, there is a certain degree of correlation. Rough grounds, like F and B provide good 'tooth', so most of the paints remained adhered, even if they contained some zinc. Smooth grounds, like D and E, do not have enough crevices that allow the paint to interlock, so more samples are delaminating. The exception is ground H, which has a mean roughness of 558 nm, yet has a relatively high percent of samples delaminating from it. A closer look at the AFM images (figure 7), shows that ground H actually has very smooth areas and very large lumps, and not the more homogenous texture of the other grounds. An image of ground G is provided as an example of a normal ground texture.

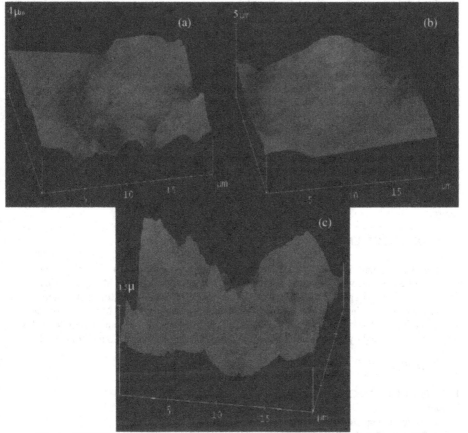

Figure 7. AFM images show: (a) and (b) different areas of ground H, (c) ground G.

<u>The paintings</u>

The actual paintings have only been examined with XRF and FTIR because of the amount of sample available. The 'blue painting' contains mostly light blue and green areas. The light blue was made with zinc white and ultramarine blue oil paints. The zinc soaps, indicated by the sharp peak in the FTIR spectra, explain why that paint has failed to the point that almost all of it had flaked off. The green paint in that painting has much less zinc in it, and is better adhered. Zinc was identified in the ground of this painting as well. The medium in the ground does not match any of the FTIR reference spectra examined and is probably some experimental material that did not provide good adhesion.

The grey/brown painting has a very high percentage of zinc soaps in both the paint and the ground, causing delamination. In this case the paint medium appears to be alkyd, while the ground is probably oil. It is interesting that the alkyd paint is delaminating, since none of the alkyd paints in the 1999 samples delaminated; however, none of those samples contained zinc, only lead. This indicates that lead soaps alone will not cause delamination, or not as quickly.

CONCLUSIONS

The naturally aged composite samples, prepared in a lab, provided the opportunity to narrow down the possible causes of delamination. The samples showed clearly that even when using high quality acrylic grounds, oil paint may delaminate from them. The determining factor is the metals present in the oil paint. The analysis found that all delaminating samples contain metal soaps and particularly zinc soaps. The presence of lead, in addition to zinc, increased the problem. Two paintings where the paint separated entirely from the ground were found to contain higher levels of zinc soaps than any of the composite samples. The AFM results confirmed that rough grounds provide better adhesion than smooth ones and that if the ground is rough, paint may adhere despite some zinc in the paint. One factor that may be important, the quantities of surfactants that migrated to the surface of the ground, will be examined in the near future.

ACKNOWLEDGMENTS

The authors would like to acknowledge the following people for their advice, supply of samples, or assistance with technical examinations: Marion Mecklenburg, Michael Skalka, Bronwyn Ormsby, Gregory Smith, Mark D. Gottsegen, Bill Berthel, Christina Young and Jay Krueger, Jie (Jessie) Sui, Jane Wang, Kevin McKenna, Allison Rutter and Mary K. Andrews. We would like to thank the faculty at the Queen's Art Conservation Program. Funding for this project was provided by the Natural Sciences and Engineering Research Council of Canada.

REFERENCES

1. L. Robinet and M. C. Corbeil, Studies in Conservation 48, 23-40 (2003).
2. Ibid.
3. H. Wexler, Chemical Reviews 64 (6), 591-611 (1964).
4. J.J. Boon, F. Hoogland and K. Keune in AIC paintings specialty group postprints, compiled by H. M. Parkin (Washington, AIC, 2007), pp. 16-23.
5. A. Stewart, Official Digest, Fed. Paint & Varnish Prod. Clubs 311, 1100–1113 (1950).
6. J. van der Weerd, PhD dissertation, University of Amsterdam/AMOLF, 2002.

Mater. Res. Soc. Symp. Proc. Vol. 1047 © 2008 Materials Research Society   1047-Y03-01

The Relation Between the Fine Structural Change and Color Fading in the Natural Mineral Pigments Azurite and Malachite

Ryoichi Nishimura, and Ari Ide-Ektessabi

Graduate School of Engineering, Kyoto University, Yoshida Honmachi, Sakyo-ku, Kyoto, Japan

ABSTRACT

Many ancient Asian and Japanese paintings have been drawn with natural mineral pigments. The discoloring mechanism of these pigments has been a real concern for the characterization, restoration and preservation of the ancient cultural properties. The authors expect that the color fading is deeply related to the chemical composition and the fine structural change of the major elements. Therefore the purpose of this paper is to make clear the relation between the fine structural change and color fading.

We analyzed several representative pigments of Japan, including copper carbonate hydroxide pigments (azurite and malachite, "gunjo" and "ryokusho" called in Japanese) by x-ray fluorescence analysis (XRF) and x-ray absorption fine structure (XAFS). In order to examine the deterioration of pigments, some of them were exposed to a heated condition of 260°C under oxidizing condition, as is a common practice in the preparation of pigments in Japan. The spectral reflectivity was collected on spectrophotometer system at 10 minutes intervals.

Here we propose to compare the results obtained from XRF and XAFS with the spectral reflectivity. The results demonstrate that the chemical composition and the fine structural changes can provide valuable information for revealing the discoloring mechanism, and this then leads to an original color deduction of the ancient cultural property.

INTRODUCTION

Natural mineral pigments are widely used in many ancient Japanese paintings [1]. The natural pigments are made from the dry powder produced by breaking up the minerals or rocks into fine particles by grinding. In order to subtly modify the color, heating of the fine powders often occurs, according to modern painters. Then water and a glue binder are mixed with the pigment prior to use. By analyzing the color of these paintings, much information about pigments can be obtained. In our research group, deduction of pigments is aimed. It is performed by comparing the color of paintings to a reference database. But the colors of pigments are fading over the centuries. The colors of existing paintings are the results of color fading. In addition, there is a quick discoloring that is caused by chemical reaction such as burning. These color changes of the pigments prevent us from estimating the pigments.

In this research, the relation between the chemical composition and color fading due to heating was studied. This will lead to improvement of pigments deduction. The pattern of discoloring depends on each pigment. Therefore, it is necessary to investigate the patterns separately.

Azurite and malachite, well-known Japanese pigments, were exposed to a heat for varying times. The spectral reflectivity was compared with the results obtained from XRF and XAFS. X-ray fluorescence analysis using Synchrotron Radiation (SR-XRF) analysis is a powerful method for investigating the distribution of trace elements with high spatial resolution. X-ray Absorption Fine Structure analysis using Synchrotron Radiation (SR-XAFS) delivers the fine structural change of each element. Additionally, XRF and XAFS analyses are non-destructive techniques and the samples can be retained after the analyses.

EXPERIMENT

The traditional Japanese pigments, azurite and malachite, (respectively, "gunjo" and "ryokusho" in Japanese) were selected for study. The azurite, basic copper carbonate, has the chemical formula $2CuCO_3 \cdot Cu(OH)_2$. It is well known that azurite transforms to malachite, $CuCO_3 \cdot Cu(OH)_2$, as a result of hydration to a less thermally state[2].Ten pigment samples of each powdered mineral were prepared. All of them have been grinded and sieved to a standard particle size. For one sample no heating was applied to the control. Nine samples were heated with nine different holding times from 10 min to 90 min in increments of 10 minutes.

Japanese-style painters often roast the natural mineral pigments to modify the shade of colors before they draw their works, but the temperature is not controlled. In order to treat the data more quantitatively, the pigments in our experiments were heated in a drying oven (Iuchiseieido, DO-300) at a constant temperature oxidizing condition. The temperature was kept at $260 \pm 1°C$. Then, the pigments were dissolved in a bit of water with glue, and applied to Japanese papers made from hemp. All of our experiments were conducted on the painted papers.

At first, the spectral reflectance was measured by a spectrophotometer (HITACHI, UV2550/ geometric consideration: integrating sphere measurement). Spectral reflectance from 350-800 nm with 1 nm interval was measured in order to observe the color variation caused by heat.

At next, XRF and XAFS analyses were performed at the Photon Factory in the High Energy Accelerator Research Organization, Tsukuba, Japan. The synchrotron radiation was monochromated by a multilayered reflecting mirror. The monochromated X-rays were focused by slits. The incident and transmitted X-rays were monitored by ionization chambers that were set in front of and behind the samples. The fluorescent X-rays were collected in a solid-state detector at 90° to the incident beam. The incident X-ray energy was 15 keV for XRF analysis, whereas the energy was scanned at the Cu K absorption edge from 8.90 to 9.09 keV for XAFS analysis. The cross-sectional area where the beam contacted the pigment samples was 0.84 mm square. Measurements were performed in the air. When XRF and XAFS analyses were conducted, the samples were fixed to acrylic plates with kapton tapes.

RESULTS AND DISCUSSION

The spectral reflectivity was obtained. The Figure 1 and Figure 2 show the spectral reflectance of the samples. The graphs show that the longer the holding time was, the lower the value of maximum reflectance became, whereas the wavelength of the peak position changed very little. This means that the luminosity decreased, but the color wavelength did not change [3]. In fact, the samples appear darker along with the heating time.

Additionally, it was found that after a heat treatment of around 40—60 min at 260 °C, the pigment color changed abruptly.

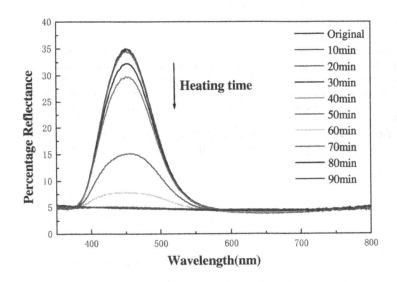

Figure 1. The spectral reflectance of azurite is lowered by discoloring.

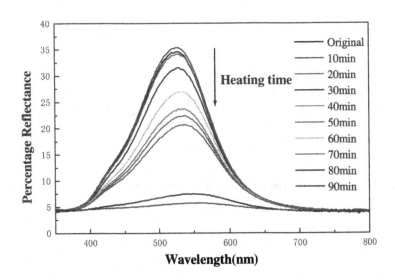

Figure 2. The spectral reflectance of malachite is lowered by discoloring just as in Figure 1.

XRF ANALYSIS

XRF analysis delivered the distribution of trace element included in the azurite and malachite. XRF spectra of samples made from azurite and malachite are shown in Figure 3 and Figure 4 respectively. The spectra indicate that almost same elements, Cu , Fe and Pb were included in these two pigments. However, their colors were quite different, because the peaks of the spectral reflectance of azurite were to the short-wavelength side, compared with malachite in Figure 1 and Figure 2.

In addition, the spectra of the heat discolored samples were almost same as the as-received samples. A change in chemical-bonding state of the main element, Cu, caused the color difference.

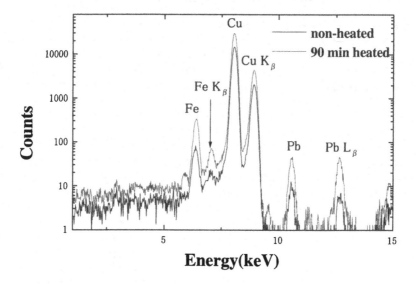

Figure 3. XRF spectra of non-heated and heated samples made from azurite.

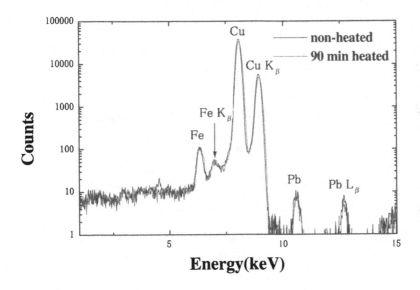

Figure 4. XRF spectra of non- heated and heated samples made from malachite.

XAFS ANALYSIS

Figure 5 and Figure 6 show the XAFS spectra of heat discolored samples made from azurite and malachite. The spectra were normalized to the maximum intensity of each spectrum. It was found that the absorption edge of the samples heated for 80 and 90 min shifted to lower energy. Figure 1 and Figure 2 show that the colors of long time heated pigments were quite different from as-received samples. It was demonstrated that there was a relation between chemical-bonding state of Cu and the color fading.

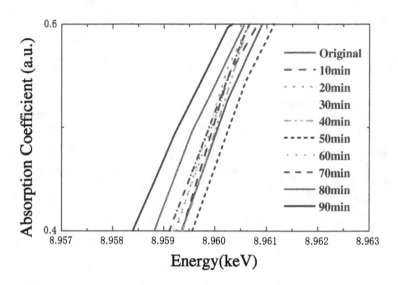

Figure 5. Position of Cu K absorption edge of heat discolored samples made from azurite.

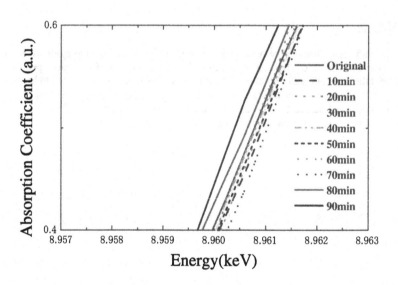

Figure 6. Position of Cu K absorption edge of heat discolored samples made from malachite.

CONCLUSIONS

In this research, the relationship between the fine structural change and color fading of natural mineral pigments was studied. The spectral reflectivity was obtained. The data show that the color of pigments abruptly changed at a certain point of 40 to 60 minutes at 260 °C.

Then, the spectral reflectance was compared with the results of materials analyses, XRF and XAFS. As a result of XRF analysis, it was found that the elemental distribution of azurite were similar to that of malachite. However, the spectral reflectance of azurite and malachite were quite different. In addition, the XRF spectra of the heat discolored samples were almost same as the as-received samples. These results and XAFS spectra show that the spectral reflectance is altered by chemical bonding states of pigments.

ACKNOWLEDGMENTS

The XRF and XAFS analyses were performed at the Photon Factory in the High Energy Accelerator Research Organization, Tsukuba, Japan. The authors would like to thank Professor A. Iida, for his valuable advice. We would like to express our gratitude to Mr. M. Kim and Miss. Y. Nakao, Japanese-style painters, for their helpful suggestions and comments about the old Japanese pigments.

REFERENCES

1. T.Katayama, and A.Ide-Ektessabi, JSME - Kansai Branch Annual Conference Vol.2005, No. 80 (20050318) pp. "9-43"-"9-44".
2. M.Odlyhaa, N.S.Cohena, G.M.Foster and R.H.West, "Dosimetry of Paintings : Determination of the Chemical Change in Museum Exposed Test Paintings (Azurite Tempera) by Thermal and Spectroscopic Analysis"Thermochimica Acta 365, pp. 53-63 (2000).
3. N.Sasaki, Y.Manabe and K.Chihara, "Analysis of Spectral Distribution on Discoloring Process of Dyed Fabric" The Color Science Association of Japan, Vol. 31, No. 3, pp. 172-182(2005).
4. A.Iida, "X-ray Analysis by Synchrotron Radiation – X-ray Fluorescence Analysis and XAFS", Biomedical Research on Trace Elements, vol. 14 (3), pp. 188-195 (2003).

Archaeological Science

Mater. Res. Soc. Symp. Proc. Vol. 1047 © 2008 Materials Research Society   1047-Y01-02

Prehistoric Ceramics of Northern Afghanistan: Neolithic Through the Iron Age

Charles C. Kolb
Division of Preservation and Access, National Endowment for the Humanities, 1100
Pennsylvania Avenue, NW, Washington, DC, 20506

ABSTRACT

For nearly four millennia, Afghanistan has been at the crossroads of Eurasian commerce and remains ethnically and linguistically diverse, a mosaic of cultures and languages, especially in the north, where the Turkestan Plain is a conduit for the so-called Silk Route, a series of "roads" that connected far-flung towns and urban centers and facilitated the transfer of goods and services. The research reported herein involves the comparative analysis of archaeological ceramics from a series of archaeological sites excavated in northern Afghanistan in the mid-1960s by the late Louis Dupree and me. I served as the field director (1965-1966) and analyzed the ceramics excavated from all six archaeological sites. These were Aq Kupruk I, II, III, and IV located in Balkh Province (north-central Afghanistan) and Darra-i-Kur and Hazar Gusfand situated on the border between Badakshan and Tarkar Provinces (extreme northeastern Afghanistan). Ten of the 72 ceramic types from the Aq Kupruk area have been published [1, 2, 3] but none of the 53 wares from northeastern Afghanistan have been described. The majority of the Aq Kupruk materials are undecorated (plain ware) ceramics but there is a unique series of red-painted decorated ceramics (Red/Buff, numbered types 45 through 52) with early first millennium BCE designs but the pottery dates to the BCE-CE period. The results of ceramic typological, macroscopic, binocular and petrographic microscopy (thin-section analysis and point counting) are reported.

INTRODUCTION

Afghanistan, "Land of the Afghans," formerly the Republic of Afghanistan, and now the Islamic Republic of Afghanistan (Jomhuri-ye Eslami-ye Afghanestan in Dari Persian, Da Afghanistan-ye Eslami Jamhawriyat in Pashto) has undergone dramatic sociopolitical changes during the past 150 years, particularly since 1989. As a crossroads of Eurasian commerce for nearly four millennia, the country has served as conduit for invaders from east and west, and was recipient of major religious ideologies including Buddhism and Islam. Hence, Afghanistan is ethnically and linguistically diverse, a mosaic of cultures and languages, especially in the north, where the Turkestan Plain is a conduit for the so-called Silk Route, a series of "roads" that connected urban centers and facilitated the transfer of goods and services. The current study focuses on two sets of archaeological sites in the Hindu Kush mountain range (the western extension of the Himalayas) adjacent to the Turkestan Plain, four in Balkh Province (the Aq Kupruk sites)

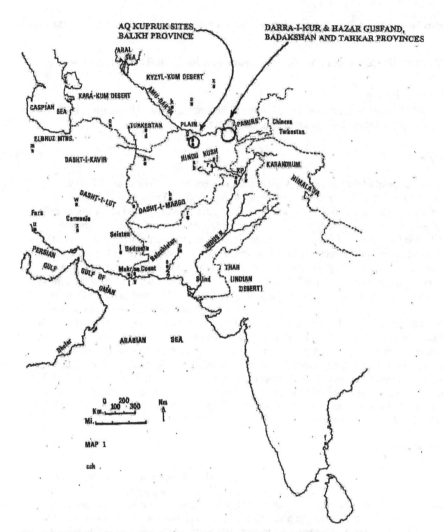

Map 1 - *South Central Asia.* Major physiographic features appear in capital letters, and major cultural regions and sites appear in capital and lower case letters. Abbreviations: KP indicates the Khyber Pass; a. Alexandria Arion (Herat), b. Alexandropolis (Kandahar), c. Anau, d. Antiochia Margiana (Merv), e. Aq Kupruk, f. Arikamedu (Pondicherry), g. Balkh (Bactria), h. Bamiyan, i. Bampur, j. Begram, k. Bokhara, l. Dambi-Koh, m. Ecbatana, n. Kabul, o. Kulli-Damb, p. Nal, q. Peshawur, r. Quetta, s. Samarkand (Marcanda), t. Shar-i-Sokhta, u. Siraf, v. Sutkagen-dor, w. Tal-i-Iblis, x. Tashkent, y. Taxila, z. Tepe Yahya. Modern political boundaries are approximate.

Figure 1: Map of South Central Asia.

and two on the border between Badakshan and Tarkar Provinces (the Badakshan sites), see Figure 1. The Aq Kupruk sites archaeologically span all periods from Upper Palaeolithic through Contemporary Nomadic cultures and are situated on a major north-south commercial route through the Hindu Kush. Aq Kupruk I (Ghar-i-Mar or "Snake" Cave) has the most comprehensive set of stratified assemblages for all chronological periods and is unique in Afghan prehistory for that reason [2, 3]. The prehistory and history of the country has been well-documented by Louis Dupree [4, 5].

LOCATION, ENVIRONMENT, AND ECOLOGY

Afghanistan is a landlocked Central Asian nation bounded on the north by three former Soviet Socialist Republics, now members of the Commonwealth of Independent States (Turkmenistan, Uzbekistan, and Tajikistan), on the northeast by the People's Republic of China, to the east by a portion of the disputed territory of Jammu and Kashmir controlled by Pakistan, on the south by Pakistan, and on the west by Iran. Afghanistan's current boundaries were established during the late 19th century in the context of the "Great Game," a rivalry between the British then occupying India in the Asian Subcontinent and the Russians in Central Asia; Afghanistan was called the "Northwest Frontier" by the British. The country covers an area of 652,223 sq km (251,825 sq mi) -- approximately the size of Texas or the United Kingdom -- and has a configuration similar to a clenched right fist with the thumb extended to the northeast, the narrow Wakhan Corridor leading to China. The Hindu Kush mountains run northeast to southwest and divide the northern provinces from the rest of the country [6].

Major physical features include the Amu Darya (the ancient Oxus River) forming much of the northern international border; steppes, the Turkestan Plains in the north; a mountainous center and northeast dominated by the Hindu Kush mountain range (the western extension of the Himalayas) and the Pamirs; the Seistan Basin with plains in the southwest through which flows the Helmand River; and deserts (Dasht-i Margo) to the far west and south (Registan). Except for the Amu Darya which flows westward into the Caspian Sea and the eastward flowing Kabul River which joins the Indus, other rivers end in inland seas, swamps, or salt flats. The geology of Afghanistan is largely controlled by the topography and is highly complex. The fan-shaped, central highlands are comprised of old, resistant bedrock; and the younger, soft, and erodable sediments of Cenozoic age (Paleogene, Neogene, Quaternary) that wrap around the edges, starting in the northern Turkestan Plains, proceeding through the western lowlands and the Sistan Basin to the southeastern mountains and foothills. The highest, northeastern part of the country is characterized mainly by Prepaleozoic and Paleozoic metamorphosed sediments and granitic intrusions. West of Kabul this group divides into two zones, the widest band of which strikes southwest toward Kandahar in the south with the narrow zone heading towards Herat in the northwest. To the north, Cretaceous and Paleocene limestones and red sandstones dominate; to the south are older Jurassic to Cretaceous limestones and sandstones. Tertiary (Paleogene, Neogene) sedimentary rocks are especially dominant along the border with Pakistan. Soils clearly show the impact of overuse. Soils in river valley alluvial plains are predominantly sedimentary due to repeated floodplain depositions. Saline, alkali, and salt marsh soils are found in poorly drained areas of the Turkestan Plains where soluble salts (sodium, calcium, and magnesium) are concentrated through high evaporation. Desert soils are predominantly sand and occur in the desert wastes of southwestern Afghanistan and on the Turkestan Plains [7, 8, 9, 10, 11].

In Balkh Province, the Hindu Kush consists of faulted and folded ranges separated by valleys, some of which are the open, terraced floodplain type, as at the village of Aq Kupruk, but the majority of these valleys are narrow and deeply incised, as in the Balkh River Gorge immediately north of the town where the four archaeological sites are located. The bedrock, especially the limestone, is strongly jointed, while the northern flank of the Hindu Kush is composed of a broad northern extending alluvial and loess-covered apron. Two large intermittent tributaries (Qala Gundug to the west and Darra-i-Archa to the southeast) join the Balk River north of the town where the river has created a broad, open alluvial plain. Three Aq Kupruk caves (I, II, IV) were formed by solution in limestone at ca. 50,000 BP [12]. Excavation at these sites yielded 21,446 sherds. Badakshan is a mountainous, rugged portion of northeastern Afghanistan characterized by peaks rising 4,000-6,000 m with granite and dolomitic limestone outcrops and some sandstone. Excavations at Darra-i-Kur and Hazar Gusfand produced 2,464 ceramic specimens. Earthquakes and heavy glacial outwashes are characteristics of northeastern Afghanistan.

CHRONOLOGY

Situated at the intersection of three geographically and culturally different worlds -- India with its monsoons, Central Asia with its steppes, and the Iranian plateau -- Afghanistan has seen a succession of invaders and colonizers, as shown in Table 1: Archaeological Sites in Afghanistan: Periods. Archaeologically, Afghanistan's prehistory dates earliest to 50,000 year ago and includes Middle and Upper Paleolithic sites (particularly in the northern foothills of the Hindu Kush), and abundant Bronze and Iron Age sites in the north, south, and east. Modern Afghanistan, located at the crossroads of Central, West, and South Asia, incorporates partially or wholly the ancient regions of Aria, Bactria, Sogdiana, Arachosia, and Drangiana -- part of the Achaemenid Empire (ca. 559 BC–330 BCE), and more recently the Alexandrian Empire of Alexander the Great and his Hellenic successors (330-150 CE). The territory was also a part of the Parthian, Yüeh-chih/Tokharian, Saka, and Kushana polities, the Sasanid Empire (ca. CE 226–651), and was incorporated into the lands conquered by Arab Muslims ca. 700. Portions of Afghanistan remained under the Abbasid Caliphate into the 9th century and then under the temporary control of a series of polities, the Samanids, Kerts, Ghaznavids, Ghurids, Mongols (1220-1506), and the Timurids and successor states. In the 16th century the Mughal Empire of India and Safavids (Persians) were political rivals with Afghanistan as a frontier until the emergence of the native Durrani dynasty that began rule in 1747 and continued nominally under Pashtun Abdali tribe rulers until the communist coup in April 1978. Afshar Persian sociopolitical influence remained into the 19th century when Afghanistan was thrust into the geopolitical arena of Britain versus Russia (initially through the British East India Company), the so-called "Great Game." Three Anglo-Afghan wars (1838-1842, 1878-1879, and 1919) served to help establish Afghan independence in the 20th century [5, 6].

PREVIOUS ARCHAEOLOGICAL RESEARCH

French and British archaeological scholars have focused on the excavation of monuments and portions of ancient cities, while Russian research has focused on burial kurgans and spectacular finds such as the Tillia tepe Bactrian gold. American research has emphasized culture history, chronology, material culture (stone and metal tools and ceramics) and defining inter-regional contacts. Scientific exploration in Afghanistan began after September, 1922, when the French government negotiated a diplomatic treaty with Afghanistan. In it was recognized the establishment of the Délégation Archéologique Française en Afghanistan (DAFA). French research concentrated on pinpointing evidence for the spread of Hellenism, tracing the silk route, and studying the relationship of Gandharan art to the Buddhist art of the Afghan area. Balkh was probed (1924) for evidence of Hellenism but without results; and work shifted to the major ancient sites of Bamiyan, Begram, and Hadda. DAFA returned to Balkh in 1947 but failed to find any pre-Kushan evidence. In 1949 Daniel Schlumberger transferred operations to Lashkari Bazaar in the south, thus beginning DAFA's first large-scale study of Islamic ruins. J. M. Casal directed its first Bronze Age excavations at Mundigak in 1951. In 1952 Schlumberger went to Surkh Kotal, north of the Hindu Kush, where excavations revealed the absolute evidence for an indigenous Bactrian art and shed new light on the development of Gandharan art. In 1963, excavations began at Ai Khanoum on the Afghanistan-Soviet Central Asian northern boundary, and subsequent research revealed the easternmost city of Greek culture yet known but one with many distinctly oriental traits and syncretism in architecture, art, and religion.

Excavations by archaeologists from any country other than France did not occur until after World War Two. Surface collections were made by a British team led by Beatrice de Cardi in 1950 in Kandahar and Farah provinces. Site excavations began in 1950 during an expedition of the American Museum of Natural History (AMNH) at Deh Morasi Ghundai directed Louis Dupree. This work provided the first Bronze Age data for the territory of Afghanistan. Subsequent American research dealt principally with the prehistoric period. In 1954 Carleton S. Coon discovered an Aurignacian Upper Paleolithic blade industry (ca. 32,000 BCE) and a Mesolithic one (ca. 10,500 BCE) in the rock-shelter at Kara Kamar. Dupree made an extensive survey in northern Afghanistan in 1959 and conducted excavations at Aq Kupruk (1962 and 1965), where findings included a large and elaborate Upper Paleolithic assemblage (ca. 15,000-10,000 BCE), and in Badakshan Province on 1966. Other Americans investigated in the Seistan deserts, where results were especially important for the Islamic period: George Dales led work for the University of Pennsylvania's University Museum (1968-1971); and William Trousdale explored for the Smithsonian Institution (1971-1977). From the beginning of systematic archeology in Afghanistan, some Afghan scholars, such as Ali A. Kohzad, were closely involved. Although archeology is not regarded as a prestigious career in Afghanistan, a small number of Afghan students have studied the subject in France and Italy. C. C. Mustamandy opened the first Afghan-directed excavations in 1965 at Hadda, and when the Afghan Institute of Archaeology (AIA) was established in 1966, he became its first director-general. The Kyoto University Archaeological Mission to Central Asia has emphasized Buddhist sites, while the Istituto Italiano per il Medio ed Estremo Oriente (IsMEO) has focused on Islamic period sites and monuments. Archeological research and restoration inside Afghanistan virtually ceased after the Soviet intervention in 1979 and resumed only in 2005. An entire generation of scholars – Afghan and foreign – have been denied the opportunity to pursue archaeological research in Afghanistan and to study collections housed in the Kabul Museum since 1919 [15]. The AMNH-excavated collections that were deposited in the Kabul Museum in 1966 have not yet been located, although the complete vessels and other objects were saved and have been cataloged to international standards (2003-2005) [16, 17].

Table 1: Archaeological Sites in Afghanistan by Period
(After Kolb 2001)

Period or Culture (n = 36)	Location in Afghanistan	Chronology	Number of Sites (known and reported) for that period)	

Periods or cultures in **bold** are represented in the Aq Kupruk (AK) and Badakshan (B) sites

Period or Culture	Location	Chronology	Number of Sites	AK/B
Lower Paleolithic		<50,000 BCE	1	-
Middle Paleolithic		50,000 - 30,000 BCE	10	B
Upper Paleolithic		15,000 - 10,000 BCE	7	AK
Epi-Paleolithic		10,000 - 8,000 BCE	15	AK
Neolithic		8000 - 4000 BCE	20	AK, B
Chalcolithic/Bronze Age		4000 - 1500 BCE	43	AK
Iron Age		1500 - 700 BCE	45	AK
Early		1500 -> BCE		AK, B
Late		<- 700 CE		AK
Achaemenid		530 - 330 BCE	89	AK
Seleucid		330 - 185 BCE	39	
Mauryan	southeast	275 - 185 BCE	6	
Greco-Bactrian	north	BCE 250 - 160 CE	79	AK
Parthian	west	CE 160 - 225 CE	36	AK
Indo-Greek	east	CE 155 - 90 BCE	7	
Saka	east+southeast	BCE 90 - 20 CE	9	
Indo-Parthian	south	CE 20 - 225 CE)	64	
	east	CE 20 - 75 CE)		
Greater Kushan	north+east	75 - 225	169	AK
Early Sasanian	north+east	275 - 345)	266	AK
	south+west	275 - 450)		
Kushano-Sasanian	north	345 - 425)	111	AK
	east	345 - 450)		
Hephtalite	north	475 - 565)	166	AK
	east	450 - 535)		
Later Sasanian	west	450 - 650)	255	AK

	north	565 - 650	)	
	south+east	535 - 650	)	
Turkish Khanate	north	650- 875	)	129 AK
	central	650 - 800	)	
Arabs [Islam]	southwest	650 - 800	)	32
	northwest	650 - 820	)	
Turki Shahi	east	650 - 850	)	71
	southeast	650 - 800	)	
Tang [China] eastern Wakhan		-		1
Tahirid northwest		820 - 875		6
Samanid	north	875 - 1000	)	146 AK
	south	900 - 980	)	
	southwest	910 - 1000	)	
Hindu Shahi	east-central	875 - 1000	)	26
	central-east	875 - 980	)	
	southeast	875 - 960	)	
Ghaznavid	southeast	960 - 1150	)	231 AK
	east-central+south	980 - 1150	)	
	central-west+west	1100 - 1150	)	
	north	1100 - 1150	)	
Seljuk	north	1050 - 1150		108
Ghurid		1150 - 1220		235
Kart	northwest	1245 - 1350		7
Kayani	southwest	1260 - 1380		26
Chaghati	north+east	1330 - 1380		12
Timurid		1380 - 1500		162 AK, B
Late Islamic				
Contemporary Nomadic				AK, B

THE AMNH EXPEDITIONS TO NORTHERN AFGHANISTAN

Four major American expeditions to northern Afghanistan (1962, 1965, 1966, 1967) were sponsored by the American Museum of Natural History to Afghanistan and funded by the National Science Foundation, American Philosophical Society and The Pennsylvania State University [4]. These expeditions were organized by Louis Dupree, who had engaged in archaeological research in Afghanistan since 1950. I served as the field director for two of these projects (1965-1966) and analyzed the ceramics excavated from the excavations at six archaeological sites: Aq Kupruk I, II, III, and IV in Balkh Province and Darra-i-Kur and Hazar Gusfand in Badakshan and Tarkar Provinces. The fourth year of the field research was devoted to non-ceramic Paleolithic archaeological sites. The 1965-1966 field team included Dupree (expedition organizer and stone tool specialist). Kolb (field director, ceramicist, medical officer, and paymaster), and in 1966 Dexter Perkins, Jr., a paleontologist who analyzed animal remains, notably *Ovis-capra* (sheep-goat) and *Bos* (cattle), and assisted in the excavation of a Neanderthal (*Homo sapiens neanderthalensis*) occipital skull fragment at the site of Darra-i-Kur in 1966. The chronologies of these six sites span 50,000 years, from Middle Paleolithic (Mousterian) to the present. The longest stratified sequence in Afghanistan occurs at Aq Kupruk I: Upper Paleolithic, Epi-Paleolithic, Neolithic, Bronze Age, Iron Age, Achaemenid, Early Kushan, Great Kushan, Early Sasanian, Kushano-Sasanian, Hephalites, "Early" Islamic, "Late" Islamic, and Contemporary Nomadic [13, 14]. Nineteen radiocarbon samples (all from charcoal) were collected and processed (16 for Aq Kupruk I, and 3 for Aq Kupruk II) and the results reported [2, 3, 4].

Four sites situated north of the bazaar town of Aq Kupruk ("White Bridge" in Uzbeki Turkic, 35°05' N, 66°51' E) were investigated. A major trench (2.5 m x 42.0 m) at Aq Kupruk I (Ghar-i-Mar or "Snake" Cave), located on the east side of the Balk River, was excavated to a depth of 14.5 m (the water table occurred at 12.5 m) and yielded 19,765 sherds. This cave, actually a rock shelter with a slight overhang, is situated on the caravan route from the Turkestan Plain to the Bamiyan Valley to the far south, and on to Kabul and eventually the Subcontinent. On the west side of the river opposite this site, a second set of "T-shaped" trenches was excavated at Aq Kupruk II Ghar-i-Asb or "Horse" Cave) to bedrock at a depth of 6.3 m (the main trench was 2.5 m x 24 m), producing mostly lithic assemblages but also 1,436 sherds. Aq Kupruk III was an open-air Upper Palaeolithic campsite with 13 sherds. On what was to be the final day of the 1965 field season, a test pit was excavated in Aq Kupruk IV, yielding 232 sherds most of which fit three vessels (two Red Streak Burnished bowl-dishes, and a Black/Cream Ware round base, double strap-handle amphora-like jug) plus ten disarticulated secondary burials dating to the Early to Late Iron Age (1st century BCE to 5th-6th centuries CE). Elaborate grave offerings including a bronze Han dynasty mirror, iron projectile points, an iron dagger and horse trappings were also recovered in the subsurface 1.5 m x 1.8 m stone-lined burial chamber [2, 4]. In the decades of the 1950s and 1960s, pottery was made at Aq Kupruk by itinerant, traveling artisans who moved from village to village, used local clays, and produced wares "on demand." The silty alluvial clays at Aq Kupruk are conducive to pottery manufacture but collected samples were, unfortunately, not exportable to the United States for analysis.

In pursuit of early Paleolithic materials, the 1966 expedition undertook excavations in two caves near the small village of Chenar-i-Baba Darwesh (36°44' N, 69°59' E), located west of Kishm. Three trenches were excavated in Darra-i-Kur ("Cave of the Valley") with the main one (2.5 m x 24 m.) taken to a massive roof fall at 7.5 m. A total of 2,334 sherds were recovered, dating primarily from the Neolithic and Late Iron Age, and a *Homo sapiens neanderthalensis* occipital skull fragment was recovered from a lower level. The second site, Hazar Gusfand ("Cave of a Thousand Sheep"), was tested but a similar massive rock fall was located at 2.0-3.0 m; 130 sherds were recovered. The ceramics from the Aq Kupruk area are distinct from those recovered in Badakshan. The latter have tempers derived mostly from granite and sandstones, often with sand and micaceous inclusions and with some types tempered by crushed slate, or calcite, and crushed sherds (grog). Both the Aq Kupruk and Badakshan sites have vegetable fiber tempered ceramics and crushed limestone aplastics.

Table 2: Ceramics by Site

Aq Kupruk I (Ghar-i-Mar, "Snake" Cave)
  Chronology: 15,000 BP through all periods to the present
  Ceramic Types Identified: n = 72 of 72 (Aq Kupruk Series)
  Specimens: n = 19,765 sherds (1,849 rims, 17,025 bodies, 599 bases, and 192 handles)

Aq Kupruk II (Ghar-i-Asb, "Horse" Cave)
  Chronology: 20,000 BP through all periods except Late Islamic to the present
  Ceramic Types Identified: n = 22 of 72 (Aq Kupruk Series)
  Specimens: n = 1,436 (128 rims, 1,285 bodies, 11 bases, 1 handle)

Aq Kupruk III (Upper Paleolithic Open Air Site, no Persian name)
  Chronology: 20,000-15,000 BP and Contemporary Nomadic
  Ceramic Types Identified: n = 4 of 72 (Aq Kupruk Series)
  Specimens: n = 13 (2 rims, 8 bodies, 3 bases)

Aq Kupruk IV ("Skull" Cave, no Persian name)
  Chronology: 500 CE (Late Iron Age) Early Islamic and Contemporary Nomadic
  Ceramic Types: n = 17 of 72 (Aq Kupruk Series)
  Specimens: n = 232 (20 rims, 203 bodies, 7 bases, 2 handles)

Darra-i-Kur ("Cave of the Valley")
  Chronology: 35,000 BP Middle Paleolithic, Neolithic, "Goat Cult" Neolithic, Early Iron
        Age, Late Islamic, Contemporary Nomadic
  Ceramic Types: n = 47 of 53 (Badakshan Series)
  Specimens: n = 2,334 (1,215 rims, 2,071 bodies, 32 bases, 16 handles)

Hazar Gusfand ("Cave of a Thousand Sheep")
  Chronology: BCE/CE Early Iron Age, Neolithic, Late Islamic, Contemporary Nomadic
  Ceramic Types: 13 of 53 (Badakshan Series)
  Specimens: n = 130 (1 rim, 128 bodies, 1 base)

METHODOLOGY

The American Museum of Natural History contractual agreement with the
Republic of Afghanistan's Ministry of Culture and Kabul Museum called for a basic
analysis of the excavated materials prior to dividing the collection between AMNH and
the Kabul Museum. I processed all six lithic and ceramic collections and made the
division of objects. This analysis and division was overseen by Mohammed Ibrahim
Khan of the Kabul Museum (who had been trained in Italy) and Philippe Gouin of
DAFA. All of the whole ceramic vessels and metal artifacts (silver rings, bronze artifacts
and mirrors, horse trappings, etc.), semiprecious stones, unique and decorated ceramic
objects, osteology (human and non-human), textiles, and stone carvings were retained for
the Kabul Museum collections. With the permission of A.A. Motamedi, Director of the
Kabul Museum, some materials were transferred temporarily to the United States for

analysis by scholars and were subsequently returned to Afghanistan. Major laboratory analyses were undertaken by J. Lawrence Angel (Smithsonian Institution) on the human osteology, Lawrence Lattman (The Pennsylvania State University) geological specimens, and Earle Caley (Ohio State University) metals. Approximately 12% of the ceramics was retained for the AMNH collections and a select sample of pottery specimens (90 sherds) was sent for petrographic analysis at Penn State [1, 2, 3, 13, 14]. In 2002, the collection deposited at the American Museum was transferred from New York to a centralized repository at the Peabody Museum, Harvard University to join other excavated artifact type collections (notably those of George Dales and Walter Fairservis). The samples analyzed at Penn State were deposited in Frederick Matson's ceramic laboratory.

Petrographic microscopy is a flexible technique and has the ability to assist in characterizing textural aspects of clays, vessel, and/or sherds from the analysis of thin sections [18, 19]. In addition, petrographic microscopy is the only technique that can describe individual mineral grains within a vessel or sherd. On the basis of cost effectiveness, ease of sample preparation, and simplicity of analytical procedure, the petrographic microscope still offers the best method for identifying and classifying mineralogical compositions of ceramic pastes, and relating these to their respective potential aplastics, especially those of inorganic origins. The thin sectioning of sherds and the study of the prepared specimens by petrographic microscopy has been demonstrated to be especially useful when the clay minerals and aplastics of a limited geological or geographical distribution are identifiable within the clay body. In these instances it is a relatively easier task to locate an area or areas from which, based on clay and/or aplastic and/or mineral temper components, a particular ceramic artifact is most likely to have originated. Some methodological questions continue to arise about the validity and efficacy of point counting [20, 21] but these have been addressed [22, 23, 24], and the technique remains reliable. The preparation of ceramic thin sections of the Aq Kupruk and Badakshan sherds, the use of a polarizing microscope, and guidance were provided by Frederick Matson in the Penn State ceramic laboratory.

The long-term plan developed by Dupree and Kolb was to prepare a major monograph on the results of the field research. An interim report was published [4] and I agreed to process and publish on the ceramics with financial support provided by Dupree (but support was not forthcoming). Based on my analysis in Afghanistan's Kabul Museum (three weeks in September 1966) and later at Penn State (three months in the spring of 1967), several publications were prepared and papers presented [1, 2, 3, 13, 14]. One concerned numerical ceramic types 40 through 43 from Aq Kupruk: "A Red Slipped Pseudo-Arretine Ceramic from South Central Asia" [1, 2] which was related to ceramics produced in southeast India that imitated a Roman form known in the Levant and Mesopotamia, and was also recovered at Shamshir Ghar in southern Afghanistan. A Chinese bronze mirror, carnelian beads, from Iran, and semi-precious stone (lapis lazuli) were directly associated with two complete bowls providing evidence of cultural contact and amalgamation along the Silk Route. A second study concerned Aq Kupruk ceramic types 20 through 25, published as "Kushano-Sasanian Storage Jars from Central Asia: A Study in Ceramic Ecology and Ethnoarchaeology" [3]; based on ceramic pastes and temper inclusions, these vessels were made locally in northern Afghanistan in northeastern Iran, or southern Turkmenistan. The pastes and aplastics were consistent with the geology and clays from the Turkestan Plains.

The results of ceramic typological, macroscopic, binocular and petrographic microscopy (thin-section analysis and point counting) are next reported. One major problem is the lack of comparative ceramic analyses by thin-section study or by any physico-chemical technique. Italian, French, and Russian archaeologists working in Central Asia do not used petrographic microscopy to define ceramic wares, relying instead on similarities in vessel shape and decorations. Geographically, the closest comparable thin section investigations of Central Asian ceramics were undertaken by scholars from the International Merv Project at Merv, Turkmenistan (1992-1995) [25]. Merv served for millennia as a gateway for travelers and traders along the Silk Road between east and west. The city's origins date to the 3rd millennium BCE, and it is included in the Bactria-Margiana Archaeological Complex (BMAC) and there are substantial Parthian and Sasanian occupations through CE 650. Plain wares dominated the Parthian and Sasanian assemblages at Merv, and three coarse ware fabrics were defined: vitrified clay or "slag," calcite and grog, and coarse organic material. In the total assemblage, Puschnigg found the following percentages in the Merv assemblage: fine ware (74.5%), unidentified (20.2%), "slag" tempered [overfired or vitrified] (3.5%), grog tempered (1.4%), and calcite tempered (0.4%). Puschnigg selected only 17 sherds from five fabric groups for petrographic assessments by Louise Joyner [26]. The results of that unpublished study have been shared with me with the stipulation that is not be further shared or published. These fabric groups were: slag-tempered, calcite tempered, grog-tempered, silty quartz and biotite mica, and red fabric (a fine slipped bowl). All five groups were petrographically consistent with the local deltaic alluvial geology.

ANALYSES AND RESULTS

Prior to the thin section analysis, I completed basic typological, macroscopic, and binocular petrographic analyses following methodologies and guidelines provided by Shepard and Rice [27, 28] and previously used in my other assessments [2, 3]. Although of limited value, paste hardness was expressed in terms of Mohs' Scale (differentiating Hard, Semi-hard, Semi-soft, and Soft) and paste texture using the Wentworth Classification (Silt through Coarse) was used to describe aplastics and fabrics, ceramic color designations were quantified using the Munsell Soil Color system, although field designations have been retained for convenience (e.g., Buff and Cream). Laboratory analyses resulted in the reclassification of some fabrics, notably on the basis of color variants related to firing.

The first data set, **Aq Kupruk Ceramics**, characterizes and provides salient information on the petrofabric groups or wares (n = 11) and types (n = 73 types) and quantities of pottery recovered from the four Aq Kupruk sites (I, II, III, IV) in north-central Afghanistan. A second separate and distinct data set, **Badakshan Ceramics**, is dissimilar to the Aq Kupruk materials but, like the first, similarly documents the ceramics (n = 8 petrofabric groups or wares and n = 53 types) from the two Badakshan sites (Darra-i-Kur = D-i-K, and Hazar Gusfand = HG) in northeastern Afghanistan. The two sets of tabulations are similarly organized: letter designation (A, B, etc.) define major **ceramic groups or wares** based upon petrographic assessments. Numbers (1, 2, etc.) designate **ceramic types** based on pastes, aplastics, decoration, and vessel form(s) as defined in the field and in the laboratory. The quantities of sherds (rims, bodies, bases, and appendages) for each ceramic type are listed by archaeological site at the end of the narrative descriptions of each group or ware.

Aq Kupruk Ceramics

For the Aq Kupruk pottery, ceramic types are designated numerically; i.e. # 1-3); Badakshan types (considered in the subsequent section of this analysis) are *italicized*; i.e. *# 4-5*.

A Glazed Wares (n = 3 types, # 1-3): 71 specimens

These fragmentary specimens are recent historic pieces, either Late Islamic (Type 1) from the 9th-12th centuries or 20th century contemporary wares all recovered from the top humus layers at the two major sites (Aq Kupruk I and II). The blue and green glazed specimens are bowl or dish fragments identical to contemporary pottery produced in the modern village of Istalif located 38 km north of Kabul. The fabrics are silt to fine with very fine vegetable fibers (0.03 mm in length -- likely cattail fuzz with a few seeds from swamp or riverine sources) and occasional calcined limestone fragments noted in fabric cross-sections. The vegetable material is too fine to be chaff or fibers derived from dung. Matson observed the use of cattail fuzz, locally called *luq*, in his ceramic ethnoarchaeological study of Istalif potters in 1968 [20]. Aplastics are sparse in the pastes (<15% in relation to the paste in cross-section). Firing was in an oxidizing atmosphere.

1 Glazed Islamic-Timurid Polychromes (white, black, blue, purple, green)
    Aq Kupruk I: 12   Aq Kupruk II: 2
2 Glazed Cobalt Blue (with green design)
    Aq Kupruk I: 37   Aq Kupruk II: 6
3 Glazed Green
    Aq Kupruk I: 12   Aq Kupruk IV: 2

B Black/Cream Wares (n = 5 types, # 4-6): 251 specimens (3 thin sections)

Black/Cream vegetable fiber temper in a silt past texture is characteristic of this ware group, although a few specimens (n = 10) have calcareous inclusions or fine-grained quartz fragments (<0.02 mm in diameter). Very fine vegetable fibers (0.03 mm in length -- likely cattail fuzz with a few seed exocasts) characterize types 4 and 5. The aplastics are sparse (<15 % in relation to the paste in cross-section). Vessel forms include small handled jars or amphoras and several simple, non-hemispherical bowl form. Seven sherds in type 6 have quartzite, calcite, hematite, micaceous schist, sandstone, and unidentified fine-grained inclusions. The latter are from basins or small jars (apparently without appendages). The black paint on cream surface typically appears on the jar forms. The pottery was highly fired in an oxidizing atmosphere so that some sherds exhibit a greenish cast; there are no black cores. Chronologically, the ware is Late Iron Age Kushano-Sasanian in date and the locus of production could be in the Turkestan Plains and its riverine tributaries wherever silts alluvium is present, including Aq Kupruk.

4 Black/Cream (vegetable fiber temper) plain

Aq Kupruk I: 168   Aq Kupruk II: 3   Aq Kupruk IV: 57
5 Black/Cream (vegetable fiber temper) stamped decoration
    Aq Kupruk I: 13
6 Black/Cream (crushed limestone temper) plain
    Aq Kupruk I: 10

C Cream Wares Group (n = 5 types, # 7-11): 1,407 specimens (5 thin sections)
    Cream-Hard Red vegetable fiber temper sherds generally have a silt paste but
some specimens have a fine paste with fine paste with fine vegetable fibers (0.03 mm in
length -- likely cattail fuzz with a few seed exocasts). The aplastics are very sparse (<5
% in relation to the paste in cross-section). The 15 sherds with quartz and sand temper
(likely derived from sandstone) also have hematite fragments, iron sulfide in quartz
grains, and calcite. Vessel forms include small jars, bowls (apparently hemispherical) ,
vase-goblets with pedestal bases, and spouted water jars. The lighter exterior color is
achieved by salt washing during the oxidizing firing process and some specimens have
appliquéd decoration. This group of ceramics dates to the Late Iron Age Kushano-
Sasanian era and was likely produced in northern Afghanistan or the Turkestan Plains.
Sasanian Creamwares (type 11) are typically fired in a reducing atmosphere, but type 11
exhibits oxidation firing.

7 Cream-Hard Red (vegetable fiber temper) plain
    Aq Kupruk I: 657   Aq Kupruk II: 479   Aq Kupruk III: 11
8 Cream-Hard Red (vegetable fiber temper) linear striation decoration
    Aq Kupruk I: 14
9 Cream-Hard Red (vegetable fiber temper) appliquéd design decoration
    Aq Kupruk I: 6
10 Cream-Hard Red (fine crushed quartz and sand temper) red stripe decoration
    Aq Kupruk I: 15
11 Cream Ware (vegetable fiber temper) plain
    Aq Kupruk I: 215   Aq Kupruk II: 10

D Coarse Wares Group (n = types, # 12-19): 11,510 specimens (10 thin sections)
    Coarse Hard Buff is one of the predominant wares at Aq Kupruk I (4,757 sherds)
and has crushed stone or crushed limestone temper (15-30% in relation to the paste in
cross-section) with coarse and medium texture paste. The other aplastics (all <0.5 mm in
diameter) may include grey and red quartzite fragments, crushed river pebbles, calcined
calcite, chert fragments, hematite, and some fine chaff or grasses (sometimes with
unidentified seeds). These are thick-walled vessels low-fired in oxidizing atmospheres
and exhibit black coring. Typical forms include large, medium, and small-size storage
jars with a wide variety of rim forms, some basins and bowls, and a few dishes. This ware
dates from the Kuprukian Ceramic Neolithic to Kushano-Sasanian era, ca. 4000 BCE to
CE 400, but are more frequent during the Early Iron Age. Coarse Soft Buff, types 16-18
(1,410 sherds), are similar in terms of vessel forms and dates to types 12-15, but are fired
at lower temperatures and have larger dark cores. The ware is most common in the
Kuprukian Ceramic Neolithic. Type 19, Coarse Soft Buff (3,936 examples) with crushed
stone and pebble temper (> 0.3-1.5 cm and up to 20% of the paste in cross section) has

similar paste characteristics to types 12-18 but is friable and easily crumbles. The specimens range in thickness to 4.5 cm. and appear to be the remains of *tanur-e-nanpazi* (bread ovens) constructed in-situ in the Aq Kupruk caves. Locally available materials transported over short distances would likely be used and the fact that the types 12-18 are similar is pastes (rather than tempers) may suggest a local manufacture for this group of ceramics. However, all of these pastes and tempers are consistent with the geology of northern Afghanistan, the Turkestan Plains, and (less certain) the Iranian Plateau, as well as the alluvial plain at the village of Aq Kupruk.

12 Coarse Hard Buff (crushed stone temper) plain
     Aq Kupruk I: 2,238   Aq Kupruk II: 203   Aq Kupruk IV: 3
13 Coarse Hard Buff (crushed stone temper) appliquéd design decoration
     Aq Kupruk I: 46
14 Coarse Hard Buff (crushed limestone temper) plain (see also Figure 2)
     Aq Kupruk I: 2,173   Aq Kupruk II: 23   Aq Kupruk IV: 39
15 Coarse Hard Buff (crushed limestone temper) straight-line linear decoration
     Aq Kupruk I: 32
16 Coarse Soft Buff (crushed limestone temper) plain
     Aq Kupruk I: 1,259   Aq Kupruk II: 25   Aq Kupruk IV: 71
17 Coarse Soft Buff (crushed limestone temper) textile impressed interior
     Aq Kupruk I: 24   Aq Kupruk II: 2
18 Coarse Soft Buff (crushed limestone temper) impressed chevron design decoration
     Aq Kupruk I: 22   Aq Kupruk II: 7
19 Coarse Soft Buff (crushed stone and pebble temper) plain
     Aq Kupruk I: 3,421   Aq Kupruk II: 515

E Hard Buff Wares Group: inorganic temper (n = 6 types, # 20-27): 3,543 specimens (6 thin sections)

Hard Buff with crushed limestone temper (3,412 sherds) is the norm but 112 have crushed quartz and sand temper (likely a variant) and another 20 sherds exhibit micaceous temper [3]. All specimens exhibit medium to fine paste textures and the temper is predominantly crushed dolomitic limestone (45-85% of all aplastics) with very coarse particles (1.0-2.0 mm) most common (30-50%) but with coarse (0.5-1.0 mm) represented 15-25% and occasional granule and pebble (2.0-4.0 and 4.0+ mm) examples in the paste. Calcined calcite, clear-rounded quartz, and rounded plagioclase feldspar are found in the silt pastes. Minor constituents include ferruginous claystone, siltstone, rounded orthoclase feldspar, Muscovite mica (in natural clay), gypsum, and occasional seeds. Two seed exocasts came from *Triticum compactum* (club wheat) or *T. aestivum* (bread wheat), both known from sites in the Iranian Plateau and Turkestan Plain since 3500 BCE and used in the production of *nan* (an unleavened bread). The vessel forms typically include massive, large, and medium-size storage jars (from 155.0 to 45.0 cm in height) either plain (undecorated) or less frequently decorated with appliqué (representations of goat heads/horns, lunates, and buttons), wavy line, punctation, straight line, or chevron decorations on the vessel necks or shoulders, and a grey wash at the shoulder. There are 15 jar rim forms, but rarer vessel forms include a basin and several vase and bowl forms. These ceramics date to the Parthian/Greater Kushan/Kushano-

Sasanian/Later Sasanian eras, ca. 1 BCE-CE 700, with a majority of the specimens related to the Kushano-Sasanian period, CE 75-400. The pastes are consistent with the geology of northern Afghanistan, the Turkestan Plains, and the alluvial plain at the village of Aq Kupruk.

20 Hard Buff (crushed limestone temper) plain (see also Figure 2)
      Aq Kupruk I: 2,932   Aq Kupruk II: 12   Aq Kupruk IV: 11
21 Hard Buff (crushed limestone temper) appliquéd design decoration
      Aq Kupruk I: 9
22 Hard Buff (crushed limestone temper) wavy-line striation decoration
      Aq Kupruk I: 202   Aq Kupruk IV: 2
23 Hard Buff (crushed limestone temper) punctuation decoration
      Aq Kupruk I: 100
24 Hard Buff (crushed limestone temper) straight-line striation decoration
      Aq Kupruk I: 82
25 Hard Buff (crushed limestone temper) impressed chevron design decoration
      Aq Kupruk I: 62
26 Hard Buff (micaceous temper) plain
      Aq Kupruk I: 9   Aq Kupruk IV: 11
27 Hard Buff (crushed quartz and sand temper) plain
      Aq Kupruk I: 108   Aq Kupruk II: 3   Aq Kupruk IV: 1

F Hard Buff Ware: organic temper (n = 3 types, # 28-30): 1,149 specimens (3 thin sections)
      Hard Buff vegetable fiber temper ceramics are similar to the group previously characterized, e.g. Hard Buff Wares Group: inorganic temper," with the exception that the temper was chaff or vegetable fiber (0.04-10.5 mm in length). Large and medium storage jars were produced and there are 6 rim forms associated with the ware and these overlap in configuration with the storage jar rims in types 20-25. These ceramics also date to the Parthian/Greater Kushan/Kushano-Sasanian/Later Sasanian eras, ca. 1 BCE-CE 700, with a majority of the specimens related to the Kushano-Sasanian period, CE 75-400. The pastes are consistent with the geology of northern Afghanistan, the Turkestan Plains, and the alluvial plain at the village of Aq Kupruk. Type 31 (micaceous temper) may be an import of Sasanian date and origin.

28 Hard Buff (vegetable fiber temper) plain
      Aq Kupruk I; 1,089   Aq Kupruk II: 34   Aq Kupruk III: 1
29 Hard Buff (vegetable fiber temper) impressed chevron design decoration
      Aq Kupruk I: 4
30 Hard Buff (vegetable fiber temper) wavy-line striation decoration
      Aq Kupruk I: 21

G Hard Red Wares Group (n = 9 types, # 31-39): 1,380 specimens (4 thin sections)
      The Hard Red group includes sherds with vegetable fiber temper (types 34-36 and 36, n = 562 specimens), finely crushed quartz and sand temper (derived from sandstone) was noted in 71 sherds, crushed limestone temper was noted in 29 specimens, and

Muscovite mica micaceous temper occurs in 11 examples. All of the type 34-36 vegetable fiber specimens were chaff or dung tempered (10-20% temper in relation to the paste in cross-section) and had fine textures while the others exhibited coarse to medium paste textures with aplastics representing 20-30% of the paste in cross-section. All 9 types seem to be medium or large storage jars dating to the Early and Late Iron Age.

31 Hard Red (micaceous temper) plain
    Aq Kupruk I: 10   Aq Kupruk II: 1
32 Hard Red (fine crushed quartz and sand temper) plain
    Aq Kupruk I: 60   Aq Kupruk II: 5   Aq Kupruk III: 1
    Aq Kupruk IV: 2
33 Hard Red (fine crushed quartz and sand temper) stamped decoration
    Aq Kupruk I: 3
34 Hard Red (vegetable fiber temper) plain
    Aq Kupruk I: 338   Aq Kupruk II: 80
35 Hard Red (vegetable fiber temper) textile impressed
    Aq Kupruk I: 4
36 Orange Slipped Hard Orange (vegetable fiber temper) plain
    Aq Kupruk I: 6
37 Red Slipped Hard Red (fine crushed quartz and sand temper)
    Aq Kupruk I: 95
38 Red Slipped Hard Red (vegetable fiber temper) plain
    Aq Kupruk I: 132   Aq Kupruk II: 2
39 Red/Orange-Buff (crushed limestone temper) plain
    Aq Kupruk I: 27   Aq Kupruk II: 2

H Red Streak Burnished Ware Group (n = 6 types, # 40-44, 72): 1,079 specimens (5 thin sections)
    Red Streak Burnished vegetable fiber temper, bears a striking resemblance to Roman Arretine ceramics which were imitated in Europe, North Africa, southwest Asia, and recovered as far east as Arikamedu (Pondicherry), southeastern India [1, 2]. At Aq Kupruk the 1,079 sherds included two fragmented but restored, nearly complete vessels from Aq Kupruk IV (deposited in the Kabul Museum and recatalogued to international standards in 2005). Vessel forms include a plate-bowl with direct upright rim, and several small jar forms, including one miniature (unguent?) vessel with a restricted neck. The ware dates to the Late Iron Age, likely 150-400 CE [2], but is most likely up to 200 years earlier given the association with a Han dynasty bronze mirror. The clay was levigated and the paste texture is consistently silt; aplastic inclusions by percentage volume is sparse (<15%). The temper is an extremely fine vegetable fiber <0.7 mm in length and 0.1-0.03 mm in diameter in cross-section, and is likely cattail fuzz; a few seed exocasts (neither wheat nor barley) were rarely encountered and additional aplastics (likely introduced during the processing of the clay include crushed limestone (<0.5 mm in diameter), crushed granitic and basaltic particles (<0.3 mm in diameter), and occasional ferric oxide particles occurring naturally in the ferruginous clays. Characteristic of the ware is its light red-colored slip (normally 10 YR 6/6, Munsell 1954) thickly applied over a pink paste (commonly 5 YR 8/4); the sherds exhibit

complete oxidation. Some specimens exhibit a separation of the thick slip from the fabric. The production loci have not been identified with certainty but would be anywhere in the Turkestan Plains and its riverine tributaries wherever silty alluvium is present (including the area of Aq Kupruk). Joyner's Group 5: Fine Red Fabric [26] is dissimilar to petrofabric H specimens from Aq Kupruk.

40 Red Streak Burnished (vegetable fiber temper) plate-bowl, plain (see also Figure 2)
    Aq Kupruk I: 807   Aq Kupruk II: 1   Aq Kupruk IV: 18
41 Red Streak Burnished (vegetable fiber temper) plate-bowl, linear striations
    Aq Kupruk I: 13
42 Red Streak Burnished (vegetable fiber temper) jar, plain
    Aq Kupruk I: 212   Aq Kupruk IV: 1
43 Red Streak Burnished (vegetable fiber temper) jar, punctuation decoration
    Aq Kupruk I: 4
44 Red Streak Burnished (vegetable fiber temper) jar, stamped decoration
    Aq Kupruk I: 23
72 Red Streak Burnished (vegetable fiber temper) dish (?), fugitive gold/purple-blue
    decoration
    Aq Kupruk I: 1

I Red-on-Buff Wares Group (n = 8 types, # 45-52): 1,083 specimens (15 thin sections)

The Red/Buff wares (1,070 sherds at Aq Kupruk I; 10 at Aq Kupruk II, and 3 at Aq Kupruk IV) characteristically have a monochrome red (and variants) on a buff or natural surface (natural or occasionally self slipped) have never been reported from any other site in Afghanistan [13]. All of the paste textures are silt or very fine. Temper variants include vegetable fiber temper derived from fine chaff or dung, and cattail fluff (38.0 % of the rims and 34.5% of the bodies); fine crushed limestone (14.1 and 3.4% of rims and bodies, respectively); and combined fine crushed limestone and vegetable fiber (15.1% of the rims and 24.9% of the bodies). Crushed stone with lesser amounts of crushed limestone, and vegetable fiber (fine chaff or dung) account for the remainder. The aplastic distinctions were dependent on whether the artisan was manufacturing a jar or storage vessel (crush limestone preferred but sometimes mixtures of crushed limestone with vegetable fiber) as opposed to a dish or bowl forms (fine vegetable fiber temper). Low neck jar forms (likely for storage or serving food) and hemispherical bowls and other bowl forms used as serving vessels were the predominant shapes.

This ceramic is associated with five radiocarbon dates of the Early Iron Age and four from the Late Iron Age (there are 91 direct associations of sherds with charcoal specimens), e.g., 1500 BCE to CE 700, but with a vast majority of the specimens from the more recent period, 100 BCE to CE 700. There are 36 Munsell red color variants in the rim sherd collection and 41 associated with the bodies. The red paint is often applied thickly on the natural or buff colored paste. A total of 69 design motifs -- mostly geometric -- have been defined and include (in frequency): bands, parallel line series, groups of spirals ("tazi" tails), groups of equilateral triangles, groups of solid or outlined diamonds, and series of chevrons [4, 13]. A perplexing issue is that similar painted decorations date from 5000 BCE into the first millennium BCE, suggesting Chalcolithic and Bronze Age rather than an Iron Age chronology. However, similar motifs occur on Black/Red painted ceramics from Bronze Age Baluchistan, notably in Gedrosia [30]. More recently, de Cardi [31] has redefined Parthian period Londo Ware to include a "Late Londo" Ware that is chronologically Scytho-Parthian and Kushan in date that also resembles the Aq Kupruk series. A potential remnant of the Red/Buff painted pottery tradition may be seen in ethnographic ceramics produced in the Northwest Frontier Province of Pakistan during the 20th century [32]. Specimens of Types 46, 48, and 52 are illustrated in Figure 2.

45 Red/Buff (fine crushed stone temper) plain
    Aq Kupruk I: 176
46 Red/Buff (fine crushed limestone temper) plain
    Aq Kupruk I: 238
47 Red/Buff (fine crushed limestone temper) textile impressed, painted exterior design
    Aq Kupruk I: 5
48 Red/Buff (vegetable fiber temper) dish, painted exterior or interior designs
    Aq Kupruk I: 89
49 Red/Buff (vegetable fiber temper) dish, painted exterior and interior designs
    Aq Kupruk I: 73
50 Red/Buff (vegetable fiber temper) bowl, painted exterior or interior designs
    Aq Kupruk I: 36   Aq Kupruk IV: 1
51 Red/Buff (vegetable fiber temper) bowl, painted exterior and interior designs
    Aq Kupruk I: 60
52 Red/Buff (vegetable fiber temper) jar, painted exterior designs
    Aq Kupruk I: 393   Aq Kupruk II: 3   Aq Kupruk IV: 9

J Semi-hard Wares (n = 4 types, # 53-56): 1,558 specimens (5 thin sections)
    The Semi-hard Buff ceramics have fine crushed limestone (19 examples) or vegetable fiber tempers (54 sherds); Semi-hard Red-Orange with fine crushed limestone temper is a firing variant. Semi-hard Red sherds with crushed stone temper is a major type at Aq Kupruk I (type 56, 1,465 specimens). Paste textures are coarse to medium in these medium and large-size jar forms with aplastic inclusions by percentage volume is 15-20%. The chronology is Early Iron Age with some specimens perhaps earlier. The pastes and tempers are consistent with the geology of northern Afghanistan, the Turkestan Plains, and the region surrounding the village of Aq Kupruk.

53 Semi-hard Buff (fine crushed limestone temper) plain
    Aq Kupruk I: 18   Aq Kupruk II: 1
54 Semi-hard Buff (vegetable fiber temper) plain
    Aq Kupruk I: 52   Aq Kupruk II: 2
55 Semi-hard Red-Orange (fine crushed limestone temper) plain
    Aq Kupruk I: 4
56 Semi-hard Red (crushed stone temper) plain
    Aq Kupruk I: 1,465   Aq Kupruk II: 15   Aq Kupruk IV: 1

K Soft Buff Ware Group (n = 3 types, # 57-59): 158 specimens (2 thin sections)
Soft Buff ceramics have crushed stone temper in a coarse to medium texture paste with aplastics >10% in fabric cross-sections; 7 sherds have textile impressed interiors. The ware seems to be associated exclusively with small and medium-size globular-shaped storage jars and dates to the Ceramic Neolithic. The provenance cannot be determined with any degree of accuracy but pastes and tempers are generally consistent with the geology of northern Afghanistan, the Turkestan Plain, and the region around the village of Aq Kupruk.

57 Soft Buff (crushed stone temper) plain
     Aq Kupruk I: 144   Aq Kupruk IV: 2
58 Soft Buff (crushed stone temper) wavy-line striation decoration
     Aq Kupruk I: 5
59 Soft Buff (crushed stone temper) textile impressed
     Aq Kupruk I: 7

Unidentified: 200 specimens
60 Undefined and Uncategorized Ceramics (mostly fragments from *tanur-e-nanpazi* [bread ovens])

Miscellaneous Minor Ceramics: 73 specimens
61 Cream slipped Hard Red (vegetable fiber temper) orange painted
     Aq Kupruk I: 1
62 Tan Slipped Hard Red (Crushed stone temper) plain
     Aq Kupruk I: 1
63 Coarse Hard Buff (crushed stone temper) straight-line striations
     Aq Kupruk I: 8
64 Coarse Soft Buff (crushed stone temper) reddish surface
     Aq Kupruk I: 7
65 Tan-Brown (crushed limestone and vegetable temper) plain and burnished
     Aq Kupruk I: 18
66 Black/Red/Natural Buff (vegetable fiber temper) polychrome decorations
     Aq Kupruk I: 8
67 Natural Clay or Water-worn Sherd (vegetable fiber temper)
     Aq Kupruk I 16
68 Soft Buff (crushed shell temper) plain
     Aq Kupruk I: 4
69 Blackware (calcite/quartzite temper) plain
     Aq Kupruk I: 4
70 Red slipped Cream (vegetable fiber temper) plain
     Aq Kupruk I: 3
71 Hard Black (crushed stone tamper) plain and burnished
     Aq Kupruk I: 3

Badakshan Ceramics: Darra-i-Kur and Hazar Gusfand

The *italicized* numbers are the type designations for the Badakshan pottery; i.e. *# 1-3*.

A Glazed Wares (n = 3 types, *# 1-3*): 29 specimens
The majority of the fragments are Late Islamic polychromes with white, black, blue, purple, green, red, and yellow colors. The blue and green glazed specimens are identical to or close variants of glazed pottery produced in the modern village of Istalif located near Kabul (see the description for Glazed Wares from Aq Kupruk). Very fine vegetable fibers (cattail fuzz from swamp or riverine sources are likely) and occasional calcined limestone fragments occur in fabric cross-sections.

1 Glazed Blue
      D-i-K: 1
2 Glazed Green
      D-i-K: 6
3 Glazed Islamic Polychromes (white, black, blue, purple, green, red, and yellow)
      D-i-K: 17   HG: 5

B Black Ware (n = 2 types, *# 4, 29*): 40 specimens (1 thin section)
Baba Darwesh Black Ware, represented by 40 sherds, with crushed calcite temper is unique and easily discerned with the temper very visible in the dark fabric. It occurs in globular jar forms with medium necks and either slightly flaring or direct rims. Some sherds are reddish grey in color due to firing variations. These vessels are likely cooking wares since calcite aplastics serve to more evenly dissipate heat in the ceramic fabric. Decorations include simple wet-grooved striations, incision, punctation, and channeled geometric decorations in the form of chevrons, outlined equilateral triangles, multiple parallel lines, cross hatching, zigzags, and ladder-like motifs. Type *29* was conflated into type *4*. This pottery is directly associated with Dupree's "Goat Cult Neolithic" which has a radiocarbon date of 3800-3300 BP [4], and the ware may be related to the Burzahom Neolithic.

4 Baba Darwesh Black Ware (crushed calcite temper) various decorations
      D-i-K: 40
29 Soft Black Ware (crushed stone temper) plain
      D-i-K: None (reclassified)

C Buff Wares (n = 13 types, *# 5-17*): 307 specimens
Hard Buff sherds with crushed stone temper (230 specimens, type's *# 9-17*) have a variety of decorative techniques. The crushed stone includes granites, dolomitic limestone, and various tiny river pebbles all of which are consistent with the local geology. Hard Buff micaceous and sand temper (38 sherds) and Hard Buff vegetable fiber temper (27 sherds) are less common in the assemblage. Medium-size jars are predominant in the sample, but medium-size and small storage jars and what appear to be basins also appear. The chronology may include the late Neolithic, but specimens are mostly associated with the Early and Late Iron Age. The pastes and tempers are

consistent with the geology of northern Afghanistan (less so the Turkestan Plains), especially Badakshan and the Wakhan Corridor.

5 Hard Gray (vegetable fiber temper) firing variant, plain
   D-i-K: 2
6 Hard Buff (micaceous sand temper) plain
   D-i-K: 10   HG: 28
7 Hard Buff (vegetable fiber temper) plain
   D-i-K: 2   HG: 18
8 Hard Buff (vegetable fiber temper) decorated
   D-i-K: None   HG: 7
9 Hard Buff (crushed stone temper) plain
   D-i-K: 191   HG: 30
10 Hard Buff (crushed stone temper) straight-line striation decoration
   D-i-K: 7
11 Hard Buff (crushed stone temper) wavy-line striation decoration
   D-i-K: 4
12 Hard Buff (crushed stone temper) wavy-line striations and punctuation decorations
   D-i-K: None
13 Hard Buff (crushed stone temper) wavy-line decoration
   D-i-K: 1
14 Hard Buff (crushed stone temper) punctuation decoration
   D-i-K: 3
15 Hard Buff (crushed stone temper) medial ridge on body
   D-i-K: 1
16 Hard Buff (crushed stone temper) grooved decoration
   D-i-K: 4   HG: 1
17 Hard Buff (crushed stone temper) stamped decoration (palmate)
   D-i-K: 2

D Cream Slipped Wares (n = 4 types, # *18-21*): 29 specimens
   This is a small group of sherds that seem inconsistent with the local geology and may be imports. Hard Cream Slipped Buff Ware with micaceous-sandy temper (7 sherds) vary from Hard Cream Slipped Red Ware (types *19-20*, 19 specimens) which are vegetable fiber tempered with chaff and/or dung (0.05-4.5 mm in length) and occur in the form of bowls and small jars. Unique to the group are 3 sherds of Hard Creamware, self-slipped Cream slipped on Cream, with vegetable fiber temper. All appear to date to the Late Iron Age and do not appear to be a locally-produced. Type *21* is likely Sasanian in chronology and origin.

18 Hard Cream Slipped Buff Ware (micaceous-sandy temper) plain
   D-i-K: 2   HG: 5
19 Hard Cream Slipped Red Ware (vegetable fiber temper) plain
   D-i-K: 16
20 Hard Cream Slipped Red Wares (vegetable fiber temper) straight-line striation decoration

D-i-K: 3

21 Hard Creamware [Cream-on-Cream slipped] (vegetable fiber temper) plain
D-i-K: 3

E Red Wares (n = 5 types, # *22-26*): 108 specimens (1 thin section)

Hard Red vegetable fiber temper ceramics (91 specimens) is a plain ceramic (one sherd is incised) with a silt to fine texture that is chaff or dung tempered (0.05-4.4 mm in length) and has dark grey or black cores. A small group (17 sherds) exhibits evidence of a red wash on a buff surface which is likely derived from salt washing, Aplastics are 15+% of the paste in cross-section in what appear to be small and medium-size storage jars. These ceramics are Late Iron Age in date and possibly Kushan or Kushano-Sasanian and likely made of clays from the alluvia of the Turkestan Plains.

22 Hard Red Wash/Buff (vegetable fiber temper) plain
D-i-K: 17
23 Hard Red (vegetable fiber temper) plain
D-i-K: 77   HG: 13
24 Hard Red (vegetable fiber temper) incised decoration
D-i-K: 1
25 Hard Red/Buff (vegetable fiber temper) painted exterior designs
D-i-K: None
26 Hard Red/Buff (vegetable fiber temper) painted exterior and interior designs
D-i-K: None

F Semi-hard Buff Wares (n = 2 types, # *27-28*): 40 specimens

Semi-hard Buff occurs in two temper types -- crushed stone (35 sherds) and chaff vegetable fiber temper (5 specimens) in jar or basin forms probably dating to the Neolithic. The pastes and tempers are consistent with the geology of northern Afghanistan.

27 Semi-hard Buff (crushed stone temper) plain
D-i-K: 35
28 Semi-hard Buff (vegetable fiber temper) plain
D-i-K: 5

G Soft Buff Wares (n = 17 types, # *30-46*): 1,866 specimens (3 thin sections)

This is a heterogeneous cluster with Soft Buff crushed stone temper, with a variety of decorative techniques, predominating (types *32-42*, 1,793 sherds) Impressions, incision, punctation, appliqué and painting occur on various specimens; reddish-brown paint occurs on only two specimens(reclassified from types *25-26*). Soft Buff micaceous-sandy temper likely derived from crushed sandstone (9 sherds), Soft Black/Buff sherds (3 specimens) and Soft Red (7 examples) are minor variants of the vegetable fiber temper cluster. The Soft Buff vegetable fiber temper sherds (types *42-44*, 39 examples) are low fired and exhibit black coring. Specimens are mostly associated with Neolithic and the pastes and tempers are consistent with the geology of northern Afghanistan (less so the Turkestan Plains), especially Badakshan and the Wakhan Corridor. Fourteen examples of **crushed slate** tempered pottery are unique in the assemblage, are non-local in origin, and also date to the Neolithic.

30 Soft Buff (micaceous-sandy temper) plain
    D-i-K: 9
31 Soft Buff (crushed slate temper) plain
    D-i-K: 14
32 Soft Buff (crushed stone temper) plain
    D-i-K: 1,741   HG: 2
33 Soft Buff (crushed stone temper) textile impressed interior
    D-i-K: 10   HG: 3
34 Soft Buff (crushed stone temper) straight-line striation decoration
    D-i-K: 11
35 Soft Buff (crushed stone temper) punctuation decoration
    D-i-K: 8
36 Soft Buff (crushed stone temper) impressed triangle motif
    D-i-K: 1
37 Soft Buff (crushed stone temper) impressed triangle and ladder motifs
    D-i-K: 1
38 Soft Buff (crushed stone temper) impressed chevron motif
    D-i-K: 4
39 Soft Buff (crushed stone temper) horizontal channel decoration
    D-i-K: 3
40 Soft Buff (crushed stone temper) appliquéd node decoration
    D-i-K: 2
41 Soft Buff (crushed stone temper) medial ridge decoration
    D-i-K: 5
42 Soft Buff (crushed stone temper) painted reddish-brown
    D-i-K: 2
43 Soft Buff (vegetable fiber temper) plain
    D-i-K: 36
44 Soft Buff (vegetable fiber temper) grooved decoration
    D-i-K: 1
45 Soft Black/Buff (vegetable fiber temper) plain
    D-i-K: 3
46 Soft Red (crushed stone temper) plain
    D-i-K: 3   HG: 4

H Semi-soft Buff Wares (n = 5 types, # *47-49, 53*): 73 specimens
    Semi-soft Buff vegetable fiber temper (types *47-49*) have mostly chaff as an aplastic but some specimens also have small amounts of fine vegetable fiber in fine past textures; black coring is common. Semi-soft Black is a variant and has heavy black coring. Six sherds (type *53*) have crushed clay or **grog** tempers which unique to this region. The grog is essentially the same as the matrix of the sherds. Joyner's Group 3: Grog Tempered Fabric, which contains feldspars and biotite and Muscovite micas, is similar [26]. What may be a variant of type *47*, overfired Semi-soft Red (type *50*, 8

sherds) has crushed **calcite** temper. The latter is not related to the distinctive **B Black Ware** described previously and seems to approximate Joyner's Group 2: Calcite Tempered Fabric with feldspars and micas [26]. All of the Badakshan sherds are associated with Neolithic levels and the provenance might be anywhere in northern Afghanistan (less so the Turkestan Plain) and especially Badakshan and the Wakhan Corridor.

47 Semi-soft Buff (vegetable fiver temper) plain
     D-i-K: 39   HG: 8
48 Semi-soft Buff (vegetable fiber temper) impressed circular decoration
     D-i-K: 1
53 Semi-soft Buff (crushed clay temper)
     HG: 6
49 Semi-soft Black (vegetable fiber temper) plain
     D-i-K: 11
50 Semi-soft Red (crushed calcite temper) plain
     D-i-K: 8

Unidentified and Miscellaneous: 2 specimens
51 Undefined and Uncategorized Ceramics
     None
52 Clay "lumps," unfired
     D-i-K: 2

CONCLUSIONS

Central Asia, or more correctly Middle Asia, situated between Mesopotamia and the Indus Valley, has only recently been characterized as a region that is an unrecognized and independent region of urbanism and civilization rather than a "backwater" between the Tigris and Euphrates rivers and the Indus [33]. Afghanistan is central to the concept of Middle Asia which geographically covers the "traditional" area of Central Asia and the Turkestan Plains, the Iranian Plateau and a portion of the Arabian Peninsula. The concept was discussed and refined at the International Association for the Study of Early Civilizations in the Middle Asian Intercultural Sphere held in Ravenna in July 2007. There is much yet to learn about the nature of urbanism, village-city dynamics, commerce, and technological exchange in Middle Asia. For example, the rise and demise of the major urban center of Jiroft (3000-1800 BCE) in southeastern Iran and other centers such as Tepe Yahya (5500 BCE-CE 300) and Shahr-i Sokhta, "The Burnt City," (3200-2100 BCE) in Sistan-Baluchistan. Northern Afghanistan is also within the Bactrian Margiana Archaeological Complex (BMAC) also known as the Bactrian or Oxus civilization. Climatic change from a wetter to a desiccated environment is but one proposed cause of the decline of these urban centers and the migration of peoples to regions that provided subsistence, including foothills and river valleys. The role played by Afghanistan in these changes is not adequately understood because of a lack of a comprehensive archaeological settlement analysis, especially for the Turkestan Plains, that would include GPS and GIS.

Badakshan: B Black Ware, *Type 4*, "Baba Darwesh Black Ware" (crushed calcite temper), "Goat Cult Neolithic."

Aq Kupruk: D Coarse Wares Group, Type 14, Coarse Hard Buff (crushed limestone temper), jars, Kuprukian Ceramic Neolithic levels.

Aq Kupruk: I Red-on-Buff Wares Group, Types 46, 48, 52 (fine crushed limestone, vegetable fiber tempers), examples of bowls and jars, Early Iron Age.

Aq Kupruk: E Hard Buff Wares Group, Type 20, (crushed limestone temper). jar in situ, Late Iron Age.

Figure 2: Examples of Major Ceramic Wares and Types

A primary goal of this research is to make available basic ceramic data, analyses, and interpretations of pottery from archaeological sites at Aq Kupruk in north-central Afghanistan and Badakshan in northeastern Afghanistan. The assemblages from these two regions are distinct and there is overlap only in Late Islamic and modern glazed specimens. A total of 21,446 sherds were recovered from four excavated sites at Aq Kupruk and subsequent analysis identified 72 ceramic types that were categorized into 11 petrofabric groups or wares; 58 thin sections were prepared and studied. Excavations at two sites in Badakshan yielded 2,464 specimens that represented 53 ceramic types; 8 thin sections were analyzed. The pastes of the thin-sectioned sherds reflect a great deal of diversity and the mineral assemblages in the sherds from these two distinctive regions reflect the differences in regional geology. There is a good correlation between the ceramic typologies and the petrofabric groupings, indicating that certain vessel types were fabricated using specific fabrics/pastes and, to some extent, the tempers.

I expect that a number of the generic wares and types described herein are common to other archaeological sites located in northern Afghanistan, Turkmenistan, Uzbekistan, Tajikistan, Kyrgyzstan, and southern Kazakhstan – notably the drainages of the Amu Darya, Syr Darya, and Zeravshan Rivers, and the Turkestan Plains. However, the descriptions and illustrations of ceramics from the Neolithic through Late Iron Ages found in the archaeological literature of this vast region are insufficient for adequate typological correlations and chronological associations. The ceramics are only generally described in terms of vessel forms (and sometimes by decorative techniques) rather than by ware or type designations that are defined on the basis of tempers. Hence, the work by Joyner and her colleagues is a much needed and long awaited contribution to the understanding of Central Asia [25, 26]. Notably, none of the ceramics from the four Aq Kupruk sites and the two from Badakshan are fired to stoneware temperatures and no specimens can be associated with metallurgy. Indeed, no slag from metallurgical production has been found in any of the six sites. Some ceramics from Aq Kupruk such as the Glazed Wares, Red Streak Burnished Ware Group, and Red-on-Buff Wares Group are distinctive and easily recognizable in the field as well as in the laboratory. The Glazed Wares, Black Ware, and some Soft Buff Wares (crushed slate temper) from Badakshan are readily discerned.

The production of basic cooking wares, serving and storage ceramics appears most commonly to be at the household or village level within both regions. Six petrofabric groups or wares at Aq Kupruk were produced by skilled artisans: B Black/Cream Wares (n = 5 types, # 4-6); C Cream Wares Group (n = 5 types, # 7-11); E Hard Buff Wares Group: inorganic temper (n = 6 types, # 20-27); F Hard Buff Ware: organic temper (n = 3 types, # 28-30); H Red Streak Burnished Ware Group (n = 6 types, # 40-44, 72), and I Red-on-Buff Wares Group (n = 8 types, # 45-52). The provenance of the Red Streak Burnished Ware Group and the Red-on-Buff Wares Group is especially intriguing. Among the Badakshan ceramics, petrofabric D Cream Slipped Wares (n = 4 types, # 18-21), appears to be a Sasanian import from an undefined location. It is hoped that this report may shed some light on commerce and interregional economic and sociopolitical interactions in Middle Asia.

ACKNOWLEDGMENTS

I want to especially acknowledge guidance provided in the field by the late Louis Dupree and advice and counsel by the late Frederick R. Matson during the laboratory assessment of these ceramics. A special thanks to Louise Joyner and the British Museum Department of Scientific Research for sharing unpublished petrographic data. Any errors in analysis and interpretation are mine. I am also grateful for the comments of two external reviewers.

REFERENCES

1. C. C. Kolb, *Current Anthropology* **18**, 536-538 (1977).
2. C. C. Kolb, *East and West* **33**, 57-103 (1983).
3. C. C. Kolb, *Ceramic Ecology, 1988: Current Research on Ceramic Materials*, ed. C C. Kolb, (Br. Archaeo. Reports, S-513, Oxford, 1989), pp.175-259.
4. L. Dupree et al., *Prehistoric Research in Afghanistan (1959-1966)*, ed. L. Dupree (Trans. Am. Phil. Soc. 62:4, Philadelphia, 1972).
5. L. Dupree, *Afghanistan*, rev. ed. (Princeton, 1990).
6. C. C. Kolb, *Encyclopedia of Modern Asia*, Vol. 1, ed. D. Levinson (Scribner's, 2002), pp. 19-23.
7   S. H. Abdullah and V. M. Chmyriov, *Geology and Mineral Resources of Afghanistan*, 2 vols. (Ministry of Mines and Industries, Kabul, 1980).
8. Islamic Republic of Afghanistan, *Afghanistan Geological Survey* (Islamic Republic of Afghanistan, Ministry of Mines and Industry, Kabul, 2007). http://www.bgs.ac.uk/afghanminerals/Index.htm [hosted by the Br. Geol. Survey].)
9. J. F. Shroder, Jr., "Geography," *Encyc. Iranica*, Vol. 1, ed. E. Yarshater (Routledge, 1983), pp. 486-491.
10. Y. Smirnova, *Terrain Analysis of Afghanistan* (East View Cartographic, Minneapolis, 2003).
11. R.Wolfart and H. Wirrekindt, *Geologie von Afghanistan* (Gebrüder Borntraeger, Berlin and Stuttgart, 1980).
12. L. Lattman, "Provisional Comments on the Geology of Cave and Terrace Sites in Northern Afghanistan," (Dept. of Geosciences, The Pennsylvania State Univ., 1969).
13. C. C. Kolb, "Painted Pottery from Baluchistan: An Ethnoarchaeological Study," (Am. Anth. Assn., 1987).
14. C. C. Kolb, "The Ceramic Ecology of Pottery from Aq Kupruk Afghanistan: The Petrographic Evidence for External Relationships," (Arch. Inst. Am., 1993).
15. C. C. Kolb, "Beyond the Bamiyan Buddhas: The Fate of Ceramic Collections in the Kabul Museum," (Am. Anth. Assn., 2001).
16. C. C. Kolb, "The Kabul Museum Collections Revisited," (Am. Anth. Assn., 2005).
17. N. H. Dupree, "Prehistoric Afghanistan: Status of Sites, Artefacts and Challenges of Preservation," *Art and Archaeology of Afghanistan: Its Fall and Survival: A Multi-Disciplinary Approach*, ed. J. van Kreiken-Peters, Juliette (Brill, 2006), pp. 79-93.
18. R. E. Carver, *Procedures in Sedimentary Petrology* (Wiley, 1971).
19. F. Chayes, *Petrographic Modal Analysis* (Wiley, 1956).
20. M. Frangipane and R. Schmid, *Schweizerische Mineralogische und Petrographische Mitteilungen* **54**, 19-31 (1974).

21. J. F. W. Negandank, *Neue Jahrbücher fuer Mineralogie*, Abhandlungen **116**, 308-320 (1972); **117**, 183-195 (1972).
22. H .W. Catling and A. Millett, *Archaeometry* **9**, 92-97 (1966).
23. J. S. Galehouse, *Journal of Sedimentary Petrology* **39**, 812-885 (1969).
24. L. van der Plas, and A.C. Tobi, *American Journal of Science* **263**, 87-90 (1965).
25. G. Puschnigg, *Ceramics of the Merv Oasis: Recycling the City* (Univ. College London, Inst. Arch. Pub., Left Coast Press, dist. by Univ. of Arizona Press, 2006).
26. L. Joyner, "Petrographic Analysis of Parthian and Sasanian Ceramics from Merv, Turkmenistan," (Br. Mus. Dept. of Sci. Res., DSR Project 7081, 1999).
27. A. O. Shepard, *Ceramics for the Archaeologist*, Pub. 609 (Carnegie Inst. of Washington, 1968).
28. P. M. Rice, *Pottery Analysis: A Sourcebook* (Chicago, 1987).
29. F. R. Matson, "Ceramic Aspects," The National Geographic Society/Smithsonian Institution Reconnaissance Expedition to Afghanistan, Iran and Turkey, August-September 1968" (Dept, of Anthropology, The Pennsylvania State Univ., 1968).
30. M. A. Stein, *An Archaeological Tour to Gedrosia* (Mem. of the Archaeo. Survey of India, 43, Indian Archaeological Survey, New Delhi, 1931).
31. B. de Cardi, *Archaeological Surveys in Baluchistan: 1948 and 1957* (Occ. Pub. 8, Institute of Archaeology, University of London, London, 1983).
32. O. S. Rye and C. Evans, *Traditional Pottery Techniques of Pakistan* (Smithsonian Contrib. to Anthro 21, Smithsonian Inst, Press, Washington, 1976).
33. A. Lawler, "Middle Asia Takes Center Stage," *Science* **317**(5838), 586-590 (3 August 2007).

Mater. Res. Soc. Symp. Proc. Vol. 1047 © 2008 Materials Research Society   1047-Y01-06

Obsidian Subsources Utilized at Sites in Southern Sardinia (Italy)

Robert H. Tykot[1], Michael D. Glascock[2], Robert J. Speakman[3], and Enrico Atzeni[4]

[1]Anthropology, University of South Florida, 4202 E. Fowler Ave., SOC107, Tampa, FL, 33620
[2]Research Reactor Center, University of Missouri-Columbia, Columbia, MO, 65211
[3]Museum Conservation Institute, Smithsonian Institution, Suitland, MD, 20746
[4]Dipartimento di Scienze Archeologiche e Storico-Artistiche, Universita di Cagliari, Cagliari, Italy

ABSTRACT

While geochemical analysis of obsidian artifacts is now widely applied around the world, both new instrumental methods and new research questions continue to be applied in archaeology. In the Mediterranean, many analytical methods have been employed and proven successful in distinguishing all of the island sources. In this study, results are presented from the virtually non-destructive, LA-ICP-MS multi-element analysis of 95 carefully selected obsidian artifacts from four neolithic period sites in southwestern Sardinia. The patterns of exploitation of specific Monte Arci obsidian subsources revealed in this study support a down-the-line model of obsidian trade during the neolithic period, but with chronological changes that are best explained by increased socioeconomic complexity.

INTRODUCTION

The existence of obsidian sources on Sardinia (Italy), and their usage by prehistoric cultures for a variety of stone tools, has been widely known at least since the 19th century when early geological and archaeological studies were conducted [1, 2]. While it was generally thought that obsidian artifacts found in Sardinia would have actually come from the natural sources at Monte Arci, and not from one of the other central Mediterranean island sources, chemical analyses to fingerprint sources and match artifacts to them were not successful until the 1960s, and very little was done in Sardinia until the 1980s, when it became clear that there were multiple subsources on Monte Arci that could be chemically distinguished [3, 4]. Detailed survey and analysis of geological samples from Monte Arci have since then identified at least seven subsources of usable obsidian, and demonstrated that different chronological and geographic patterns of their usage made it necessary to assign artifacts to specific subsources and not just to Monte Arci [5-9]. Possible explanations for this include differences in the quantity and physical properties of obsidian from each subsource; the accessibility of the subsources and the modes of transport; and changes in the socioeconomic system involved in production and exchange [10]. Type SA obsidian exists *in situ* in a concentrated area near Conca Cannas on the southwestern slopes of Monte Arci, and is black-gray, fairly glassy, with a high level of transparency. Types SB1(a,b,c) and SB2 may be found in multiple outcrops on the western slopes, but in less concentrated quantities, and varies from transparent to opaque, sometimes with many phenocrysts. Type SC obsidian, which is black, opaque, and less glassy, is abundant in mainly secondary deposits on the east side of Monte Arci [7, 11, 12].

It is generally thought that earlier neolithic societies were egalitarian in nature, while specialization in labor activities and the emergence of social/economic ranking are thought to have emerged during the late neolithic period. This is examined here by establishing whether there was consistency or change over time in the use of obsidian subsources in Sardinia, as well as any geographic variation in their usage. Obsidian artifacts were selected from four archaeological sites in southern Sardinia (Figure 1) to complement the previous studies done on a significant number of sites and artifacts in northern Sardinia and in Corsica [6, 11, 13, 14]. Those studies suggested that there was a major chronological change, mainly with a shift from all three major Sardinia subsource groups (SA, SB, SC) being used during the Early Neolithic (ca. 5700-4700 BC), to negligible use of SB and greater use of SC by the Late Neolithic (ca. 4000-3200 BC). This study tested whether or not the same patterns existed in southern Sardinia, and therefore the hypothesis that socioeconomic systems rather than subsource location/access were the main factors in obsidian artifact production and exchange.

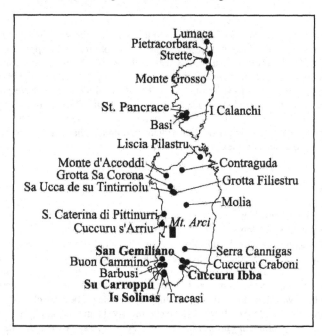

Figure 1. Neolithic sites in Sardinia and Corsica mentioned in this paper. Sites with obsidian tested in this study are in bold.

ARCHAEOLOGICAL SAMPLES

Archaeological samples were selected in 2002, from four neolithic sites in southern Sardinia, with the specific intention of expanding upon the limited data on obsidian source patterns in the region south of the Monte Arci sources. The only previous study, by J. Michels et al. [15], used atomic absorption spectroscopy to analyze a total of 53 artifacts from 5 neolithic period sites, all in the southwestern most part of Sardinia, in the Iglesias and Carbonia regions. The selection of 95 samples from these four sites would significantly expand the data available for comparing subsource patterns in different regions.

Su Carroppu (Sirri, Carbonia) is an Early Neolithic site located in southwestern Sardinia. In the form of a rock-shelter, excavations were done in the 1960s and 1970s by Enrico Atzeni, producing at least a few hundred obsidian artifacts [16-18]. A set of 15 obsidian samples were initially analyzed by Michels et al. [15], while another 20 artifacts were tested in this study (Figure 2a). A separate project, on 63 samples, was recently done by Lugliè et al. [19], using particle induced X-ray emission (PIXE).

Is Solinas (Giba) is a Late Neolithic open-air site located in the southwestern corner of Sardinia, very close to the Golfo di Palmas. A surface collection of obsidian, numbering in the thousands, and other lithic and ceramic materials were made by Atzeni [20]. A set of 24 samples were selected for analysis.

Cuccuru Ibba (Capoterra) is another Late Neolithic open-air site, located on the western side of the Stagno di Cagliari, not far from the Nuragic (Bronze Age) settlement of the same name, in an area now submerged. A large quantity of Ozieri (Late Neolithic) ceramic materials were collected on the surface by farmers, along with flint and obsidian tools, and marine shells [21-22]. Twenty-nine obsidian artifacts were selected for analysis.

San Gemiliano (Sestu) is a Late Neolithic village settlement located inland, due north of Cagliari. The extensive site consists of the remains of hut dwellings spread over several hectares, and large numbers of lithic and ceramic artifacts have been collected from surface deposits [23]. Twenty-two obsidian artifacts were selected for analysis (Figure 2b).

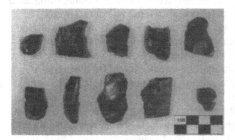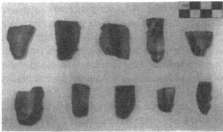

Figure 2. Examples of obsidian artifacts tested, from (a) Su Carroppu (left) and (b) San Gemiliano (right).

METHODS

It has been widely demonstrated that many different analytical methods have the capability of distinguishing between obsidian subsources, if geological samples from known localities are also analyzed to compare with. The most widely used methods for Mediterranean obsidian have been instrumental neutron activation analysis (INAA), X-ray fluorescence (XRF), the electron microprobe, and more recently, particle induced X-ray emission (PIXE) and laser ablation ICP mass spectrometry (LA-ICP-MS) [11, 24-25]. The main issues for archaeologists selecting one of these methods have been cost, availability, and whether they are destructive to artifacts. In this study, LA-ICP-MS was specifically selected because of its low per-sample cost and negligible destructive effect.

ICP mass spectrometry was applied to obsidian sourcing in the Mediterranean nearly two decades ago, but the initial laser ablation units and software available at that time were not user-friendly, while producing liquid solutions from volcanic obsidian required the use of potentially harmful acids (e.g. HF) at high temperatures and pressures [26]. So even though the quadrupole ICP-MS system used at the time was successful in distinguishing obsidian subsources and analyzing other archaeological materials [27], other low-cost, minimally destructive analytical methods (mainly the electron microprobe) were used by the first author for his dissertation research [2, 6-7].

Since that time, newer ICP-MS instruments and accessories have been developed, and LA-ICP-MS was chosen for this study as part of a larger central Mediterranean project investigating obsidian subsources on other islands (Lipari, Palmarola, Pantelleria) that would not be differentiable using the electron microprobe, which is limited to detecting major and minor elements. Details on the instrument and methods used for the LA-ICP-MS instrument at the Missouri University Research Reactor follow, while the results obtained for these other island sources as well as archaeological sites in various parts of Italy, have been published elsewhere [25, 28-34].

The obsidian samples in this study were analyzed using a Thermo Elemental Axiom high resolution magnetic sector ICP mass spectrometer capable of resolving masses as close as 0.001 atomic mass units apart. The ICP-MS was coupled to a Merchantek Nd-YAG 213-nanometer laser ablation unit, and up to ten samples were mounted in the laser ablation chamber at any one time. The laser was operated at 80% power (~1.5 mJ) using a 200 Fm diameter beam, firing at 20 times per second. A rectangular raster pattern of approximately 4 mm2 was drawn over a relatively flat spot on each sample. The laser scanned across the raster area at 70 Fm per second. The laser beam was allowed to pass over the ablation area one time prior to data acquisition in order to remove possible contaminants from the surface of the sample, and to permit time for sample uptake and for the argon plasma to stabilize after the introduction of fresh material. Analytes of interest were scanned three times and averaged. In most cases, the %RSD was 5-10%.

Standardization was accomplished by calibrating the instrument with the NIST SRM-610 and SRM-612 glass wafers doped with 61 elements. Two obsidian glasses calibrated in a round-robin exercise by the International Association of Obsidian Studies were also used. Monitoring the amount of material removed by the laser and transported to the ICP is complicated by several factors making normalization difficult. Conditions such as the texture of the sample, hardness of the sample, location of the sample in the laser chamber, laser energy, and other factors affect the amount of material introduced to the torch. A normalization method described by Gratuze et al. [35] was employed here.

In the current study, about 40 elements were measured using a resolution of 6000. The high resolution was necessary to reduce the number of ions striking the multiplier caused by several of the high concentration elements (esp. Na, Al, Si, K, and Fe). Relative concentrations for all elements were determined by comparing the unknowns to the NIST glass and obsidian standards. To convert the relative concentrations into absolute values, normalization was accomplished converting the relative concentrations to oxides and then normalizing the total to 100%. The method yields satisfactory concentrations for all major and trace elements.

RESULTS AND DISCUSSION

The results of the LA-ICP-MS analyses are clearly grouped into the four primary Monte Arci subsources (SA, SB1, SB2, SC) as illustrated in Figure 3. While it is possible to discriminate between SC subsources (SC1, SC2) and SB1 sub-subsources (SB1a, SB1b, SB1c), such distinctions are not significant for this study (see [11]). A simple bi-plot of element ratios Fe/Cs vs. Sm/Sr is sufficient to separate these subsources, while multivariate statistical analyses provide even clearer distinctions. The numeric results for each of the four sites tested in this study are provided in Table 1.

A comparison by chronological phases of the percentages of the three main subsources (SA, SB, SC) used at all studied sites in southern Sardinia [this study; 15, 19] indicates that there were general changes from the Early to Middle to Late Neolithic time periods, with a big drop in the presence of type SB obsidian, and a major increase in the use of type SC (Figure 4).

When the subsource percentages for Su Carroppu are compared with the data for other Early Neolithic sites in Sardinia (Santa Caterina, Filiestru, Sa Corona) and in Corsica (Strette, Pietracorbara, Lumaca), there are clearly very similar patterns of usage, especially of type SB obsidian [36] (Figure 5). A similar comparison, for the three Late Neolithic sites tested in this study, with others in southern (Barbusa, Tracasi) and northern (Molia, Sa Ucca, Filiestru, Monte d'Accoddi, Contraguda, Li Muri, Liscia Pilastro) Sardinia, and in Corsica (I Calanchi, Basi, Saint Pancrace, Monte Grosso, Strette), again shows similarity in obsidian subsource usage across these major regions (Figure 6). For these Late Neolithic sites it is also clear that type SB obsidian has become of negligible importance, while type SC obsidian dominates most of the assemblages tested.

The presence of obsidian from two or three subsources at each archaeological site tested in this study, with similarity in patterning with sites in northern Sardinia and Corsica, strongly suggests broad geographic similarity both in the purpose of obsidian usage, and in the socio-economic circumstances in which it occurred. The absence of strong outliers supports a down-the-line model of obsidian trade during the neolithic period, in which the quantity/frequency of obsidian in lithic assemblages decreases with distance from Monte Arci, but with similar representation of the multiple obsidian subsources. The geographically consistent decrease in the use of the SB subsource by the Late Neolithic period is likely to result from the modest geological quantity accessible on the west side of Monte Arci, at a time when increased production and exchange of the greater quantity of accessible SC obsidian is supported by the high levels of primary reduction revealed by survey and excavation in the Sennixeddu area on the east side of Monte Arci [37].

Table 1. Obsidian sourcing results for the archaeological sites tested in this study.

Site	Time Period	No.	SA	SB	SC
Su Carroppu (Sirri, Carbonia)	Early Neolithic	20	8	5	5
S. Gemiliano (Sestu)	Late Neolithic	22	4		17
Cuccuru Ibba (Capoterra)	Late Neolithic	29	16		13
Is Solinas (Carbonia)	Late Neolithic	24	3	1	20
	Total tested	95	31	6	55
			33.7%	6.5%	59.8%

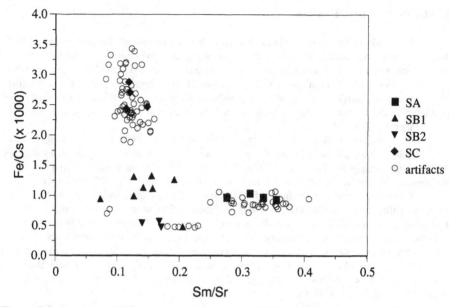

Figure 3. Laser ablation ICP mass spectrometry (LA-ICP-MS) data discriminating Monte Arci subsources (solid symbols) and the archaeological obsidian artifacts tested (open circles).

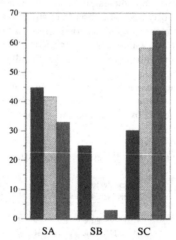

Figure 4. Bar chart of obsidian usage in southern Sardinia during three Neolithic time periods (Early Neolithic = blue, Middle Neolithic = green, Late Neolithic = red)

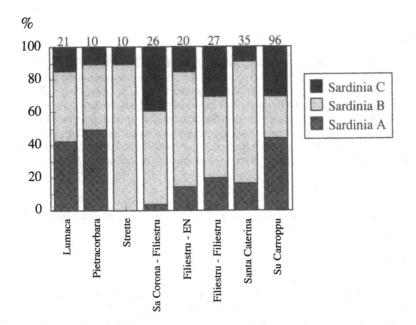

Figure 5. Bar chart of obsidian sources used at Early Neolithic sites in Sardinia and Corsica. Number of samples tested for each site shown at the top.

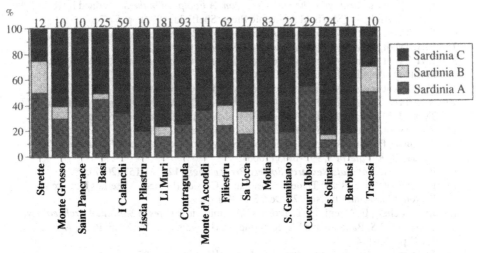

Figure 6. Bar chart of obsidian sources used at Late Neolithic sites in Sardinia and Corsica. Number of samples tested for each site shown at the top.

CONCLUSION

The results obtained using virtually non-destructive, LA-ICP-MS analysis of 95 selected obsidian artifacts from four neolithic period sites in southern Sardinia parallel the patterns of usage of Monte Arci subsources found for sites in northern Sardinia and in Corsica. These patterns of exploitation support a down-the-line model of obsidian trade during the neolithic period, but with chronological changes that are best explained by increased socioeconomic complexity. Further analyses and integration of sourcing data with lithic typology [19], usewear studies [38], and investigation and excavation of source localities and production workshops [12, 39] will lead to better understanding of the role of obsidian in the prehistoric Mediterranean.

ACKNOWLEDGMENTS

The research presented here was supported by a grant to the first author from the National Science Foundation (BCS-0075535).

REFERENCES

1. C. Puxeddu, *Studi Sardi* **14-15**, 10-66 (1958).
2. R.H. Tykot, *Prehistoric Trade in the Western Mediterranean: The Sources and Distribution of Sardinian Obsidian* (PhD dissertation, Harvard University, 1995).
3. B.R. Hallam, S.E. Warren, C. Renfrew, *Proc. Prehistoric Society* **42**, 85-110 (1976).
4. M.P. Mackey and S. E. Warren, *Proceedings of the 22nd Symposium on Archaeometry* (1982) pp. 420-431.
5. R.H. Tykot, in *Sardinia in the Mediterranean: A Footprint in the Sea*, edited by R.H. Tykot and T.K. Andrews (Sheffield Academic Press, Sheffield, 1992) pp. 57-70.
6. R.H. Tykot, *J. Mediterranean Archaeology* **9**, 39-82 (1996).
7. R.H. Tykot, *J. Archaeological Science* **24**, 467-479 (1997).
8. R.H. Tykot, in *Archaeological Obsidian Studies: Method and Theory,* edited by M.S. Shackley (Plenum, New York, 1998) pp. 67-82.
9. R.H. Tykot and A.J. Ammerman, Antiquity **274**, 1000-1006 (1997).
10. R.H. Tykot, in *Social Dynamics of the Prehistoric Central Mediterranean*, edited by R.H. Tykot, J. Morter, J.E. Robb (Accordia Research Institute, London, 1999) pp. 67-82.
11. R.H. Tykot, *Accounts of Chemical Research* **35**, 618-627 (2002).
12. C. Luglie, F.-X. Le Bourdonnec, G. Poupeau, M. Bohn, S. Meloni, M. Oddone and G. Tanda, *C.R. Palevol* **5**, 995-1003 (2006).
13. R.H. Tykot, *Materials Research Society Proceedings* **712**, 143-157 (2002).
14. R.H. Tykot, Sources and Trade of Obsidian in Corsica (France), Materials Research Society, Boston, Massachusetts, Nov. 29-Dec. 3 (2004).
15. J.W. Michels, E. Atzeni, I.S.T. Tsong, G.A. Smith, in *Studies in Sardinian Archaeology*, edited by M.S. Balmuth and R.J. Rowland, Jr. (University of Michigan Press, Ann Arbor, 1984) pp. 83-114.
16. E. Atzeni, *Rivista di Scienze Preistoriche* **XXVII**, 478-479 (1972).
17. E. Atzeni, *Rivista di Scienze Preistoriche* **XXXII**, 357-358 (1978).
18. E. Atzeni, C. Lugliè, M.V.G. Migaleddu, in *L'ossidiana del Monte Arci nel Mediterraneo, La ricerca archeologica e la salvaguardia del paesaggio per lo sviluppo delle zone interne della Sardegna, Atti del 2º Convegno Internazionale, Pau, 28-30 novembre 2003* (Cagliari, 2004) pp. 185-200.

19. C. Lugliè, C., F.-X. Le Bourdonnec, G. Poupeau, E. Atzeni, S. Dubernet, P. Moretto, L. Serani, *J. Archaeological Science* **34**, 428-439 (2007).
20. E. Atzeni, *La preistoria del Sulcis Iglesiente* (Viale Elmas, Cagliari, 1987).
21. E. Atzeni, *Cagliari preistorica* (ETS Editrice, Pisa, 1986).
22. E. Atzeni, *Cagliari Preistorica* (University Press Archeologia, Cagliari, 2003).
23. E. Atzeni, *Studi Sardi* **XVII**, 5-93 (1962).
24. R.H. Tykot, in *Physics Methods in Archaeometry. Proceedings of the International School of Physics "Enrico Fermi"*, edited by M. Martini, M. Milazzo and M. Piacentini (Società Italiana di Fisica, Bologna, 2004) pp. 407-432.
25. R.H. Tykot, in *Acts of the XIVth UISPP Congress, University of Liège, Belgium, 2-8 September 2001. Section 9: The Neolithic in the Near East and Europe*. BAR International Series 1303:25-35 (Archaeopress, Oxford, 2004).
26. R.H. Tykot, *International Association for Obsidian Studies Newsletter* **5**, 9 (1991).
27. R.H. Tykot and S.M.M. Young, Archaeological Chemistry Symposium Series **625**, 116-130 (1996).
28. L. Lai, R.H. Tykot and C. Tozzi, in *Atti del XXXIX Riunione Scientifica dell'Istituto Italiano di Preistoria e Protostoria: Materie prime e scambi nella preistoria italiana, Firenze, 25-27 November 2004* (Firenze, 2006) pp. 598-602.
29. R.H. Tykot, in *L'ossidiana del Monte Arci nel Mediterraneo: recupero dei valori di un territorio*, a cura di P. Castelli, B. Cauli, F. Di Gregorio, C. Lugliè, G. Tanda and C. Usai (Tipografia Ghilarzese, Ghilarza, 2004) pp. 118-132.
30. R.H. Tykot, in *Préhistoire et protohistoire de l'aire tyrrhénienne/Preistoria e protostoria dell'area tirrenica*, a cura di C. Tozzi and M.C. Weiss (Felici Editori, 2007) pp. 217-220.
31. R.H. Tykot, B.A. Vargo, C. Tozzi and A. Ammerman, in *Atti del XXXV Riunione Scientifica, Le Comunità della Preistoria Italiana. Studi e Ricerche sul Neolitico e le Età dei Metalli* (Istituto Italiano di Preistoria e Protostoria, Firenze, 2003) pp. 1009-1112.
32. R.H. Tykot, A.J. Ammerman, M. Bernabò Brea, M.D. Glascock and R.J. Speakman, *Geoarchaeological and Bioarchaeological Studies* **3**, 103-106 (2005).
33. R.H. Tykot, T. Setzer, M.D. Glascock and R.J. Speakman, *Geoarchaeological and Bioarchaeological Studies* **3**, 107-111 (2005).
34. R.J. Speakman, M.D. Glascock, R.H. Tykot, C. Descantes, J.J. Thatcher, C.E. Skinner, K.M. Lienhop, in *ACS Symposium Series* **968**, 275-296 (2007).
35. B. Gratuze, M. Blet-Lemarqu and J.-N. Barrandon, *Journal of Radioanalytical and Nuclear Chemistry* **247**, 645-565 (2001).
36. R.H. Tykot, *American Chemical Society Symposium Series* **831**, 169-184 (2002).
37. R.H. Tykot, C. Lugliè, T. Setzer, G. Tanda and R.W. Webb, *International Association of Obsidian Studies Bulletin* **35**, 9 (2006).
38. T. Setzer, R.H. Tykot and C. Tozzi, *International Association for Obsidian Studies Bulletin* **31**, 8 (2004).
39. R.H. Tykot, M.R. Iovino, M.C. Martinelli and L. Beyer, in *Atti del XXXIX Riunione Scientifica dell'Istituto Italiano di Preistoria e Protostoria: Materie prime e scambi nella preistoria italiana, Firenze, 25-27 November 2004* ((Istituto Italiano di Preistoria e Protostoria, Firenze, 2006) pp. 592-597.

Mater. Res. Soc. Symp. Proc. Vol. 1047 © 2008 Materials Research Society   1047-Y01-05

Analysis of Modern and Ancient Artifacts for the Presence of Corn Beer; Dynamic Headspace Testing of Pottery Sherds from Mexico and New Mexico

Theodore Borek[1], Curtis Mowry[1], and Glenna Dean[2]

[1]Materials Characterization, Sandia National Laboratories, 1515 Eubank Blvd. SE, MS 0886, Albuquerque, NM, 87123
[2]Archeobotanical Services, PO Box 658, 21581 Hwy 84, Abiquiu, NM, 87510

Abstract A large volume-headspace apparatus that permits the heating of pottery fragments for direct analysis by gas chromatography/mass spectrometry (GC/MS) is described here. A series of fermented-corn beverages were produced in modern clay pots and the pots were analyzed to develop organic-species profiles for comparison with fragments of ancient pottery. Brewing pots from the Tarahumara of northern Mexico, a tribe that produces a corn-based fermented beverage, were also examined for volatile residues and the organic-species profiles were generated. Finally, organic species were generated from ancient potsherds from an archeological site and compared with the modern spectra. The datasets yielded similar organic species, many of which were identified by computer matching of the resulting mass spectra with the NIST mass spectral library. Additional analyses are now underway to highlight patterns of organic species common to all the spectra. This presentation demonstrates the utility of thermal desorption coupled with GC/MS for detecting fermentation residues in the fabric of unglazed archaeological ceramics after centuries of burial.

Introduction This work was instigated by an inquiry to perform a spot analysis for furfural on pottery sherds.[1] Since the spot test involves the use of hydrochloric acid, and we did not want use this potentially destructive test on historical specimens, an alternative analytical scheme that utilized dynamic large-volume headspace sampling was proposed.

Dynamic large-volume headspace sampling with gas chromatographic separation with mass spectrometric identification, as used in our laboratory, is a technique that permits the collection of trace organic species from materials that are undergoing thermal treatment in an inert atmosphere. We have developed this technique because:

1. It does not utilize solvents; there is no need to be concerned with selectivity, dilution, or loss of analytes.

2. Ultimate temperature may be selected to minimize thermal stress on articles tested.

3. If necessary, further testing on the same article may be performed without too much concern for this method altering nonvolatile residues.

4. It is nondestructive to the article examined at the temperatures used here.

A test specimen is placed in a heating apparatus that has been previously demonstrated to be free of organic species. The test specimen is then slowly heated in a flowing stream of filtered ultra-high purity nitrogen, and the offgases collected using a cryogenically cooled, 3-trap environmental air sampling system (Entech Instruments, Simi Valley, CA). This system removes most of the water and carbon dioxide from the analytical stream, then concentrates the organic species, permitting part per billion detection limits. The system is represented schematically in Figure 1. This scheme can be reduced to 1) heating to evolve volatile organic species, 2) concentration of organics, 3) separation of organics, and 4) identification of species by mass spectrometry.

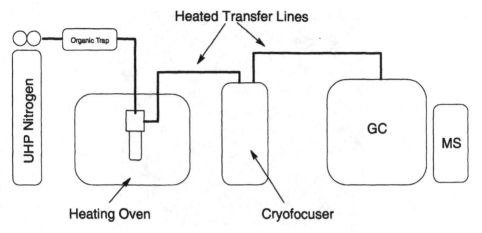

Figure 1. Schematic of Analytical System for Pot Sherd Volatile Analysis

Separation of the focused organic species occurs in the gas chromatograph. Each separated species is then analyzed by mass spectrometry; this mass spectrometric information may then be computer-matched to a database, and a tentative identification of the organic species may be made. Confirmation of species identification may be made later through analysis of analytical standards.

To determine if the analytical system would work as proposed and provide a baseline set of organic species response, intentional fermentation of a corn-based liquid was conducted. This intentional fermentation was carried out in clay pots produced by Archeobotanical Services; the liquid was produced using standard fermentation practices. Corn was purchased at a local feed store, allowed to germinate in warm water, and then chopped. This chopped material was strained, and the liquid placed into the clay pot and either allowed to ferment using wild yeast or was intentionally inoculated with Brewer's yeast. A typical fermentation is shown in Figure 2.

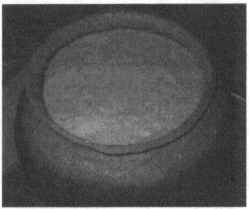

Figure 2. Corn liquid fermentation in modern clay pot

After the fermentation was judged to be complete, the liquid was drained from the pot, the pot allowed to air-dry, the pot was sectioned for analysis as shown in Figure 3. Samples for analysis could be taken from anywhere along the section profile.

Figure 3. Modern pot preparation for volatile analysis

We also obtained brewing vessels from the Tarahumara Nation in the Republic of Mexico. The Tarahumara brew 'tiswin' or 'tiswino' from corn for social and ceremonial purposes. Two brewing vessels were obtained in order to search for chemical markers of corn-based brewing for comparison to archaeological samples. One vessel, next to an 18 inch ruler, is shown in Figure 4.

Figure 4. Tarahumara Brew pot

Finally, a collection of recently excavated sherds was obtained from an archeological dig conducted in New Mexico. These field samples had not been pre-treated in any manner prior to analysis.

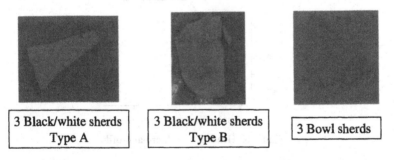

| 3 Black/white sherds Type A | 3 Black/white sherds Type B | 3 Bowl sherds |

Experimental For each analysis, a piece of sherd no larger than 20mm by 30mm was placed in a 25mm diameter by 150mm long glass test tube and this tube was attached to the headspace gas sampling apparatus as shown in Figure 5. This fixture was then placed in a heating oven for thermal treatment of the sample.

Figure 5. Pot Sherd Heating Fixture

Ultra high purity nitrogen, conditioned by an organic gas trap, was passed through this fixture at 40cc/min. The heating oven warmed the samples from 50°C to 190°C over a 27 minute period using the profile shown in Figure 6.

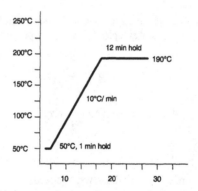

Figure 6. Heating oven thermal profile

Organic gases evolved from the sample were concentrated in a commercial cryofocusing inlet system. The cryofocused gases were then separated using gas chromatography, using an Agilent DB-624 column, 60meter x 0.25mm x 1.4μm film thickness; GC heating conditions were 35°C for 8 minutes, ramp at 7°C/min to 150°C, ramp at 12°C/min to 255°C, and hold at 255°C for 10 minutes. The transfer line to the mass spectrometer temperature was maintained at 260°C. The GC column flow was set to 1.4mL/min at 35°C; the column was maintained at a constant pressure of 21.5 psig. The mass spectrometer was operated in the full scan mode, 33 to 340 amu, 2.4 scans/sec, peak threshold at 150 counts, and a 3 minute solvent delay. The mass spectrometer source was heated to 230°C; the quadrupole to 150°C. Organic gases such as methane, ethane, ethylene, acetylene, propane, and methanol are not observed under the conditions used for this study.

Discussion A variety of chemical species was detected in most samples. In some samples it was clear that the technique was limited by the sample size. These species ranged from simple hydrocarbons (pentane, octane) to aldehydes (acetaldehyde, decanal), chlorinated hydrocarbons (chlorobenzene), and oxygenated compounds (furfural, furan, 2-butenol).

A typical result for a modern pot used for fermentation is shown in Figure 7. Each peak is a distinct chemical species.

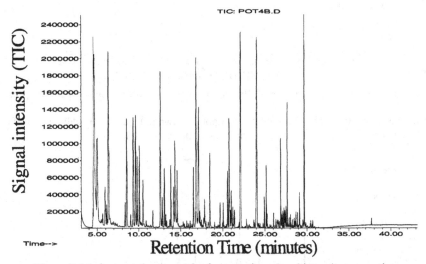

Figure 7: Modern pot sherd used for fermentation, signal intensity versus time

A closer examination of this chromatogram with some of the peaks identified is shown in Figure 8. In addition to furfural, which was the original species of interest, many other species are observed and identified.

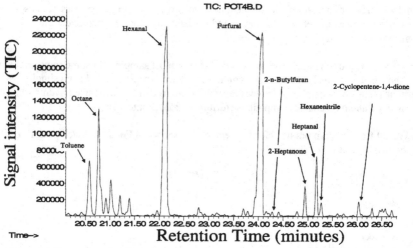

Figure 8: Closer examination of previous chromatogram with identification of several peaks.

The analysis of a portion of a Tarahumara brew pot is shown in Figure 9. In addition to furfural, which is indicated on the figure, many other organic species are observed.

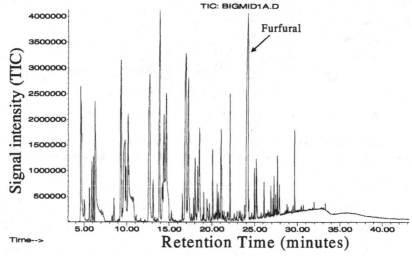

Figure 9: Tarahumara brew pot sherd, signal intensity versus retention time

The analysis of one of the ancient sherds is shown in Figure 10. This example shows just a small portion of the chromatographic result. The peak intensities for this sample are much less than those observed in the modern samples, but are nevertheless present in the analysis. The species observed are indicated on the figure.

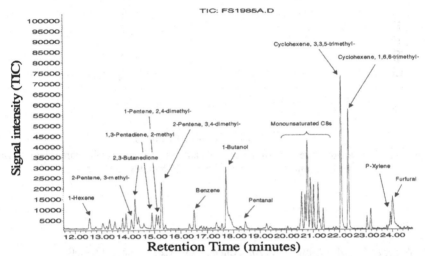

Figure 10: Black/white sherd type B analysis

Further evaluation of the data was then performed; it is necessary to confirm that species with similar retention times have the same mass spectral signature, and then which species may be found across some or the entire sample set. While chromatograms can be compared strictly as x,y data, it is important to remember that with complex samples such as these the mass spectral data must be taken into account as well.

For example, an overlay of the total ion chromatogram plots from several analyses is shown in Figure 11. Almost all of the chromatograms have a peak detected at 22.0 minutes, and the species present has a mass fragment of mass to charge (m/z) of 109 amu. Upon closer examination of the mass spectral data, however, it is revealed that the there are 2 distinct chemical species desorbing from the archeological and Tarahumara Nation samples.

Additional differences can be observed in the selected ion plot, such as the peaks at 22.4 and 22.9 minutes, only observed in Tarahumara Nation pots. Additional data analysis is aimed at finding such differences and similarities that may indicate the type of usage for the archaeological sherds.

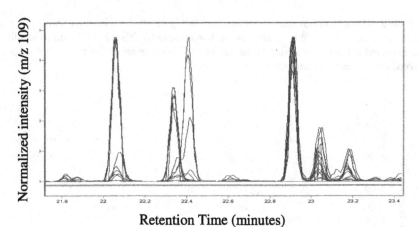

Retention Time (minutes)

Figure 11: Comparison of results: Archeological samples (red) and Tarahumara (blue) overlaid chromatograms region from 21.5 to 23.4 minutes retention time

References:

1. V. Anger and S. Ofri, *Fresenius Z. Anal Chem*; 1964; v.203, no.6, p.422-430

Mater. Res. Soc. Symp. Proc. Vol. 1047 © 2008 Materials Research Society   1047-Y02-02

"For Whom the Bell Tolls" Mexican Copper Bells From the Templo Mayor Offerings:
Analysis of the Production Process and Its Cultural Context

Niklas Schulze
FFyL-IIA, UNAM, Mexico City, Mexico

ABSTRACT

The 3389 copper (alloy) bells from offerings included in successive building phases of
Late Postclassic Templo Mayor (A.D. 1325 – 1520) of Tenochtitlan (Mexico City) are the results
of production processes influenced by social, economic, ideological and technological factors.
The compositional and morphological variability of the bells in the earlier construction phases of
the Templo Mayor suggests the presence of several workshops in or around Tenochtitlan, while
the reduction of this spectrum to one bell type made of copper-tin bronze, points towards a
standardization of the production process and a decrease in the number of workshops that
supplied the Templo Mayor in later phases. The compositional and morphological information,
as well as contextual analysis and comparison with other Mexican bells, give insights into the
bells' symbolism, the mechanisms used to supply the temple with offerings, the organization of
metalwork and the rationale behind some of the technological choices of the artisans. The
detected changes through time (seem to) point to important shifts in the ideological, economic,
social and technological influences on the artisans' choices in the latter half of Aztec rule.

INTRODUCTION

During the past 30 years the excavations at the Templo Mayor of Tenochtitlan have
brought to light more than 160 offerings with a wide array of different materials and object
types. These finds are of special importance and interest, not only because the Templo Mayor
was the real and symbolic center of power in the capital of the Aztec Empire, but because they
offer a rare insight into the culture and cosmovision of a society of the past [1, 2]. One group of
objects frequently included in these offerings that has not yet received much attention is the large
collection of copper bells.

All the bells found before 2003 - a total of 3389 bells coming from 48 offerings - were
included in the present investigation. The majority of these artifacts are *Spherical* or *Pear-
shaped* and measure between 1 and 4 centimeters in height. While many Mesoamerican bells do
not come from controlled excavations, the Templo Mayor collection allows studying the contexts
of the bells and even shows a clear relative and absolute chronology. The offerings with their
bells can be attributed to one of the seven distinct building phases of the Main Temple, which
allows correlating the observable trends in bell forms and compositions with the expansion of the
Aztec Empire. While the rule of the Aztec Triple Alliance of Tenochtitlan, Texcoco y Tlacopan
covered nearly 100 years before being interrupted by the Spanish conquest in 1520, most of the
bells were found in building phases IVb to VII – a period covering only 51 years.

The aim of the present investigation was to understand some of the ideological,
economic, social and technological factors that influenced the production process of the bells and
show that their study might provide information about the society that produced them. On the
following pages the analysis of the bells and the main results of the investigation will be
presented. The objective of the XRF analysis of the bells was to confirm or reject the hypothesis

of a correspondence between morphological, chronological and compositional trends. The large amounts of data generated during the investigation were presented in a different context [3] and for lack of space will only be repeated here in a summarized form.

RESULTS OF THE MORPHOLOGICAL AND XRF ANALYSIS OF THE BELLS

The technological decisions [4] that artisans have to take during the elaboration of the bells using the lost wax casting process [5], are influenced by ideological, economic, social and technological considerations, making the objects the result of a compromise between the cultural and physical opportunities and limitations. The large variety of different bell forms and types - not to forget the diverse alloys, mainly copper or gold based - in all of prehispanic America illustrates the many different solutions societies found in response to the cultural and technical challenges of making these metallic sound-artifacts. The variety of bell forms at the Templo Mayor, however, is rather limited. Four main forms can be identified: As the size-distribution graph clearly shows, the *Pear-shaped*, *Spherical*, *Oval* and *Tubular* bells form dense clusters that point towards a relatively small internal variability (Figures 1 and 2).

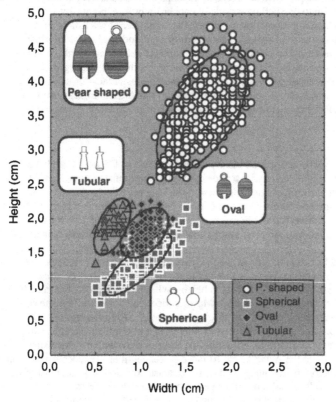

Figure 1. Size distribution of the four basic bell shapes found at the Templo Mayor

196

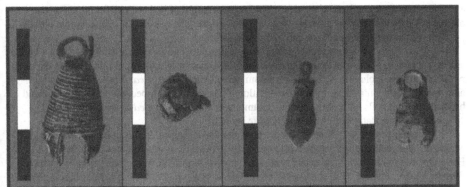

Figure 2. The four basic bell shapes found at the Templo Mayor. The *Spherical* bell still contains its casting core

Two basic forms, *Pear-shaped* and *Spherical* (Figure 2), dominate the collection and in combination represent more than 80 % of all the bells. While the *Spherical* bells are extremely uniform, the *Pear-shaped* ones allow a differentiation in types and sub-types based on the differences in the false wirework and ornamentation [6].

The distribution of the offerings that contain bells in the temple area shows that more offerings containing larger quantities of bells are located in the southern half of the building, which is dedicated to Huitzilopochtli, the main god of the Mexicas, associated with war and tribute (Figure 3).

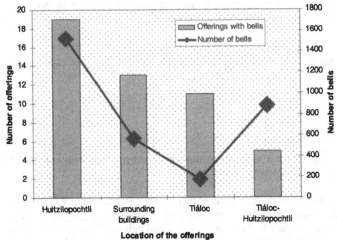

Figure 3. The distribution of offerings with bells in the Templo Mayor.

Fewer offerings, with fewer bells on average, are located in the northern half of the temple, dedicated to Tlaloc, the Mexica god of fertility and agriculture. Matos Moctezuma [7] suggests that the two halves of the pyramid could represent the mountains that clash at the entrance to *mictlan*, the land of the dead. The largest amount of bells can be found in the offerings located exactly on this east-west axis that divides the two halves of the temple pyramid. These trends seem to be maintained in all building stages.

Apart from the distributional differences of absolute and relative amounts of bells, one can detect a general trend over time in the use of the different bell forms. While the first building phases (until and including phase V) were dominated by the use of *Pear-shaped* bells – while also the other three forms were used – the building phases VI and especially VII are dominated by the *Spherical* bells with only very few examples of the *Pear-shaped* types and a nearly complete discontinuation of *Oval* and *Tubular* bells [6].

In order to verify that the morphological and chronological trends coincide with the compositional groupings, it was necessary to analyze the largest possible sample of bells. However, due to curatorial and security reasons, it was impossible to prepare the bells or move them to a laboratory for analysis. On these grounds, it was decided to analyze the objects using a portable XRF prototype [8], made available by the Institute of Physics at the UNAM in Mexico, without removing the corrosion layer. Experiments have shown that the mechanical elimination of the oxide layer in these cases has only a limited potential of improving the analytical results [6]. This is due to the natural heterogeneity of the archaeological metals and the thin walls of the bells (many of them less than 1 mm) that make a complete removal of the corrosion – which attacks the metal from the interior and exterior – impossible, except in very few cases (for a detailed description of the analytical procedure see [3]).

The results of the analysis proved that the composition provides a valid grouping criterion on the one hand, while, on the other, an individual result probably does not reflect the real composition of the object due to its heterogeneity and the processes of element depletion and enrichment taking place in the oxide layer [9]. The general trends, however, are clear. Three elements (arsenic, lead and tin) could be identified as the main alloying metals (with mean values above 2 %) while the three other elements commonly detected (iron, silver and antimony) normally were present only in trace amounts (with mean values below 0.5 %) (Table 1).

Element	Min. (%)	Max. (%)	Mean	Range
Fe	0.03	0.91	0.14	0.88
As	0.00	36.65	2.92	36.65
Ag	0.00	8.86	0.44	8.86
Sn	0.00	28.37	2.34	28.37
Sb	0.00	1.11	0.12	1.11
Pb	0.00	37.82	3.86	37.82

Table 1. The mean values and the range of the three minor and three major elements identified in the copper based bells from the Templo Mayor (n = 567)

The four bell forms cannot be distinguished by their minor elements composition. However, especially for the *Pear-shaped* bells it seems possible to detect two groups (Figure 4), differentiated by the presence or absence of antimony. This could indicate the exploitation of at least two sources of copper minerals. More metal objects and material from possible source areas will have to be studied for a better understanding of the provenance of the raw materials.

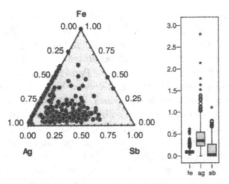

Figure 4. Minor elements (iron, silver and antimony) in the *Pear-shaped* bells

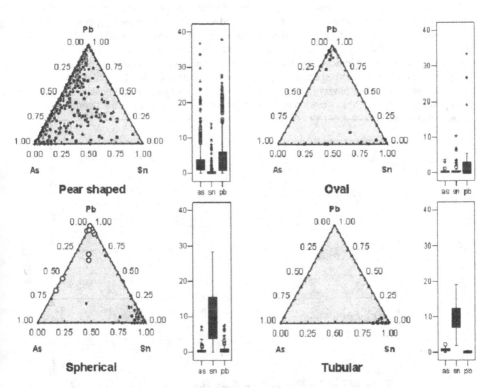

Figure 5. The major element composition of the bells (lead, tin and arsenic). The *Spherical* bells show a clear chronological differentiation between the early lead containing bells (building phase IVb, white dots) and the later tin containing ones (building phases V – VII, black dots)

199

The major elements distribution, on the other hand, offers rather clear patterns when looking at the ternary and the box plots together (Figure 5). The relatively high amounts of tin (average of around 10 %) used in the *Spherical* and *Tubular* bells clearly distinguishes them from the *Oval* and *Pear-shaped* ones. The latter two are mainly separated from each other by the higher arsenic content of the *Pear-shaped* bells. However, while the *Pear-shaped* bells seem to cover a large spectrum of different compositions, the *Oval* and *Globular* bells each form two distinct groups, based on differential lead and tin contents. The reason for this is not clear in the case of the *Oval* bells (that have generally very low amounts of both metals), but the *Spherical* bells show a chronological division: While these bells contained appreciable amounts of lead in building phase IVb, the composition changes to copper-tin in the later phases. This means that the *Tubular* (that were only identified in building phase IVb) and the *Spherical* bells containing tin (from building phases V-VII), although similar in composition, were not contemporaneous.

In spite of these rather obvious differences in alloy use, it is impossible to speak of 'recipes' that the prehispanic metallurgists followed, considering that the alloy composition covers an ample range of concentrations, as especially the example of tin in *Spherical* bells shows (Figure 6). While arsenic was generally mixed as a mineral with the copper minerals or the copper metal melt, which makes it difficult to control its quantity, tin could be added as metal in order to form relatively constant compositions. However, the wide range of alloy concentrations seems to indicate that the metal workers did not attempt – or did not have the ability – to create alloys with specific physical or mechanical properties, like color or strength. That the results are not arbitrary due to difficulties with the analytical instruments or the composition changes in the corrosion layer is shown by the existence of distinct compositional groups that coincide with sub-types of the *Pear-shaped* bells [6].

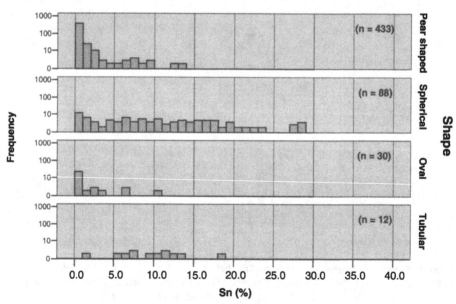

Figure 6. The tin concentration in the differently shaped bells from the Templo Mayor

THE TEMPLO MAYOR BELLS IN THEIR MESOAMERICAN CONTEXT

While in South America pure metals and alloys were sometimes used to create contrasting gold and silver objects, associated, for example, with the right and left side of the body [10, 11], the Templo Mayor bell metals had a large spectrum of colors: Their different compositions indicate that colors ranged from copper red through orange and yellow to nearly white. It was probably this wide spectrum of different colors that was sought by the artisans, rather than specific colors. During the production process of these bells the metalworkers had to solve several technological problems, mainly related to mould design, alloy behavior (for example, gas uptake, fluidity, solidification rates) and process organization. Every step in this process is connected to the others and a change in one parameter, for example composition, had to be responded to by changing one or several of the others in order to assure consistently high quality products. The accomplishment of the prehispanic metallurgists is not the consistent elaboration of one or two specific alloys, but rather the ability to 'make the process work' with the huge number of very different alloy compositions.

The meaning of the bells and their metal, therefore, cannot be assumed to be the same as that of gold and silver. In ethnohistorical sources (e.g. the works of Fray Bernardino de Sahagún and Fray Diego Durán) as well as on stone statuary, bells are frequently associated with war and warriors, the female aspect of the 'Earth Monster' Tlaltecuhtli, the lunar goddess Coyolxauhqui (who even has the Aztec word for bell = *coyolli* in her name), the gods of *pulque* and drunkenness, and mortuary contexts. All of these elements can be linked to the concepts of change and transition [3]. These are also very well exemplified by the material – metal – itself. The hard, brittle mineral becomes liquid metal upon heating, and then hardens to form a malleable material that is brilliant and reflective when polished. Furthermore, in contrast to gold, copper and its alloys in contact with the environment quickly develop a green or black patina.

The composition data seems to suggest that at least some of the alloys used at the Templo Mayor, especially for the *Pear-shaped* bells that contain lead in relatively high concentrations, are unique in Mesoamerica (Figure 7). However, some areas are not very well documented yet, and future investigations will have to supply more data. Also the size data of the Templo Mayor bells, as compared to a collection of illegally traded and confiscated prehispanic bells from all over Mexico [12] and the collection of the Regional Museum of Guadalajara in West Mexico [13], shows a low degree of variability (Figure 8), which together with the clear trends of alloy use, supports an argument against multiple origins.

If we base our research on the data gained from the analysis of the Templo Mayor bells, it seems possible to argue in favor of local production. This argument is strengthened by the fact that the ethnohistoric sources state that raw materials that reached the city as tribute, gift or trade item were redistributed to local artisans that produced the objects – on some occasions specifically for ritual acts at the Templo Mayor [14]. The reduction of bell forms and the domination of small *Spherical* bells in the latest building stages suggest that less material, time and energy were invested, while the division of labor was increased and the products became more standardized. Also, the number of producing workshops seems to have been reduced, which indicates that production and the associated decision-making processes became more centralized.

The increase in production efficiency becomes evident when comparing the two main bell 'models'. By avoiding the false-wirework normally seen in the *Pear-shaped* bells and not

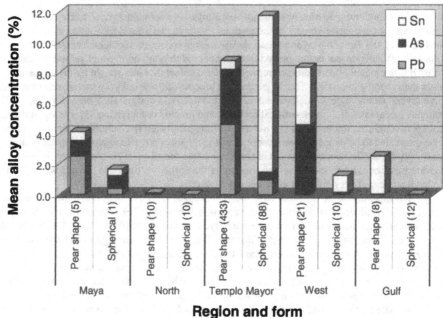

Figure 7. Mean alloy composition of pear-shaped and spherical bells from the Maya area, North Mexico, the Templo Mayor, West Mexico, and the Gulf area [15 – 18]

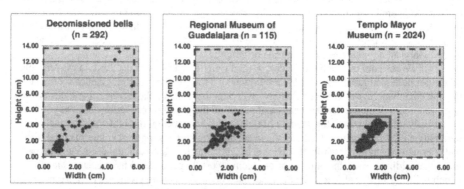

Figure 8. Height and width of the bells from different collections. Size data for the bells from the Regional Museum of Guadalajara was taken from Hosler [13]

removing the casting cores from many of the *Spherical* bells, as can be seen in Figure 2, two very labor-intensive steps were eliminated. One reason for these changes at the beginning of building phase VI can be a transformation in the general organization of the empire under the new *tlatoani* Ahuizotl. He was responsible for initiating the second cycle of large-scale expansion [19] and reached the Pacific coast with his conquering army. This added some areas of western Mexico rich in minerals to the direct zone of influence of the empire.

CONCLUSIONS

Bells made of copper and its alloys were of great importance in prehispanic Mesoamerica, as the large amount of these artifacts indicates. The use-contexts and associations of the bells allow connecting them to the concepts of change and transition. The general lack of size and shape variability of the 3389 Templo Mayor bells, as compared to other Mexican collections, indicates a small number of workshops that produced these sound artefacts. This rejects the possibility that the bells reached Tenochtitlan as tribute from different regions of the Aztec realm. The composition data, which shows the use of alloys different from those of other regions (especially leaded bronze and leaded copper), seem to suggest local production. In the later building phases of the Templo Mayor the variability of form and composition became even more reduced, which indicates standardization and centralization of decision-making processes. Although basic forms are clearly linked to the utilization of specific alloys, no 'recipes' concerning the exact alloy concentrations were used. No effort seems to have been made to create bells with special colors. In summary, it can be said that the changes through time in morphology and composition of the Templo Mayor bell offerings point towards important shifts in the social, technological, economic and ideological influences on the 'technological choices' of the artisans in the latter half of the Aztec dominion, possibly linked to the expansion of the empire under the Aztec ruler Ahuizotl.

ACKNOWLEDGMENTS

The author wishes to thank Luis Barba, José Luis Ruvalcaba and Nicolas Caretta, as well as Alicia Rosas, Hans Jürgen Schulze and Peter C. Kröfges for their guidance, input and patience. This research was supported by Mexican grants from UNAM-DGEP, UNAM-PAPIIT (IN400302) and CONACyT (U49839-R).

REFERENCES

1. L. López Luján, *The Offerings of the Templo Mayor of Tenochtitlan*, Niwot: University Press of Colorado, 1994.
2. E. Matos Moctezuma, *Tenochtitlan*, México: El Colegio de México y Fondo de Cultura Económica, 2006.
3. N. Schulze, *El proceso de producción metalúrgica en su contexto cultural: los cascabeles de cobre del Templo Mayor de Tenochtitlan*, México: PhD Thesis in Anthropology, UNAM, 2008.
4. B. Sillar and M.S. Tite, "The Challenge of 'Technological Choices' for Materials Science Approaches in Archaeology," *Archaeometry* 42 (2000) 2-20.
5. D. Easby, "Sahagún y los orfebres precolombinos," *Anales del INAH* (1955-57) 85-118.

6. N. Schulze and J. L. Ruvalcaba-Sil, "Metales Prehispánicos: el caso de la colección de cascabeles del Templo Mayor de Tenochtitlan," in *Rayos X y otras Técnicas Físicas en Arte, Arqueología e Historia*, México: Sociedad Mexicana de Cristalografía, in press.
7. E. Matos Moctezuma, *The Great Temple of the Aztecs: Treasures of Tenochtitlan*, London: Thames and Hudson, 1988.
8. F. Picazo Navarrete, J. L. Ruvalcaba, K. López and F. Jaimes, "Diseño y construcción de un dispositivo de fluorescencia de rayos X portátil," in *Mérida, XLVI Congreso Nacional de la Sociedad Mexicana de Física, 2003*, (2003) pp. 17.
9. W. Geilmann, "Chemische Untersuchungen der Patina vorgeschichtlicher Bronzen aus Niedersachsen und Auswertung ihrer Ergebnisse," in M. Levey, ed., *Archaeological Chemistry, A symposium*, Philadelphia: University of Pennsylvania Press, 1967, pp. 87-146.
10. W. Alva and C. B. Donnan, *Royal Tombs of Sipán*, Los Angeles: Regents of the University of California, 1993 (2nd ed.).
11. D. Schorsch, E.G. Howe and M.T. Wypyski, "Silvered and Gilded Copper Metalwork from Loma Negra: Manufacture and Aesthetics," *Boletín Museo del Oro* 41 (1996) 145-164.
12. N. Schulze, "The Copper Bells of the Templo Mayor of Tenochtitlan (Mexico): Cultural Influences on the Production Process," *Presentation given at the Archaeological Sciences of the Americas Symposium, 23-26 September 2004*, Tucson: Univ. of Arizona, (unpublished), 2004.
13. D. Hosler, *The Origins, Technology, and Social Construction of Ancient West Mexican Metallurgy* (2 vols.), Santa Barbara: PhD Thesis, University of California, 1986.
14. D. Durán, *Historia de las Indias de la Nueva España e Islas de la Tierra Firme* (2 vols.), introduction by A. Ma. Garibay, México: Porrúa, 1984.
15. D. Hosler and G. Stresser-Péan, "The Huasteca Region: a Second Center for the Production of Bronze Artifact in Ancient Mesoamerica," *Science* 257 (1992) 1215-1219.
16. D. Hosler, *The Sounds and Colors of Power: The Sacred Metallurgical Technology of Ancient West Mexico*, Cambridge, MA: The MIT Press, 1994.
17. J. W. Palmer; M. G. Hollander; P .S. Z. Rogers; T. M. Benjamin; C. J. Dufffy; J. B. Lambert and J. A. Brown, "Pre-Columbian Metallurgy: Technology, Manufacture, and Microprobe Analysis of Copper Bells from the Greater Southwest," *Archaeometry* 40 (1998) 361-382.
18. A. Morales Martínez, *Caracterización No Destructiva de Bronces Prehispánicos Mayas Mediante PIXE*, México: Tesis para obtener el título de Ingeniero Químico Metalúrgico, Facultad de Química, UNAM, 2003.
19. M. E. Smith, *The Aztecs*, Oxford: Blackwell Publishers, 2003 (2nd ed.).

Mater. Res. Soc. Symp. Proc. Vol. 1047 © 2008 Materials Research Society   1047-Y02-01

The Reduction Welding Technique Used in Pre-Columbian Times: Evidences from a Silver Ring from Incallajta, Bolivia, Studied by Microscopy, SEM-EDX and PIXE

Luis Torres Montes[1], Jose Luis Ruvalcaba[2], Demetrio Mendoza Anaya[3], Maria de los Angeles Muñoz Collazo[4], Francisca Franco Velázquez[5], and Francisco Sandoval Pérez[5]
[1]Instituto de Investigaciones Antropológicas, Universidad Nacional Autónoma de México, Circ. Exterior s/n, Ciudad Universitaria, Mexico DF, 04510, Mexico
[2]Instituto de Fisica, Universidad Nacional Autónoma de México, Mexico DF, Mexico
[3]Instituto Nacional de Investigaciones Nucleares, Salazar, Edo. de Mexico, Mexico
[4]Museo Arqueológico, Universidad Mayor de San Simon, Cochabamba, Bolivia
[5]Depto. de Materiales, Universidad Autónoma Metropolitana-Azcapotzalco, Mexico DF, Mexico

ABSTRACT

A pre-Columbian silver ring from Incallajta, Bolivia, recovered from an archaeological excavation is composed of a thin sheet of silver bent to form the ring. Two small wires in the shape of the infinity sign are joined to the surface of the ring. Four green stone beads were laid inside the four cavities formed by the wires. Energy dispersive spectroscopy (EDX) and Particle Induced X-rays Emission (PIXE) analyses of the beads proved that they were turquoise. Examination with a stereoscopic binocular microscope indicated that the two wires could have been soldered to the ring by reduction welding, because copper corrosion products were found in the interface of the welding, similar to those seen on two modern silver objects from Indonesia, decorated with granulation. Since reduction welding is a technique not reported before in pre-Columbian metallurgy, further analyses were carried out to prove that it was used here. Thus, the ring was analyzed with Scanning Electron Microscopy (SEM-EDX) and external beam PIXE, showing with certainty that the copper content in the area of the welding was higher than in any other part of the ring, with increasing copper amounts towards the center of the weld.

INTRODUCTION

The Incallajta Archaeological Ruins, an archaeological complex of about 30 hectares, in the Municipality of Tocoma, Province of Carrasco, Department of Cochabamba, Bolivia, were appointed a National Monument in 1929 (Figure 1). "The Incallajta Archaeological Research Project" [1] started in 1999 with an extensive systematic survey of valleys, highlands and surroundings undertaken since 2000. Several sites of different epochs were found with agricultural terraces, barns, access control points, water supplies and roads, among other structures. This development shows the great importance of the area for the Incas, whose main occupation dates from 1450 to 1532 A.. D. In 2006 the archaeological excavations in the inner and outer part of structure 52D, toward the west of the site, produced two metallic objects in a carefully protected context around a home furnace and garbage place, indicative that the site was the dwelling of a very important personage. The silver ring is composed of a thin sheet of silver bent to form the ring. Two small wires in the shape of the infinity sign or a figure eight, are joined to the surface of the ring. Four green stone beads were laid inside the four cavities formed by the wires. Figure 2 shows the ring pieces; only one wire was still attached to the ring. The second item is a common sewing needle.

In spite of research done on pre-Columbian metallurgy, still several issues have to be solved. In the spring of 2007, the two small metal artifacts, arrived at the Laboratory of Archaeological Chemistry of the *Instituto de Investigaciones Antropológicas-UNAM* for a technical examination. There were two questions to solve: First, identify the materials composition and its condition to recommend a conservation treatment, and second, determine their manufacturing techniques. The objects came from a site where the neighboring communities were protective of their heritage and extreme care needed to be taken to avoid even minimal damage to the artifacts, so analysis had to be non-destructive. Thus, microscopic examination and analytical techniques such as Scanning Electron Microscopy (SEM-EDX) and external beam Particle Induced X-ray Emission spectrometry (PIXE) were used for the characterization of the artifacts.

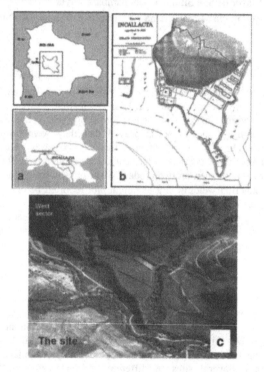

Figure 1. (a) Map of Bolivia showing Cochabamba, (b) historic map of Incallajta and (c) aerial view of the site.

Figure 2. (a) Needle and ring from Incallajta, (b) ring with incrusted green beads.

EXPERIMENT

The microscopic examination of the artifacts was carried out with a binocular stereoscopic microscope at 5 to 50 X to determine the main aspects of manufacturing. After this initial analysis, the other analytical techniques were applied.

Among the analytical techniques that use particle accelerators, Particle Induced X-ray Emission (PIXE), is a very useful tool for analyzing historic artifacts [2] and several non-destructive procedures exist that can be applied to characterize practically every type of material. The object is irradiated with a beam of protons using an external device [3], applied in this case for the analysis of the ring, the beads and needle. All the 3 MeV proton beam irradiations were performed under the observation of a micro-camera that transmits the image of the irradiation zone of the artifacts to two monitors (Figure 3).

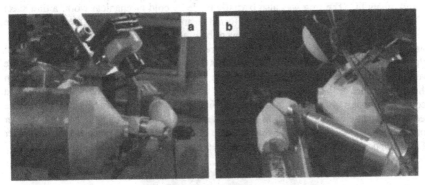

Figure 3. Analysis of the items by the external beam setup. (a) Overview of the instrument and (b) close-up of the beam on the sample.

The beam diameter was collimated to 0.5 mm. Two X-ray detectors were used in the external beam PIXE set-up: an AmpTek Si-PIN diode detector for the detection of light elements and a Canberra LEGe detector for trace element detection. In order to calibrate the external beam for quantitative analysis, standard reference materials certified by NIST of brass (SRM 1107) and homogeneous silver alloys of 0.925 and 0.720 were irradiated under the same conditions as the ring and needle. Reference materials of sediments SRM 2704 and 27011 were used as well for the analysis of the green beads.

After the PIXE analysis, the artifacts were studied in the low vacuum electron microscope of the *Instituto Nacional de Investigaciones Nucleares*. Due to their size, they fit inside the SEM chamber without physical harm. Micrographs and EDX spectra were taken using a 20 keV electron beam in regions, where under the binocular microscope, copper corrosion products were evident and also in zones that will enable determination of the original composition of the ring.

RESULTS AND DISCUSSION

Microscopic examination

The examination indicates the technique with which the items were formed: a simple hammering process. The needle was cold hammered from a wire with a technique described by Di Peso and others [4]. The ring was also partially made by cold mechanical work, a thin sheet some 0.2 mm thick and 5 mm wide, curved to form the ring of about 20 mm in diameter. At one end, two nails extended toward the sides and were bent around the other side to hold it firmly. Two small wires bent to form the infinity signs, were applied to the surface of the ring and four green-blue beads were attached in the four hollows they created. It was not clear whether the wires were joined to the ring by mechanical work or if they were welded on. One clue was seen under the microscope: at the interface of the wires and the substrate, and only in this place, a green copper corrosion product was observed (Figure 4), with properties similar to those described for malachite by Gettens [5]. Malachite is a basic copper carbonate, a common copper alteration product. The explanation for the presence of malachite in that place is that the solder was richer in copper, which corrodes preferentially because is more active than silver.

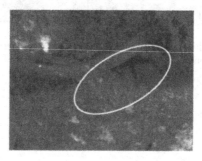

Figure 4. Wire over the ring with green copper corrosion products (inside the indicated region).

Similar localized copper corrosion was observed during microscopic examination of two modern Indonesian silver artifacts, a wrap for cigarette lighters and a container for matches, decorated with "granulations" similar to the ones in ancient gold jewelry, joined to the metal substrate (Figure 5).

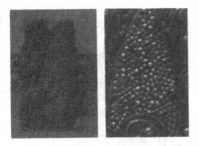

Figure 5. Indonesian silver wrap with granulations and detail.

This localized corrosion resulting in malachite was observed in the joins of the attached wires to the body of the ring. The granulation technique was used in gold artifacts, not in silver, by Egyptians, Etruscans and later other countries of Europe until the Middle Ages [6]. Jewelry of gold was decorated with small gold spheres, up to $900/cm^2$, welded to the surface of the metallic artifact. However, the granulation technique was forgotten and when goldsmiths attempted to reproduce it, heat applied to the spheres to weld them to the substrate metal formed a layer of gold on the surface because the spheres fused before the solder. In the 20th century forgeries were produced. Litteldale rediscovered in the 1940's the technique of fusion welding at the British Museum. The weld is obtained when a mixture of an organic adhesive and a copper salt is used to glue the spheres to the object. When this mixture is heated with a blowpipe, the organic adhesive burns providing C and the copper salt is reduced, depositing pure copper that diffuses and welds the spheres to the artifact, without melting them. Thus if malachite is used as copper salt $Cu_2CO_3(OH)_2$, it is reduced to metallic copper when heat is applied, taking the oxygen from the salt according to the reaction of equation 1, the copper then serves as solder:

$$CuCO_3(OH)_2 + C \rightarrow 2Cu + 2CO_2 + H_2O \qquad (1)$$

To find out if Cu was present and if its concentration is higher in places where the welding took place, without removing a sample, SEM-EDX and PIXE analytical techniques were used.

SEM-EDX analysis

Figure 6, shows some micrographs from a turquoise piece and the separated wire. The elemental composition of the ring is shown in Table I. From the EDX results it is clear that the electron beam does not penetrate deeply under the surface of the ring, since the elemental chemical composition is mainly representative of a layer with a certain amount of corrosion, however is possible to find differences between the composition at the join and that of the silver ring that can be interpreted as larger amount of copper and its corrosion products in the joins as related to the surface of the ring and wires. However, in some parts of the wire-ring interface, the evidence is confused by the presence of soil from Si, Al and O contents as well as other elements from the corrosion layers such as sulphur.

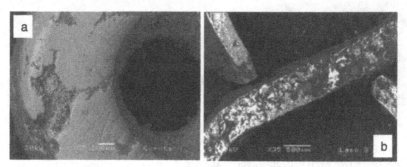

Figure 6. SEM images of the (a) turquoise piece and (b) the separated wire of the ring.

Table I. Elemental composition (weight %) of the ring obtained by EDX.

	Region				
	ring	welded joint	welded joint	attached wire	separated wire
C	16.8	23.6	28.0	13.5	13.5
O	24.9	29.6	31.3	24.4	9.33
Na	1.11	0.96	1.10	0.55	0.53
Mg	0.32	0.34	0.38	0.54	0.14
Al	2.82	3.85	4.66	7.99	1.66
Si	4.10	5.45	6.45	4.75	1.33
P	0.62	1.06	1.57	0.78	3.56
S	2.05	0.73	0.51	2.20	0.31
Cl	0.63	0.87	1.16	0.72	0.60
Ca	0.94	1.02	1.63	1.17	0.52
Fe	1.01	1.80	2.70	1.85	68.5
Cu	0.78	1.71	1.69	1.35	-
Ag	43.9	29.0	18.9	40.2	-

On the other hand, the needle presented the following elemental composition: C 29.8 %, O 29.4 %, Mg 0.29 %, Al 3.46 %, Si 5.61%, P 1.5 %, S 0.18 %, Cl 0.48 %, Fe 1.37 %, Cu 24.1 % and Sn 12.9 % indicating that it was a bronze needle but very corroded. It is also contaminated with materials from the soil, indicated by high amounts of O, Si, and Fe.

In regard with the green bead analyzed by this technique, it presents a high proportion of C: 11.6 %, O 48.6 %, Al 16.2 % P 13.0%, Fe 5.4 % and Cu 4.65 %. From the contents of Al, P and Cu contents we can identify the green stones as turquoise.

External PIXE analysis

The elemental composition of the ring determined by PIXE is shown in Table II. The composition of the ring shows an average of Ag 84.9 %, Cu 7.4 % and O 7.4%. The ratio of Cu/Ag for several parts of the ring (A, B, C) and the wire rings indicates that they were made of the same alloy. The variable proportions of Fe and O are due to soil remains, however, the surface corrosion does not seem to have significantly influenced the analytical results due to the depth penetration of the 3 MeV proton beam (about 30 μm for this metallic matrix) [7].

Table II. Elemental concentrations (weight %) of the ring determined by PIXE. Uncertainties are ±10% of the measurement.

Analyzed Region	Fe %	Cu %	Ag %	O %	Cu/Ag ratio
wire 1_1	0.08	10.0	85.4	4.5	0.12
wire 1_2	0.08	6.6	89.9	3.4	0.07
wire 1_3	0.08	4.9	89.8	5.2	0.05
wire ring 1_4	0.17	8.0	83.7	8.1	0.10
ring A	0.07	9.4	84.0	6.4	0.11
ring B	0.06	7.2	88.9	3.9	0.08
ring C	0.38	6.7	82.0	10.9	0.08
wire ring 2_1	0.28	8.4	82.9	8.4	0.10
wire ring 2_2	1.78	5.4	77.1	15.7	0.07
welded joint ext.	2.49	51.8	36.9	8.9	1.4
welded joint med.	0.32	69.4	11.9	18.4	5.8
welded joint center	0.37	70.4	12.7	16.5	5.5

In the region where the weld may be present, and as the PIXE analysis goes deeper than EDX into the inner part of the core, the concentrations of Cu increase rapidly from the exterior (51.8%) to the center of this region (70.4%), reinforcing the possibility that a reduction welding method could have been used to join the wires to the rings [6]. Soldering cannot have been used if we consider that an alloy composition of 70%Cu-30%Ag will fuse at 950°C, about 70°C higher than the fusing temperature of the silver ring alloy with a composition of 90%Ag-10%Cu [8]. This is the first report of this kind of soldering technique for pre-Columbian America.

On the other hand, the mean analytical results from PIXE for the needle agree with the EDX results and indicate that its alloy is a Cu-Sn bronze, (Fe 0.18%, Cu 50.1%, As 0.04%, Sn 31.9%, Sb 0.17%, O 17.6%). It is not possible to measure the composition accurately due to the advanced degree of corrosion and the needle is so thin that the composition of the whole metal core has been altered. The high amounts of Sn and Sb are due to surface enrichment in the patina layer.

Table III shows the elemental composition of the green beads. Due to the observed proportions of Al, Cu. and P, the beads mineral identification corresponds to turquoise. The Fe and Zn seen in the analysis is due to substitutions of Cu and Al in the turquoise crystalline structure, which could correspond to other phases such as calcosiderite and faustite [9]. The comparison of the compositions of the beads indicates that they were obtained from different sources, with the compositions of beads 1 and 2 the most similar.

Table III. Elemental concentrations (weight %) of the green beads determined by PIXE. Uncertainties are ±10% of the measurement.

Item/region	O	Al	P	Cl	K	Ca	Ti	Fe	Cu	Zn
	%	%	%	%	%	%	%	%	%	%
bead 1_1	45.6	20.0	21.0	0.046	0.069	0.083	0.013	3.63	7.34	2.21
bead 1_2	44.9	20.0	20.0	0.099	0.120	0.109	0.008	3.50	8.98	2.29
bead 2_1	37.3	26.4	27.1	0.027	0.191	0.100	0.041	1.41	6.77	0.81
bead 2_2	41.5	25.2	24.9	0.096	0.184	0.117	0.021	1.26	6.00	0.77
bead 3	46.3	20.6	18.0	0.097	0.185	0.086	0.078	1.54	12.45	0.64
bead 4	47.6	17.5	20.7	0.039	0.046	0.044	0.020	6.04	7.67	0.38

CONCLUSIONS

The microscopic examinations and the analytical results of the copper elemental concentration profile indicate that reduction welding has been used to join the wires to the silver ring discovered in Incallajta. The use of this technique has never been reported before in pre-Columbian metallurgy.

Besides the corroded surface, PIXE analysis was useful in characterizing the silver ring and the wire rings alloys and in determining the copper content of the weld due to the higher penetration of the proton beam. Unfortunately, the corrosion layer was too thick to obtain representative measurements using SEM-EDX.

The needle was manufactured using a Cu-Sn bronze. Finally, green stone beads were identified as turquoise from EDX and PIXE results. This is the first report of the use of this mineral in this region of pre-Columbian Bolivia. The compositional data of the green beads will be useful for turquoise sourcing in the future.

ACKNOWLEDGMENTS

Authors would like to thank technicians K. López and F. Jaimes for their support at the Pelletron particle accelerator during PIXE measurements. Partial financial support was provided by CONACyT Mexico grant U49839-R.

REFERENCES

1. M.A. Muñoz Collazo, 2006. Gestión Participativa del Patrimonio: Un caso Boliviano, *Cuadernos de Antropología y Patrimonio Cultural* 4. INAH-CONACULTA, Mexico, 3- 47.
2. M. A. Respaldiza & J. Gómez-Camacho (eds.), 1997. *Applications of Ion Beam Analysis Techniques to Arts and Archaeometry,* Universidad de Sevilla, Sevilla.
3. J.L. Ruvalcaba Sil. 2005. Analysis of Pre-Hispanic Items from Ancient America in *X-rays in Archaeology,* M. Uda, G. Demortier, I. Nakai, coord, Springer, Dordrecht,123-149.
4. Ch.C. Di Peso, J. B. Rinaldo and G. J. Fenner, 1974. *Casas Grandes. A Fallen Trading Center of the Gran Chichimeca.* Vol. 7 Stone and Metals. The Amerind Foundation Inc., Northland Press, 500-532.
5. R.J. Gettens, 1963. *Mineral Alteration Products in Ancient Metal Objects* in Recent Advances in Conservation, G. Thomson, Editor, Butterworths, London, 89-92.
6. G. Nestler, E. Formigli, 2004. *Granulazione Etrusca, Un'antica tecnica orafa,* Nuova Immagine Ed., Siena.
7. L. Beck, S. Bosonnet, S. Réveillon, D. Eliot, F. Pilon, 2004. Nuclear Instruments and Methods B 226, 153-162.
8. T.B. Massalsi, 1992. *Binary alloy phase diagrams,* ASM International, Ohio, 29.
9. J.L. Ruvalcaba Sil, L. Bucio, M.E. Marín & A. Velázquez, 2005. Estudio por XRD y haces de iones de teselas de un disco de turquesas del Templo Mayor de Tenochtitlán in *La Ciencia de Materiales y su Impacto en la Arqueología. Vol II*, Academia Mexicana de Ciencia de Materiales A.C., D. Mendoza, J. Arenas y V. Rodríguez coord., Ed. Lagares, México, 95-11.

Reconstruction of
Past Technologies

Mater. Res. Soc. Symp. Proc. Vol. 1047 © 2008 Materials Research Society   1047-Y02-05

Reassessing Bronze Age Manufacturing Technologies at Nuzi

Andrew Shortland[1], Katherine Eremin[2], Susanna Kirk[1], and James Armstrong[3]
[1]Department of Materials and Applied Sciences, Cranfield University, Shrivenham, United Kingdom
[2]Straus Center for Conservation, Harvard University Art Museums, 32 Quincy Street, Cambridge, MA, 02138
[3]Semitic Museum, Harvard University, Cambridge, MA, 02138

ABSTRACT

Excavations from 1925-1931 at the Hurrian city of Nuzi, in modern Iraq, yielded a large and important assemblage of glass and frit and smaller but significant assemblages of metals and ceramics. Thousands of glass, and to a lesser extent Egyptian blue and faience, beads were recovered from the site, but the vitreous assemblage also included decorated glass vessel fragments, molded figurines and amulets. Many of the ceramics are glazed and include figurines and architectural wall nails, whilst the metals include weapons, tools, jewelry and armor. The vitreous materials have been widely studied in the past but the other assemblages have received little attention. These include large quantities of glazed and un-glazed ceramics and metals, the latter being dominated by copper alloys.

The current study involves reassessment of the entire assemblage, concentrating initially on the vitreous materials, glazes and metals. Portable non-destructive X-ray fluorescence (XRF) analyses of the vitreous materials and glazes can identify the colorants even in heavily deteriorated examples which retain little indication of the original color or even material, and allow the state of preservation of the artifacts to be assessed. This has shown that many beads which do not currently appear vitreous were originally glass and that many more beads were red, yellow and white than was previously expected. Variations in preservation across the site and within individual buildings are currently being examined. Selected beads and vessels have been sampled to obtain more accurate compositional data of both altered and unaltered glass by scanning electron microscopy with energy and wavelength dispersive analysis (SEM-EDS and SEM-WDS). Portable non-destructive XRF analyses of the metals allow the different alloy types to be identified despite significant corrosion and hence the proportion of different alloys in the assemblage to be determined. Full characterization of the assemblages will allow relationships between different manufacturing technologies and the raw materials used to be investigated.

INTRODUCTION

The Site

The mound of Yorgan Tepe, which contains the Late Bronze Age city of Nuzi, is situated 13 km southwest of the town of Kirkuk in modern Iraq. It lies on a flat plain between the Kurdish mountains in the northeast and 130 km from the Tigris River to the southwest [1]. The mound was excavated between 1925 and 1931, initially for a single season by the American School of Oriental Research in Baghdad and the Iraqi Department of Antiquities, and then for four seasons by Harvard University, the last two of which were under the direction of Richard Starr. The greater part of the Late Bronze Age city was uncovered during these excavations, including the temple and palace complexes. The name Nuzi is derived from the numerous (nearly 5000) inscribed clay tablets, mostly business and legal documents, found in the upper levels at the site in both the palace and private residences [2]. The excavations were primarily concerned with remains of the Mittani Period, principally Stratum II, with earlier occupation only being uncovered in limited areas of the site [3]. The excavations showed that the site was occupied intermittently from around 5000 BCE to about 400 CE. In the mid-second millennium BCE Nuzi and the surrounding region were inhabited by people that spoke Hurrian, and the town was a regional center in the Hurrian kingdom of Arrapha. Arrapha, the remains of which lie under Kirkuk, was part of the Mittani Empire, a significant power in the Near East at that time [4, 5]. The city of Stratum II was sacked and destroyed during a period of political upheaval in the region in the mid 14th century BCE, most probably around 1340 BCE [4, 6]. The identity of the destroyers of Stratum II of Nuzi is unclear, but the nearby Assyrians have been suggested as a likely possibility based on the political situation at the time [6]. The temple and palace complexes were never rebuilt; however, a later Sassanian settlement occupied parts of the mound during the 3rd to 4th centuries CE and several storage pits from this period and later Islamic graves disrupt the stratigraphy in some areas [7]. Therefore, some care is needed in examination of the finds. The site is best known for the tablets and large numbers of glass beads found here, the majority of which came from the 1340 BCE destruction layer. However, significant quantities of other vitreous materials, glazed and unglazed ceramics and metals were also excavated. About half the excavated materials were brought to the west after a division of finds with the Iraqi Department of Antiquities. Most of these are held in the collection of the Semitic Museum, Harvard University.

The Vitreous Assemblage

The glass from Nuzi is numerically the largest assemblage of second millennium BCE glass known. Many thousands of beads and other small objects of glass were found along with numerous core-formed vessel fragments. Starr [8] notes that around 16,000 beads (of all materials but mostly glass) were found in the temple complex alone. A great variety of beads are present within the glass assemblage from very simple spherical and cylindrical forms to complex polychrome examples, molded spacer beads, inlaid beads, zooform beads, and many other types. Vandiver [9] has pointed out the technical excellence in glass working present in the assemblage and suggests that the assemblage represents a relatively advanced stage of glass technology. No complete vessels were found, possibly due to deliberate breakage and scattering by the looters of the Stratum II city [10]. However, from the fragments various forms have been suggested, including a high shouldered goblet type similar to some forms of Nuzi ware [10, 11]. The current study involves a complete re-examination of the glass and other vitreous material and the first systematic study of its distribution and weathering.

Blue was the most common color of

Figure 1: blue glass vessel fragments

218

glass reported from the Nuzi assemblage [12]. Indeed almost all of the vessel fragments, with only one or two exceptions, are noted as being predominantly blue in color with decorative inlays of white, yellow, red/brown and black, although Starr [13] notes a single yellow fragment.

No direct evidence of glass making or glass working was found on the site. However, three lumps of dark blue translucent, glass which may be fragments of raw glass ingots [11] are present in the assemblage, although not mentioned in the excavation report. Many of the beads were made by winding molten glass around a wire, some examples have been found with the wire still in place [14]. Other bead types, the sun disk pendants, and the figurines are believed to have been produced by molding molten glass.

The majority of the Nuzi glass, and certainly all of the vessel glass, comes from Stratum II which is thought to span around a century between 1440 and 1340 BCE (± 10 years) [4, 6]. It has been suggested that due to a fire earlier in the history of Temple A that the majority of the glass found there will date from its reconstruction just before the destruction of Stratum II [15]. However, given the probable status and significance of glass it is possible that objects may have been conserved for a considerable time before deposition.

There are only a few other sites in the Near East with glass finds of this period. Polychrome glass beads are reported from Level VII (1650-1540 BCE) of Alalakh in Turkey and a single glass vessel fragment is noted from Level VI of the same site [16], which is dated to c1550 BCE or slightly later. Glass vessel fragments are reported from levels V-II at Alalakh with the majority reported from Level II [16] (c1300 BCE). At Tell Brak, in northeastern Syria, a few small fragments from Level 6, which are from a similar date to Level VI at Alalakh, are mentioned [17]. Further vessel glass fragments, along with other glass objects, including glass ingots, are reported from the Mittani levels (5 to 2) at Tell Brak [18] (c1500-1200 BCE). Tell al Rimah, in modern Iraq, has also produced glass vessel fragments from the Mittani (c1550-1400 BCE) and early Middle Assyrian (1350-1200 BCE) levels of this site [19] with only a single fragment from the early Mittani levels (c1500 BCE). The Nuzi assemblage therefore dates from slightly after the first appearance of core-formed glass vessels but is contemporary with more regular production of these objects. Stratum II is also contemporary with the Amarna period in Egypt when glass was being produced in significant quantities.

The Metal Assemblage

Over 1000 metal artifacts were brought back to Harvard. The majority of these are copper or copper alloys but lesser quantities of lead, iron and rare silver and gold were also identified. Unlike the vitreous materials, which have been widely studied in the past [for example, 9, 20, 21, 22, 23, 24], the metalwork has received little previous attention. Metallographic examination of eight artifacts was carried out by Bedore Ehlers and Dixon [25]. Their results indicated use of a range of alloys: one ring of leaded brass (14.4% zinc, 4.7% lead), one ring of a mixed copper-zinc-lead-tin alloy (12.2% zinc, 6.3% tin, 3.3% lead), one ring and one pin of bronze (7.0% tin and 6.4% tin respectively and 0.4% arsenic), one pin of a copper-arsenic alloy (0.8% arsenic, 0.4% nickel and 0.4% iron), and three arrowheads and one pin termed "dirty" copper by Bedore Ehlers and Dixon [25]. The "dirty" copper had traces of arsenic, up to 0.5%, and trace to negligible amounts of tin, lead and iron, ≤0.5%. The use of zinc-bearing alloys for jewelry and "dirty" copper for arrowheads suggested deliberate selection of specific alloys for different object types but was based on analysis of only a very small proportion of the assemblage. Although the existence of early zinc-bearing alloys is now well documented from 3rd millennium onwards at sites in Southwest Asia [26], the number of such artifacts remains small. It is hence important to determine how prevalent such zinc-bearing alloys are in the Nuzi assemblage and whether these are restricted to specific typologies. Although there is no textual or archaeological evidence for primary metal or glass production at Nuzi, evidence of secondary manufacture of artifacts from glass and metal does exist and some connection between the origins of the imported and/or raw materials might be expected.

The texts from Nuzi mention a range of metals, including copper, bronze, silver, gold and lead, and provide evidence that Nuzi was a major manufacturing site for weapons, such as arrowheads, spear points, swords, chariot fittings and armor scales, which were manufactured by a guild of smiths [27]. Storehouses at Nuzi held both raw materials and manufactured items, the most abundant of which appears to have been arrowheads. Thousands of arrowheads were sent to other towns and stored in vast numbers in the palace, as well as being kept in private houses and temples. The texts also mention that bronze was kept in the palace storehouse, although the form is not specified [27]. In his excavation reports, Starr noted that arrowheads were extremely common whilst other weapons were rare and suggested that this was due to looting prior to the destruction of the city [28].

Most artifacts are described in the texts as 'copper', although silver and gold inlays are mentioned a number of times [27]. The main exception to this is arrowheads, where both copper and bronze varieties are mentioned, sometimes together. For example, a list of military equipment includes "30 copper-tipped arrows, 50 bronze-tipped arrows" [27]. A bronze helmet and the bronze arms of a bow are also mentioned and the texts describe metal workers, as a "worker of bronze" and receiving bronze [27]. From this, it is apparent that the Hurrians were well aware of the difference between bronze and copper and it seems that copper was the main metal employed, at least for military equipment. Bronze was also available but appears to have been used mainly for arrowheads.

Other Materials

The site also yielded large quantities of glazed and unglazed ceramics. The former include blue glazed lions, wall nails, jars and an offering table as well as two lion figures with yellow glaze. Little previous work has been undertaken on these ceramics although Vandiver and Paynter [9, 29] have examined some samples of glaze from the glazed ceramics. Future phases of the project plan to examine the ceramic and glaze composition, phases and manufacture in more detail. Starr notes that most of the glazed ceramics had a "a copper glaze, of a bluish green colour when first uncovered, which quickly fades in intensity in the deteriorated portions when exposed to light and air" [30]. This suggests that most of the artifacts could have looked very different when first excavated. Future phases of the project plan to examine the ceramic and glaze composition, phases, manufacture and deterioration in more detail.

METHODOLOGY

Analysis was carried out on both SEM-WDS and LA-ICPMS for over 200 samples of glass from both Egypt and Mesopotamia. They were analyzed following methodologies laid down by Norman Charnley and Julian Henderson and outlined in various publications [31, 32]. Analyses were conducted on polished sections through the samples which were examined in the Cameca SX100 microprobe in the Grant Institute of Earth Sciences, Edinburgh (all glasses from

Nuzi) and the Cambridge Microscan 9 microprobe in the Department of Earth Sciences, Oxford (all other glasses). Regular runs on a Corning glass standards 'A' and 'B' were used to check for machine drift. The two instruments allowed analysis of 22 elements, mostly major and minor, with detection limits below 0.05% for most elements.

Analyses by ICPMS were carried out using a New Wave UV-213 laser ablation system in conjunction with an Agilent 7500a ICPMS instrument. Samples were mounted in standard electron microprobe resin blocks and ablated under an atmosphere of argon. Ablation conditions were laid out in [23]. Each batch of samples included four measurements from NIST 610, throughout the duration of the session to allow for correction of instrument drift. Repeat measurements of NIST 612 (50 ppm nominal concentration) were made throughout the analytical period and the results reveal that for the majority of trace elements agreement with accepted values, as expressed by the % difference between the determined and accepted values is usually better than 10% and often better than 5% for NIST 612. Analyses indicate a 1□ precision of ~5% when concentrations are well above detection limits. Of the elements that showed the greatest deviation from accepted or consensus values, P and K are both regarded as difficult elements to determine by LA-ICPMS, particularly at the concentrations in NIST 612 (55 and 61 ppm respectively). In summary, individual analyses of most elements are subject to analytical uncertainties of <10% precision and accuracy, and detection limits for most elements are in the sub-ppm range.

Preliminary work on the alteration of the glasses from Nuzi has been carried out at the Natural History Museum, London. Mounted and polished samples were examined using the Cameca SX100 microprobe for quantified compositional analysis of the glass and alteration layers. Three mineralogical standards were used to check for machine drift and 23 elements were sought. Detection limits for most elements were below 0.05%. Back scattered electron imaging and semi-quantitative spectra of the altered glasses were obtained using a Jeol 5900 LV Scanning Electron Microscope with an Inca X-sight Oxford Instruments energy dispersive detector.

Non-destructive x-ray fluorescence (XRF) is being undertaken on all the metal artifacts to determine the range of alloys present and the relationship between typology and alloy. A portable XRF system was used, a Bruker Tracer III. This has a Rhodium tube operated at 40kV, 1.4μA and an Al-Ti filter in the path of the primary x-ray beam. Due to the corrosion of the artifacts, it is not possible to quantify the compositions of the alloys. Similarly, it may be hard to distinguish between trace and minor levels of elements such as arsenic, lead and/or tin in some examples. The same XRF system is used at 40kV, 1.4μA without a filter and with a vacuum pump giving 5-10 torr between the sample and detector to assess the vitreous materials prior to sampling. This allows identification of the main colorants and/or opacifiers, determination of the likely original color and indication of the approximate state of preservation.

Mounted and polished sample of glass and other vitreous materials were examined in a Jeol JSM-6460 LV Scanning Electron Microscope at the Museum of Fine Arts, Boston, at 20kV with an Inca X-sight Oxford Instruments energy dispersive detector. This analysis gave quantified compositions and revealed elemental distributions and different phases in the samples analyzed.

RESULTS: GLASS

Survey of the glass confirmed that blue was the most common color and accounted for nearly all vessel fragments, although a small fragment of an amber glass vessel was found during sampling in 2007. The current results show no compositional difference between the blue glass employed for beads or vessels. Similarly, the possible glass ingot fragments, the lumps of dark blue translucent glass, was found to be identical in composition to the other translucent blue glasses.

Although Vandiver [9, 20] reports the presence of opaque red glass colored with copper dendrites, little such glass has yet been identified and analyzed in this current project. Two pieces with opaque red copper-glass were found but in both cases this was combined with a lead-tin yellow. The presence of tin as a colorant would normally suggest that this glass must be intrusive. However, both these glasses are typical plant ash glass and recent studies have identified lead-tin glass beads from the Akkadian level at Nippur [33], dated to the third millennium BCE these were found in a sealed context and are therefore unlikely to be intrusive. Similarly the lead-tin yellow glass beads found at Nuzi are from contexts which are not thought to be intrusive. Further survey and sampling is required to investigate the occurrence of copper reds and the unexpected lead-tin yellows at this site.

Several examples of yellow, white and amber glass were identified and sampled during this study. However, compared to the blue glasses these are present in small numbers, only two or three examples of each color have been found so far. Examination and analysis suggests that much of the material thought to be yellow 'frit' is extremely weathered yellow glass. In contrast, all the blue 'frits' examined and analyzed to date appear to be Egyptian blue, which comprises a significant proportion of the beads. Interestingly, zinc was detected in at least one example of Egyptian blue analyzed in this study and in two examples of Egyptian blue from Nuzi analyzed by Brill [34]. In Egyptian examples of Egyptian blue, zinc is normally at or below detection levels. Brill [35] also found zinc in Egyptian blue from the later sites of Nimrud and Sardis.

Distribution of glass at Nuzi

The distribution of glass on the site is extremely interesting, particularly the glass vessel fragments. The majority of the vessel fragments noted in the excavation report come from either the temple or the palace complexes with very few from other areas on the site. The main exceptions to this are the fragments from Group 26 in the northwest of the site (Rooms A4/5; F35/29/27) which is believed to have been a storage area, possibly linked to the temple complex [36]. There is also a single fragment from Room C28, in the residential area just north of the temple, which also contained many inscribed clay tablets [37]. One fragment is known from the southwestern residential area in Room P356, the purpose of this room is unclear but a fragment of temple-type glazed pottery was found within the house group [38]. Two further fragments are known from the northeastern residential area in Rooms S157 and S104. S157 may have had a ritual function but S104 appears to have been purely residential in nature [39]. Within the museum assemblage all of the vessel fragments that can be assigned a location come from the temple or palace complexes, apart from a single fragment from the suburban dwelling of Shilwi-Teshub.

The vast majority of the glass beads excavated came from the temple complex [40]. Small numbers were found across all areas of the site but again most often in rooms that were interpreted as having a non-domestic purpose. For example, numerous glass beads were found in rooms S111 and S113 which were believed to have been ritual in nature; S111 was part of a group of rooms containing temple-type pottery and other ritual artifacts, S113 also contained an offering stand similar to that found in Temple A [41].

Opaque versus translucent glasses at Nuzi

Shortland [32] has noted that out of the 12 glass vessels known in Egypt dating to the reign of Tuthmosis III (1479-1425 BCE), 11 are opaque blue glass (the remaining vessel is of marbleized glassy faience similar to a piece found at Nuzi and has been attributed to Mesopotamia [42]. Of these, ten are opaque turquoise glass and one opaque dark blue glass. Translucent blue and other colors appear later on in Egypt and antimony opacified glasses predominate in the earliest glasses found in Egypt [32]. However, the glass assemblage from Nuzi, which is broadly contemporary with these glasses (although probably slightly later as it dates from c1440-1340 BCE), is somewhat different. No opaque vessels have been identified so far i.e. vessels where the base glass has been opacified with antimony. In the current study around 30 vessel fragments (some of which may have come from the same vessel) have been analyzed and all are translucent, with antimony only being found in the decorative inlays. This includes the devitrified samples as analyses of devitrified areas on opaque glass beads and inlays show that antimony remains even when no original glass is present.

The current analyses show that the opaque blue glasses from Nuzi have higher levels of calcium oxide and sodium oxide and lower levels of colorant (Cu) than in the translucent blue glasses. This is very different from the early Egyptian glasses analyzed by Shortland [32] which show no significant compositional difference between opaque and translucent glasses other than the presence of antimony. Shortland hence suggests that in the Egyptian opaque glasses antimony was added as a simple antimony compound rather than as calcium antimonate, with lime precipitating out of the glass as calcium antimonate crystals on cooling [32]. The observed colorant levels in the Egyptian opaque glasses also suggests that white and blue glasses are not being mixed to create the opaque glass [32]. The higher levels of calcium oxide in the opaque Nuzi glasses could suggest that calcium antimonate is being added to a translucent blue glass to create an opaque turquoise color, thus increasing the calcium oxide levels within these glasses. The Nuzi material hence appears to indicate use of a different technology to produce opaque glasses compared to that used in Egypt. The lack of opaque vessels in the Nuzi assemblage is also interesting and contrast with the other Mesopotamian site studied in detail, Tell Brak, as well as with contemporaneous Egypt. The analyzed vessel fragments from Tell Brak are predominantly opaque, rather than translucent, and all the blue opaque glasses generally follow the Egyptian model [24], although there is a single opaque turquoise example that appears identical to the Nuzi opaque material (TB7, an ingot fragment).

Alteration of the glass

Much of the glass excavated from Nuzi is in a poor state of preservation. For many of the beads and vessel fragments the original color is now lost and many objects are unstable and completely devitrified. To some extent, an indication of the original color and in some cases of the inlay can be obtained by non-destructive XRF as the colorants and/or opacifiers can often be detected analytically even where no visible evidence remains due to surface alteration and resultant loss of color. Levels of copper plus antimony, antimony plus lead, and antimony alone remain high in originally opaque blue, yellow and white glasses respectively. Glass that was originally translucent can be identified by a lack of antimony in the spectrum; an original blue being indicated by copper. The relative intensities of the silicon, Compton and Rayleigh scatter peaks in the XRF spectrum are controlled by the percentage of silicon remaining and the overall density of the glass. A hydrated, altered glass has a significantly lower silicon peak and higher scatter peaks than better preserved glass. As a result, the XRF spectrum can provide a qualitative indication of the likely state of preservation.

Starr [43] notes the poor condition and fragmentary state of many of the glass finds, noting that the blue glass has become greenish-gray or white in appearance and that the yellow and white glasses have become 'chalky' in texture. Indeed it is clear from the finds notebooks that much of the glass was not identified as such at the time of excavation. For example, mention is made of 'painted' beads and other objects whose field numbers can be linked to glass objects in the museum collection. Several glass beads were wrongly noted as being of stone or composition and many of the vessel fragments do not appear in the finds books as glass vessels, if noted at all.

Preliminary results from SEM-EDS and SEM-WDS analysis have shown that the glass objects at Nuzi can be divided into several broad groups in terms of their preservation. These range from examples that have little or no visible alteration to completely devitrified and highly unstable examples.

The first phase of alteration noted in the Nuzi glasses is the appearance of a darker colored layer on BSE images, indicating a decrease in average atomic number and/or density. Analysis of a typical blue vessel glass showed that at the surface the total had dropped to 85% as a result of hydration, the soda had been removed, the potash was reduced to around half its original value and all other oxides remained the same. This is illustrated in Figure 2 with a BSE image of surface alteration in 1930.82.55. At a slightly later stage the darker layer becomes more pronounced and enriched in silica with very

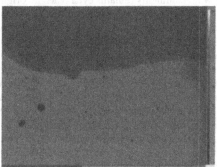

Figure 2: 1930.82.55

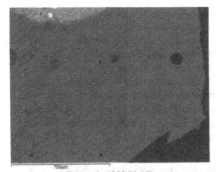

Figure 3: 1930.82.15b   **Figure 4: 1930.82.4**

low levels of alkalis and slightly enriched in colorant (copper for blue glasses such as vessel 1930.82.15b shown in Figure 3). This color change is also observed following cracks into the body of the glass and often extending well into the pristine glass, as seen in Figure 3. These darker areas are identical in composition to the outer alteration layer. This agrees well with the accepted model of initial glass alteration where ion exchange between alkalis and water at the glass surface results in a hydrated, dealkalized surface layer [44].

At the next stage of alteration, a complex system of alteration layers can develop. This frequently extends well into the body of the glass and often consists of alternating light and dark lamellae. Typical alternating lamellae are shown in Figure 4 for a copper blue glass. Digital mapping of these layers has indicated that the BSE differences are linked to variations in the levels of lime, silica and magnesium oxide and of copper in the blue glasses. For example, the paler layers are often silica-rich (up to 90% SiO_2) or have elevated copper levels (up to 25% CuO in some areas); darker layers tend to be magnesium oxide-rich (7-15% MgO compared to around 4% in the original glasses). Again this is consistent with models of glass alteration where ion exchange is accompanied by a much slower breakdown and redeposition of the silica network [44]. The lamellar structures range from layers tens to hundreds of microns across (Figure 4) to fine structures a single micron across (Figure 5). The scale has so far proved too fine to allow analysis of the individual lamellae by SEM-WDS. Whilst the exact mechanism for these structures remains debatable, they are almost certainly related to the redeposition of silica after the network has been dissolved [45]. In numerous examples the lamellar structures have begun to separate causing the alteration layers to become unstable and begin to flake away.

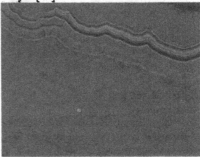

Figure 5: 1930.69.39b

Figure 6: 1930.82.62f

Water entering the glass along cracks and imperfections in the glass body can deposit material, often calcium-rich, directly from the burial environment along cracks and in voids within the glass (Figure 6). Heavy metals such as iron and manganese can also enter the glass, as can be seen in Figure 7, where manganese and iron are present in the material infiltrating between the lamellae. The colorants used in the original glass can also be affected by water. Yellow glasses show the lead being drawn along the lamellar structures within the alteration layers and the blue glasses may show enriched copper levels within the alteration layers. Antimony, however, appears to be less susceptible to alteration and in the white and opaque turquoise glasses calcium antimonate crystals continue across the alteration layers, apparently unaffected by the movement of other materials, as can be seen in Figure 8.

Figure 7: 1930.66.89b

Figure 8: 1930.66.90b

RESULTS: METALS

Non-destructive x-ray fluorescence (XRF) of the metal artifacts has enabled the assemblage to be divided into different alloy types despite considerable surface corrosion. The apparent levels of copper, lead, tin and/or zinc vary considerably within an alloy group due either to variations in corrosion or genuine compositional variations. Although there is some correlation between alloy and corrosion, with copper the most corroded and brass the least, in many cases it is not possible to deduce the likely alloy from the surface appearance.

Of the objects analyzed so far (about 25% of the assemblage), the majority are bronze or copper. Although the former dominates slightly, completion of the study will be required to determine if this is a true representation of the entire assemblage. The analyses suggest that the copper has no significant alloying elements but contains traces of one or more of arsenic, lead, tin and/or nickel. These trace elements are consistent with the "dirty" copper described by Bedore Ehlers and Dixon [25]. Only a small proportion of the assemblage analyzed consists of brass (Cu-Zn), quaternary copper alloys (Cu-Zn-Pb-Sn) and silver-copper alloys. These other alloys are restricted to jewelry and related items; all of which have now been analyzed. It is therefore considered fairly unlikely that a significant number of additional copper-zinc, copper-zinc-tin or silver will be identified. Of the zinc-bearing artifacts currently identified, twelve have a secure Hurrian provenance, three appear to come from later Sassanian graves and five have no given provenience. Similarly, of the silver alloys identified to date, two are from earlier Akkadian graves, five have a secure Hurrian provenance and one comes from a later Sassanian grave. The available quantified results indicate zinc levels of 12-14% and lead levels of 3-5% in objects from Stratum II (this study, [25]). In one ring, 6% tin was also present [25].

The current data suggests that less than 2% of the finds are likely to contain significant zinc (>1%) and less than 1% to consist of a silver-copper alloy. It is tempting to interpret this as evidence of the scarcity of these metals but some care is needed in this interpretation. The destruction layer contained few slain bodies, suggesting that most citizens either retreated to another town or surrendered prior to the sack of the city. Jewelry and related items are small, valuable and easily transported, so would be the most likely to be removed by the citizens or by subsequent looters. The remaining assemblage may hence be somewhat skewed in its typological distribution and hence in its alloy distribution if this is directly correlated to typology. Although identification of many pieces is questionable or impossible and in some cases even the true number of pieces may be debatable, examination suggests jewelry and related items comprise a minor proportion of the assemblage compared to weapons and tools. Certainly the silver and gold inlays and the larger precious metal items mentioned by texts from Nuzi are notable by their absence, undoubtedly due to being taken by the fleeing citizens or looted by the invaders.

An analytical study of metalwork from the Iranian site of Tepe Yahya [46] may provide a useful comparison, although unlike Nuzi Tepe Yahya was not destroyed by invaders. This showed that copper-arsenic alloys were the most common in both period IVB (2400-2000 BCE) and period IVA (1800-1400 BCE), with the remainder copper in the earlier period and a mixed assemblage of copper, bronze and minor brass in the later period [46]. At Tepe Yahya, zinc-bearing alloys were used for a bracelet, ribbon and a fragment [46].

Despite the textual mentions of copper and bronze arrowheads, all the arrowheads analyzed to date are copper. Of course, those recovered and analyzed represent only a small fraction of those present at the site at a given time, but the discrepancy is nevertheless surprising.

There are several distinct different arrowhead typologies [27] and it is possible that the bronze arrowheads had a distinct form, although this is not stated in the texts. Equally surprising is the fact that the various types of armor plate are all described as copper in the texts whilst all those analyzed to date are bronze.

The silver alloys were used mainly for earrings and finger rings, although silver sheet was used over the head of a large copper door stud. Most of these artifacts had not been identified as silver alloys prior to analysis. The data indicate considerable variation in silver purity but it is not possible to determine non-destructively how much of this is due to original compositional variations and how much to variations in subsequent corrosion.

Origin of the Alloys

It is tempting to speculate that the variation in alloy types and the selection of specific alloys for specific objects may have some relation to the source of these artifacts. For example, might the use of copper for arrow heads and some tools reflect local manufacture of more utilitarian tools and weapons whilst more prestigious imported tools were made of bronze? Similarly, the rarity of the zinc-bearing alloys and their restriction to jewelry, suggests this was a valuable and scarce alloy. It was probably valued for its golden color and its use limited to small decorative items. It is possible that the zinc-bearing artifacts may have been imported rather than manufactured at the site from metal stocks. Either way, the origin of primary brass and gunmetal remains a mystery. Analysis of the full set of artifacts and comparison with other contemporary sites may help us solve these questions. Early zinc-bearing alloys from south-east Asia are attributed to deliberate cementation process brass rather than accidental production from a mixed ore [26, 47]. There is no evidence, however, of primary metal production at either Tepe Yahya or Nuzi. Two finds at Nuzi were interpreted as slag by the excavators and one from Tepe Yahya is designated as 'splash?' [46], suggesting the likelihood of secondary working of imported alloys or recycling of scrap metal at these sites. In addition, the Nuzi texts show the site was responsible for production and distribution of major volumes of weapons. It seems likely that there is a common but as yet unknown source for early zinc-bearing alloys found in the near east.

LINKS BETWEEN TECHNOLOGIES

The copper used as a colorant in the blue glass generally lacks arsenic, tin or zinc at significant levels, although zinc does occur in some blue frits. This suggests relatively pure copper, or copper minerals, was used for the production of these blue glasses rather than copper alloys. Rarity and a high value to zinc-bearing alloys would explain the lack of zinc associated with copper. The presence of a large proportion of copper at the site despite the significant volume of bronze is consistent with the use of copper without scrap bronze. The lack of primary production of either glass or metal at the site means that the copper (and lead if added) used in the glass and that used for metalwork may have had a very different origin. Isotopic and trace element analysis is on-going to compare the different materials (glass, Egyptian blue, frit, glaze, metals) and different varieties of a given material (different colors or alloys). Preliminary lead isotope data from lead artifacts, blue vessel glass and yellow glaze, glass and inlays indicates that all lead isotopes are high ($Pb^{207}/Pb^{206} > 0.820$ and $Pb^{208}/Pb^{206} > 2.05$), falling outside the range for normal Egyptian materials and within and beyond that previously associated with Mesopotamian material. Hence, the same 'Mesopotamian' lead sources could have been used for the lead pieces/objects and the yellow vitreous materials found at Nuzi and these may have some connection to the copper used in the blue glass. This provides the first clear link between the different material technologies found at Nuzi and confirms their distinction from those employed for Egyptian artifacts.

CONCLUSIONS

This study has shown that the same blue glass was used for the different object types and that the possible ingots are representative of this. This may be interpreted as evidence of secondary manufacture of glass artifacts at Nuzi from an imported primary material. The data also confirms that the Nuzi glass is distinct from that found in Egypt at the same time, with greater quantities of translucent rather than opaque glass, different compositions and different technologies. Systematic study of the vitreous materials has revealed a complex pattern of alteration and allowed this to be characterized and quantified for the first time.

Examination of the metals confirms the variation in alloy types and the selection of specific alloys for specific objects. The rare zinc-bearing alloys have been shown to be largely restricted to jewelry, suggesting that this metal was probably prized for its golden color and used only for small valuable items due to its rarity. The primary origin of the various alloys, including the rare brass and gunmetal, remains debatable. It seems likely that there is a common but as yet unknown source for the zinc-bearing alloys at the various sites in the near east where such alloys are found.

Preliminary lead isotope data suggests that the same lead sources could have been used for the lead pieces/objects and the yellow vitreous materials found at Nuzi. This links the different material technologies found and confirms their distinction from those used in contemporaneous Egypt. Further analyses are required to confirm and extend these preliminary findings.

FURTHER WORK

Additional analyses of opaque and colored glasses and of the other vitreous materials are required to confirm the technological and compositional patterns identified in this study. In addition, unique beads which cannot be sampled will be examined by using non-destructive methods. Sampling of additional metal artifacts is planned although many small artifacts are intact making this difficult. Isotopic and trace element analysis is on-going to compare the different vitreous and metallic materials and their varieties.

The structural modifications and phase transitions induced by alteration of the glass will be investigated by X-ray diffraction and Raman spectroscopy. The phases produced by corrosion of the metals will be identified with the same techniques. Comparison of weathering patterns in the different vitreous materials is on-going and will be compared to pre-depositional alteration of the metals. Prior conservation examination of some metal objects [48] will provide useful information on depositional versus post-excavation metal corrosion.

REFERENCES

1. R.F.S. Starr, *Nuzi: report on the excavations at Yorgan Tepa and Kirkuk, Iraq: conducted by Harvard University in conjunction with the American Schools of Oriental Research and the University Museum of Philadelphia, 1927-1931, Volume I Text.* (Harvard University Press, Cambridge MA, 1939) p. xxix.
2. Reference 1: p. xxxv.
3. Reference 1: p. xxxiv.
4. M. Novák, in *The synchronisation of civilisations in the Eastern Mediterranean in the second millennium B.C. III. Proceedings of the SCIEM 2000 – 2nd EuroConference Vienna, 28th of May – 1st of June 2003. Contributions to the Chronology of the Eastern Mediterranean, Volume IX,* edited by M Bietak & E Czerny (Verlag der Österriechshen Akademie der Wissenschaften, Vienna, 2007) pp. 389-401.
5. J. Freu, *Histoire du Mitanni.* (Association KUBABA, Paris, 2003) p.221.
6. D.L. Stein, Zeitschrift für Assyriologie und Vorderasiatische Archäologie 79(1), 36-60 (1989).
7. Reference 1: p. 265, 294.
8. Reference 1: p. 94.
9. P. Vandiver, Journal of Glass Studies 25, 239-247 (1983).
10. Reference 1: p. 457
11. D. Barag, in: *Glass and Glassmaking in Ancient Mesopotamia: An Edition of the Cuneiform Texts which contain Instructions for Glassmakers with a Catalogue of Surviving Objects,* edited by, A.L. Oppenhein, R.H. Brill, D. Barag, A. von Saldern, (New York: Corning Museum of Glass Press, New York, 1970) pp.131-197.
12. Reference 1: p. 446
13. Reference 1: p. 458
14. Reference 1: p. 456
15. Reference 1: p. 110
16. L. Woolley, *Alalakh: An Account of the Excavations at Tell Atchana in the Hatay, 1937-1949.* (The Society of Antiquaries, Oxford, 1955) p. 301.
17. D. Oates, J. Oates and H. McDonald, *Excavations at Tell Brak: Vol I: The Mitanni and Old Bablylonian Periods.* (McDonald Institute Monographs, Cambridge; British School of Archaeology in Iraq, London, 1997) p.81.
18. Reference 1: p. 84, p. 100.
19. C. Postgate, D. Oates and J. Oates, *The Excavations at Tell al Rimah: The Pottery* (Iraq Archaeological Reports – 4. British School of Archaeology in Iraq, 1997) p.56.
20. P. Vandiver, in *Early pyrotechnology: the evolution of the first fire-using industries: papers presented at a seminar on early pyrotechnology held at the Smithsonian Institution, Washington, D.C., and the National Bureau of Standards, Gaithersburg, Maryland, April 19-20, 1979,* edited by Theodore A. Wertime, and Steven F. Wertime (Smithsonian Institution Press, 1982) pp. 73-92.
21. R.H. Brill, in *Glass and Glass Making in Ancient Mesopotamia,* edited by A.L. Oppenheim, R.H. Brill, D. Barag, and A. von Saldern, (The Corning Museum of Glass, New York, 1970) pp. 105-124.
22. R.H. Brill, *Chemical Analyses of Early Glasses, v.1 & v.2.* (The Corning Museum of Glass, New York, 1999).
23. A. Shortland, N. Rogers and K Eremin, Journal of Archaeological Science, 34, 781-789 (2007).

24.   A.J. Shortland and K Eremin, Archaeometry 48(4) 581-603 (2006).
25.   C. Bedore and C. Dixon, Context, 13, 3-4, 10-11 (1998).
26.   C.P. Thornton , in *Metals and Mines: Studies in Archaeometallurgy*, edited by, S. La Niece, D. Hook, and P. T. Craddock (Archetype Publications, London, 2007) pp. 189-201.
27.   T. Kendall, Warfare and Military Matters in the Nuzi Tablets, A Dissertation presented to the Faculty of the Graduate School of Arts and Sciences, Brandeis University Department of Mediterranean Studies, 1974.
28.   Reference 1: p.472-474.
29.   S. Paynter (forthcoming)
30.   Reference 1: p.442.
31.   J. Henderson, Archaeometry 30, 77-91 (1988)
32.   A.J. Shortland, Archaeometry 44(4) 517-530 (2002).
33.   P. Vandiver and K. Aslihan Yener 'Appendix 1. Scientific Analysis of two Akkadian glass beads' in *Nippur V The Early Dynastic to Akkadian Transition The Area WF Sounding at Nippur*, A. McMahon with contributions by M. Gibson, R.D. Briggs, D. Reese, P. Vandiver, K. Aslihan Yener, (The Oriental Institute of the University of Chicago, Chicago, 2007) pp 149-157.
34.   Reference 24: p. 468
35   Reference 24: p. 468-470
36.   Reference 1: p. 224.
37.   Reference 1: p. 234.
38.   Reference 1: p. 280.
39   Reference 1: p. 306, p. 312.
40.   Reference 1: p. 91-94.
41.   Reference 1: p. 307, p. 309.
42.   C. Lilyquist, R.H. Brill and M.T. Wypyski, *Studies in early Egyptian glass* (Metropolitan Museum of Art, New York, 1993) p. 11.
43.   Reference 1: p. 442, p. 446
44.   B.C. Bunker, J. Non-Cryst. Solid. 179, 300-308 (1994).
45.   S.D. McLoughlin, PhD Thesis, Imperial College, London, 2003. p 219
46.   C.P. Thornton, C. C. Lamberg-Karlovsky, Martin Liezers and Suzanne M. M. Young, Journal of Archaeological Science, 29, 1451–1460 (2002).
47.   C.P. Thornton and C. Ehlers, *IAMS* 23, 3-8 (2003).
48.   R.J. Gettens, Technical studies in the field of the fine arts, 1, 118-142 (1933).

Mater. Res. Soc. Symp. Proc. Vol. 1047 © 2008 Materials Research Society   1047-Y07-02

Copper Alloys Used in Barye's Hunt Scenes in the *Surtout de Table* of the Duc d'Orleans

Jennifer Giaccai[1], Julie Lauffenburger[1], and Ann Boulton[2]
[1]Conservation Division, Walters Art Museum, 600 North Charles Street, Baltimore, MD, 21201
[2]Conservation Department, Baltimore Museum of Art, 10 Art Museum Drive, Baltimore, MD, 21218

ABSTRACT

The copper alloys of the statues by Antoine-Louis Barye were examined with energy-dispersive x-ray fluorescence. Barye experimented with casting techniques in his own workshop and regularly worked with the foundry of Honoré Gonon, who re-introduced lost wax casting to 19th century Paris. Two of the technically complex Barye sculptures in the *surtout de table* of the Duc d'Orleans were difficult to cast using the more common sand casting technique. Problems with the sand casts sent to other foundries resulted in Gonon completing the casting for problematic statues in the *surtout*, after Gonon had completed lost wax casting of the three hunt scenes initially sent to his foundry. Examination of the copper alloy compositions differentiates the casts from the various foundries and determines which parts of the *surtout* were ultimately cast by Gonon.

INTRODUCTION

In 1834 the Duke of Orleans commissioned a large and grand centerpiece, a *surtout de table*, for his dining room table. The five principal sculptures, as well as some of the smaller sculptures, were commissioned from Antoine-Louis Barye.[1] For the larger works, Barye created five dramatic hunt scenes from five different areas of the world. The five principal hunt scenes, all owned by the Walters Art Museum, are the Tiger Hunt set in Mughal India (27.176), the Lion Hunt set in North Africa (27.174), the Wild Bull Hunt set in 16th century Spain (27.178), the Elk Hunt set in Central Asia (27.175), and the Bear Hunt set in 16th century Germany (27.183). See figures 1-5.

At the time the *surtout* was commissioned, Barye was regularly using the foundry of Honoré Gonon, and the majority of the casting of the five principal hunt scenes fell to Gonon. Although sand casting was dominant in 19th century Paris, Honoré Gonon worked to bring back lost-wax casting, beginning in 1829, and was the only foundry in Paris that used lost wax casting.[2] Despite Barye's successful relationship with the foundry of Gonon and his sons, the tight deadline proposed by the Duke of Orleans required that two of the principal scenes be cast by other foundries, the Elk Hunt to Quesnel and the Bear Hunt to founders Richard and Fressange.[3,4]

Figure 1. The Tiger Hunt, WAM 27.176. © The Walters Art Museum, Baltimore.

Figure 2. The Lion Hunt, WAM 27.174. © The Walters Art Museum, Baltimore.

234

Figure 3. The Wild Bull Hunt, WAM 27.178. © The Walters Art Museum, Baltimore.

Figure 4. The Elk Hunt, WAM 27.175. © The Walters Art Museum, Baltimore.

Figure 5. The Bear Hunt, WAM 27.183. © The Walters Art Museum, Baltimore.

The three hunt scenes cast by Gonon and his sons, the Tiger, Lion and Wild Bull Hunts, all have a variation on the same inscription cast into the base along with the artist's signature, *"Bronze d'un seul jet sans ciselure fondu a l'hote[l] d'Angevilliers par Honore Gonon et ses deux fils"* or "Bronze cast in only one pour without chasing at the hotel d'Angevilliers by Honore Gonon and his two sons." This signature refers to both the lost-wax casting technique and that Gonon, as well as other founders sand casting in Paris, were trying both to minimize the amount of chasing required after casting and cast the work in one piece regardless of the method of casting.

The complexity of Barye's sculptures must have been a challenge for the founders, particularly Quesnel, Richard and Fressange as archival records indicate the partially sand cast sculptures and the original models were passed to Gonon to complete casting of the Elk and Bear Hunts. The Bear Hunt has an inscription cast into the base, but it simply states that the piece was cast at Gonon's foundry and gives no mention of the bronze being "cast in one pour with no chasing". Although the only inscription on the Elk Hunt is that of Barye, most founders of the time did not sign the sculptures they cast. The accounting records from the *surtout* indicate that Édouard Quesnel sand cast part of the Elk Hunt, Gonon and Sons restored the model and completed casting and Ottin assembled and chased the statue.[4] However, Quesnel was known for his finishing work, not casting, and it is unknown if or with whom Quesnel may have contracted to initially cast the Elk Hunt.

Close examination of the underside and the casting voids of the Bear and Elk Hunts with a boroscope show that the Bear Hunt appears to be lost wax cast, with one figure, the standing hunter, being separately sand cast and attached by rivets. Viewed from the underside the Elk

Hunt appears to be primarily sand cast, raising the question of what extent Gonon was involved with the casting of the sculpture.

If the foundries used different alloy compositions, we should be able to separate and identify the contributions of Gonon, Richard, Fressange and Quesnel to the hunt scenes. In order to determine differences in the alloy composition, non-invasive X-ray fluorescence (XRF) was used to analyze all five of the hunt scenes.

EXPERIMENT

All XRF spectra were acquired with a Bruker AXS ARTAX spectrometer with Rh tube, 12.5 μm Zr filter, and 70 μm spot size polycapillary lens. Spectra were collected at 50 keV and 200 μA for 300 s of live time. Multiple locations on each sculpture were analyzed. For the Lion and Bull hunts, it was possible to analyze areas on the underside of the sculpture base that had been polished to a bare metal surface.

All XRF spectra were quantified using the Bruker AXS ARTAXCTRL software. The Bruker software uses direct comparison of the count rates for the elements of interest when quantifying. Only the spectra with an unnormalized total of all the elements between 95 and 105% were included when calculating the mean alloy composition and the uncertainty. Those spectra were normalized, re-quantified and the results averaged. The uncertainty was calculated as the standard deviation of the mean.

After a rough estimate of the copper alloy type using a general calibration file, sample spectra were evaluated with a specific calibration file developed for a copper alloy subset, e.g. quaternary alloys, tin bronzes, or high zinc brass. K_α lines were used to quantify copper, tin, and zinc. The lead L_α line was used to quantify lead. Limits of detection and quantification in a copper matrix were determined, see Table I. In general, the uncertainty obtained for each element using a quaternary bronze standard (CURM 71.33-6) is a few percent of the measured amount. However the measured lead values for the standard ranged between 4 and 6 percent, giving a standard deviation of approximately 25%. We feel this is a combined effect of lead segregation and the small spot size of the polycapillary lens. The large standard deviation for lead was considered when interpreting the results.

Table I. Limits of detection and quantification for secondary components in a copper matrix.

	Zn (K_α)	Sn (K_α)	Pb
Limit of detection	0.4%	0.4%	0.3% (L_β)
Limit of quantification	1.3%	1.5%	1.1% (L_α)

DISCUSSION

Tiger Hunt, Lion Hunt and Wild Bull Hunt

The three hunt scenes known to be wholly cast by Gonon, the Tiger, Lion and Wild Bull Hunts, showed some variation in the casting alloy used, however all were cast out of a leaded bronze of approximately 6% Sn with a range in the composition of lead (Table II). One sculpture, the Tiger Hunt, showed an addition of a few percent of zinc, forming a quaternary alloy. The Tiger Hunt was the first sculpture in the group cast by Gonon, and the absence of zinc in the later two sculptures may indicate a recipe change between 1836 and 1838. The observed range in lead is certainly partially due to the issues with lead quantification, but is large enough that it indicates a difference in the lead added to the alloys of the individual sculptures.

Table II. Mean alloy composition and uncertainty for sculptures entirely lost-wax cast by Gonon.

	# areas quantified	% Cu	% Zn	% Sn	% Pb
Tiger Hunt—1836	4	87.3 (± 0.8)	2.5 (± 0.3)	5.3 (± 0.3)	4.4 (± 0.8)
Lion Hunt—1837	3	90.1 (± 0.3)	trace	6.5 (± 0.03)	2.5 (± 0.3)
Bull Hunt—1838	3	91.8 (± 0.3)	bdl	5.9 (± 0.04)	1.5 (± 0.3)

Although Gonon predicted the pieces would be cast in one pour when inscribing his signature into the wax model prior to casting the statue, a few separate seams and patches indicate repairs made in areas of casting failure. In areas across a seam, like the tail of the elephant in the Tiger Hunt, the composition analyzed on both sizes of the join was consistent with the quaternary alloy used to cast the Tiger Hunt.

Bear Hunt

Even qualitatively, XRF spectra of fourteen spots on the Bear Hunt (Figure 6) showed two different compositions in use. The majority of the sculpture, including the lost wax cast base, was composed of a leaded tin bronze while the two hunters were cast in a high zinc brass. The separation of these two areas agrees with both the documentary and visible evidence on the statue. Records in the archives describe that part of the Bear Hunt was originally given to Richard and part to Fressange, but that the results were unsatisfactory and Gonon completed the casting with a reduced fee because the two hunters had been separately sand cast. When the sculpture was examined closely the standing hunter could easily be identified as separately cast, with visible seams indicative of sand casting, and attached by rivets to the base. However, there were no obvious indications that the hunter seated on the horse was separately cast; only the XRF results show that it was cast of the same alloy as the standing hunter.

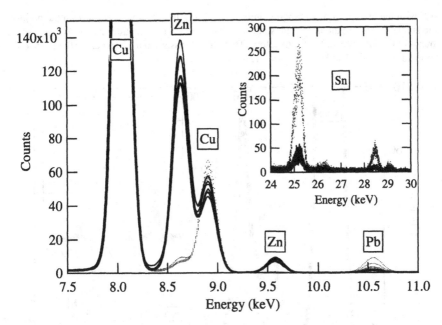

Figure 6. XRF spectra from the Bear Hunt. Dotted lines are Gonon foundry, solid lines are Richard or Fressange foundry.

When the results are quantified (Table III), the alloy used in the lost wax cast parts of the Bear Hunt roughly agrees with the alloy used in the Lion and Bull Hunts (traces of zinc in a tin bronze, with added lead). Some changes in the alloy were observed from the three hunt scenes cast entirely by Gonon; the amount of lead is greatly increased and the amount of tin increased slightly from 6 to 8 percent. In contrast, the two hunters are cast with a very different alloy, a high zinc brass with small amounts of tin and lead added to the alloy.

Table III. Mean alloy composition and uncertainty for alloys observed in Bear Hunt.

Area examined	# spots quantified	% Cu	% Zn	% Sn	% Pb	Foundry assigned
Two hunters	3	73.6 (± 0.6)	23.1 (± 0.8)	1.5 (± 0.1)	1.9 (± 0.4)	Richard or Fressange
Lost wax cast areas	4	82 (± 2)	trace	7.8 (± 0.6)	8 (± 2)	Gonon

Elk Hunt

XRF spectra of twenty-three spots on the Elk Hunt clearly indicate two different alloy compositions were used (Figure 7). The head of the most prominent figure, the largest elk, and the head of a small dog at the rear of the sculpture show only trace amounts of zinc and large tin peaks, with the amount of lead ranging widely, from 1 to 10 percent. Spectra from the majority of the statue, including the unsigned, sand cast base, show a low-medium zinc brass with small amounts of tin and lead added to the alloy. This alloy is assigned to Quesnel, or the founder contracted to cast the Elk Hunt for him (Table IV).

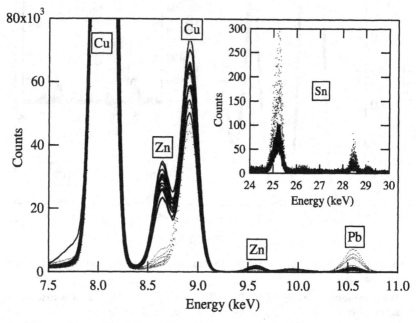

Figure 7. XRF spectra from the Elk Hunt. Dotted lines are assigned to the Gonon foundry, solid lines are assigned to Quesnel.

Table IV. Mean alloy composition and uncertainty for alloys observed in Elk Hunt.

Area Examined	# spots quantified	% Cu	% Zn	% Sn	% Pb	Foundry assigned
Sand cast areas	7	90.7 (± 0.2)	5.3 (± 0.2)	2.5 (± 0.1)	1.1 (± 0.1)	Quesnel
Large elk head, foreleg and chest; rear dog	3	87 (± 2)	trace	7 (± 2)	4 (± 3)	Gonon

The composition of the head, chest and left fore leg of the large elk and the head of a small dog at the rear of the statue is very similar to that used by the Gonon foundry in the Bear Hunt, both an eight percent tin bronze with a range in the amount of lead used, and lies within the grouping of all the Gonon foundry compositions measured (Figure 8). Although Gonon certainly was familiar with sand casting, having used the sand casting technique before switching to lost wax casting, the similarity of the alloy used for the elk and the alloy Gonon used in the Bear Hunt indicate that Gonon cast parts of the large elk and rear dog, most likely using the lost wax technique, and did not sand cast the majority of the sculpture.

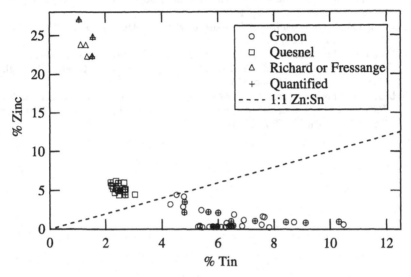

Figure 8. Alloy compositions measured from the five hunt scenes.

CONCLUSIONS

Foundries used their own recipes to mix the alloys when casting, and in this case the recipes were different enough to identify the areas cast by two different foundries in the Elk and Bear Hunts. However, the alloy composition used by the Gonon foundry changed over the few years represented here, as is demonstrated by the range in both tin and zinc used by Gonon when casting the five hunt scenes (Figure 3). This variation in alloy content underscores Gonon's reputation as an innovator who constantly experimented to improve his process.

The combination of documentary evidence from the archives and XRF analysis allowed us to assign founders to the areas of the Elk and Bear Hunt. The Bear Hunt was mostly lost wax cast by the Gonon foundry, with the two hunters sand cast by another foundry and joined to the parts of the sculpture cast by the Gonon foundry. The Elk Hunt was mostly sand cast by one foundry, with the head and chest of the largest elk, a small dog's head at the rear of the statue, and one of the front legs of the elk cast by a second foundry. Based on the use of sand casting, the lack of signature and the similarity of alloy between the large elk and the sculptures cast by the Gonon foundry, we feel that the majority of the Elk Hunt was cast by Quesnel, with the prominent large elk cast by the Gonon foundry.

ACKNOWLEDGMENTS

The authors would like to thank the Andrew W. Mellon Foundation and the Stockman Family Foundation Trust.

REFERENCES

1. Isabelle Leroy-Jay Lemaistre, "The *Surtout de Table* of the Duc d'Orléans", in *Untamed: The Art of Antoine-Louis Barye,* edited by William R. Johnston and Simon Kelly (Walters Art Museum, Baltimore, 2006), p. 26.
2. Elisabeth Lebon, *Dictionnaire des Fondeurs de Bronze d'Art: France, 1890-1950.* (Marjon, Perth, Australia, 2003), p. 168.
3. Isabelle Leroy-Jay Lemaistre, "La Chasse à l'Ours", in *Un âge d'or des arts décoratifs 1814-1848,* edited by D. Alcouffe, A. Dion-Tenebaum, and P. Ennes (Galeries Nationales du Grand Palais, Paris, 1991), p. 323.
4. Isabelle Leroy-Jay Lemaistre, "La Chasse à l'Élan", in *Un âge d'or des arts décoratifs 1814-1848,* edited by D. Alcouffe, A. Dion-Tenebaum, and P. Ennes (Galeries Nationales du Grand Palais, Paris, 1991), p. 324.

Mater. Res. Soc. Symp. Proc. Vol. 1047 © 2008 Materials Research Society   1047-Y02-07

Thermal Expansion and Residual Stress in Ancient Chinese Bronze Castings

Michelle Taube[1], and Blythe McCarthy[2]
[1]Department of Conservation, The National Museum of Denmark, I. C. Modewegsvej, Brede, 2800 Kgs. Lyngby, Denmark
[2]Freer Gallery of Art / Arthur M. Sackler Gallery, Smithsonian Institution, P.O. Box 37012, Freer Gallery of Art, MRC707, Washington, DC, 20013

ABSTRACT

Ancient Chinese bronze vessels were cast in piece molds from the Shang period (c. 1500 BCE) until at least the middle of the Eastern Zhou period (6th to 5th century BCE). Several large vessel shapes, especially those with legs, have been found to be made in several steps with the vessel body cast onto previously-made legs or other appendages. In this situation the molten metal contacts and joins to the solid pieces from the first casting. Similar interfaces are found at solid metal mold spacers or chaplets used in the casting process. In this work, we studied the thermal expansion of the bronze at cast-on joints in ancient Chinese bronzes using thermomechanical analysis (TMA). The results are compared with those from reference lead-tin bronze alloys. Modeling and microstructural analysis support TMA results that broad variations in composition still allow for successful castings.

INTRODUCTION

The Bronze Age in China began about 1650 BCE and lasted for more than 12 centuries. Bronze was used for weapons, vessels, mirrors, bells and parts of chariots. Chase details how vessels were cast using piece molds [1]. Internal cores made of clay and sand were used to keep the wall thicknesses uniform and reduce shrinkage on cooling. In some cases, molten metal would have come into contact with solid bronze during casting. In a common example, thin, rectangle-shaped mold spacers or chaplets were used to keep the inner core and mold separate and correctly positioned [2]. There is no evidence that the chaplets were trimmed in a finishing step unlike that performed on core pins in lost-wax castings.

A second example of molten metal meeting solid bronze occurs in some Zhou Dynasty vessels which have been found with legs or handles that were cast separately then attached to the vessel body using an interlocking casting method [2]. In these cases, the legs or handles were cast with the core exposed on the faces that would eventually be joined with the body. When the pieces were cool, some of the core material was carved away. Then, the first cast pieces were placed into another mold that had been designed for the entire vessel and the body was poured. The molten body metal filled the gaps carved into the core material and surrounded the metal walls of the pre-cast pieces to form an interlocking join.

The interfaces between the previously solidified legs/handles or chaplets and the body metal are well-defined. Furthermore, cracks are sometimes seen along the edge of a chaplet. It is not clear whether these cracks formed due to the stresses of differential contraction on cooling or were caused later by corrosion.

Although museum collections contain many examples of successful castings, there is evidence of failed castings. For example, chaplets decorated with cast designs that do not match the surrounding vessel have been found indicating that the chaplets were probably made from scrap, cast bronze. Gettens found that chaplets often, but not always, have higher copper content (and, therefore, a somewhat higher melting point) than the surrounding metal [2].

In order to better understand the compositional variations that could be tolerated and still result in successful castings, linear coefficients of expansion (COEs, a measure of the tendency of a material to expand with increasing temperature) were measured for reference bronze alloys and for two Chinese bronzes formed by multiple castings using TMA. An overview of the use of TMA can be found in Reference [3]. Microstructural analysis and theoretical calculations were also performed.

EXPERIMENTAL METHODS

Two types of samples were investigated in this study: standard reference alloys and Chinese bronze samples. The samples are listed in Table I. The standard reference samples consisted of three Cu-Sn-Pb alloys that were cast for the Department of Conservation and Scientific Research, Freer Gallery of Art.

Table I. Sample Description, Composition and Coefficient of Expansion

	Sample Description	Measured Composition (Cu-Sn-Pb)		Coefficient of Expansion $(x10^{-6} \cdot /°C)^b$
		EDS	bulk^a	average (measured values)
Standard	90-7-3 (Cu-Sn-Pb)	91-8-1		18.35 (18.21, 18.53, 18.30)
Reference	85-10-5	89-10-1		19.15 (20.03, 18.05, 19.36)
Alloys	75-15-10	80-19-2		19.04 (18.93, 19.11, 19.08)
Chinese Bronze Samples	FSC-B-56 (Ketel *ting*) body	82-12-6	67-8-24	16.49 (16.31, 16.93, 16.84, 16.86, 15.52)
	FSC-B-56 leg	81-14-4	70-10-19	19.15 (18.49, 18.87, 19.94, 19.29)
	FSC-B-9 (Kriger *p'an*) chaplet	88-8-3 (+2% S)		10.89 (11.01, 10.81, 10.85)
	FSC-B-9 body metal next to chaplet	86-11-3		14.11 (13.93, 14.30)
	FSC-B-9 foot rim	85-12-3		14.26 (14.25, 14.29, 13.99, 14.51)
	FSC-B-9 body metal near handle	86-12-2		14.43 (13.62, 14.70, 14.96)

^{a}Wet chemical analyses reported in [2]. ^{b}Coefficients of expansion were calculated in the temperature range 150 - 200°C.

Samples of Chinese bronzes were cut with a hacksaw from two broken vessels in the Freer Gallery Study Collection. Both vessels are described in detail by Gettens [2]. Figure 1 shows the first vessel: a three-legged *ting* (FSC-B-56, Ketel ting). This vessel was accessioned into the Study Collection when two of the legs fell off in shipment and it was discovered that the lower part of the bowl of the vessel was a modern addition made of copper sheet. The legs are attached to enough original body metal to allow for examination of the interlocking casting

Figure 1. A three-legged *ting* (FSC-B-56, Ketel ting). This vessel was accessioned into the Freer Study Collection when two of the legs fell off in shipment. The upper part of the bowl (including handles) and the legs are genuine. Courtesy of the Freer Gallery of Art, Smithsonian Institution, Washington, DC.

method in which the body was cast onto previously-made legs. Samples were taken from the body and the leg in neighboring areas.

The second vessel, shown in Figure 2, was a shallow *p'an* (FSC-B-9, Kriger p'an). This vessel was cast in one piece and provided a good opportunity to study the use of chaplets in Chinese casting. The vessel was brought into the Freer Study Collection when it broke into many pieces during shipment. It was discovered that it had been previously repaired but was a single, genuine vessel. Samples were taken of the body metal near the handle of the vessel and from the foot rim. Another piece containing a chaplet and surrounding body metal was also removed. This last piece yielded three samples: two for TMA analysis and another with both chaplet and body metal for metallographic analysis.

Thermomechanical analysis was performed with a TA Instruments TMA 2940 equipped with the macroexpansion probe. The samples were prepared with one set of parallel faces by grinding with the sample attached to a holder. The other faces were cleaned mechanically with SiC paper. The temperature was increased from room temperature to 250°C at a rate of 5°C/min. A constant force of 0.05 N was applied. The sample chamber was purged with nitrogen gas during heating and cooling to minimize oxidation. The probe was lifted during cooling to allow the sample to contract freely. A minimum of two runs was performed on each sample. The coefficient of expansion was calculated in the temperature range of 150 - 200°C using the instrument software.

Compression tests were performed on the reference alloys using the TMA to determine their elastic moduli. Samples with two parallel faces and uniform cross-sections were subjected to a compressive force that was increased at a rate of 0.05 N/min up to a maximum of 1 N. The measurements were performed at a constant temperature of 25°C. The elastic moduli were derived from the slopes of stress/strain curves constructed from the data.

Samples were prepared for metallographic examination by grinding to 600 grit SiC, followed by polishing with 6 and 1 μm diamond suspension. Final polishing was performed with 0.05 μm alumina suspension.

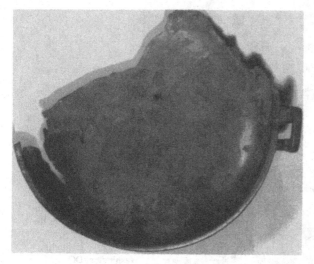

Figure 2. Shallow *p'an* (FSC-B-9, Kriger p'an). This vessel was brought into the Freer Study Collection when it broke into numerous pieces during shipment. It had been previously repaired, but was a single, genuine vessel. Courtesy of the Freer Gallery of Art, Smithsonian Institution, Washington, DC.

Standardless compositional analysis was performed by SEM-EDS to obtain an approximate composition; the corroded condition of the Chinese bronze samples and the multi-phase structure of all of the samples prevented true quantitative analysis. An FEI XL30 scanning electron microscope with EDAX x-ray analyzer was used to find the composition over the full field of view at a magnification of 200x for 200 live seconds. The analysis conditions were: 30 KV accelerating voltage, 51.2 μsec time constant and a spot size of 4.3.

RESULTS

The results of the SEM-EDS analysis are shown in Table I. Also included in the table are the results of bulk, wet chemical analyses performed on the *ting* circa 1960 [2]. The lead values differ with the two methods, but lead is difficult to analyze in part because it is almost completely insoluble in copper and tin so it accumulates in isolated areas.

Figure 3 shows an overlay plot of the TMA results for standard alloy 75-15-10. The results are linear and reproducible and are representative of the results obtained by TMA analysis. The average COEs measured for all of the samples are shown in Table I. The actual values calculated for each run are shown in parentheses. The COEs of the standard reference alloys were similar and were all approximately 18 to 19 $\times 10^{-6}$ /°C. The differences in COE in the Chinese bronze samples were more marked. In addition, the values measured for the leg of the *ting* and the chaplet of the *p'an* are significantly different from the surrounding body metal in each vessel.

Figure 4 shows an optical micrograph of the leg of *ting* FSC-B-56. The leg alloy was found to be a slightly pinker color than the body, which is an indication of higher copper content. A fair amount of $\alpha+\delta$ eutectoid is visible in the microstructure showing that the tin content is above 8 wt% [4] and closer to 10 - 15 wt% [5,6], which is in agreement with the analytical results. The interface with the body metal can be seen at the bottom of the figure. Lines of strain have been made visible by the action of corrosion.

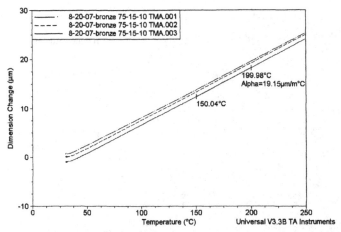

Figure 3. Overlay plot of the TMA results for standard reference alloy 75-15-10.

The body alloy of the *ting* is shown in an optical micrograph in Figure 5. α+δ eutectoid is present in this part of the casting as well. Selective corrosion of the δ phase and round areas of redeposited copper can also be seen in the figure.

The microstructures of the body metal and the chaplet of the *p'an* (FSC-B-9) are shown in Figure 6 with the chaplet visible on the right side of the figure. The color of the body metal was slightly yellower than that of the chaplet indicating lower copper content. Both the body

Figure 4. Optical micrograph of the leg alloy of the *ting* (FSC-B-56). The layer of corrosion at the bottom delineates the interface with the body metal. (brightfield, unetched)

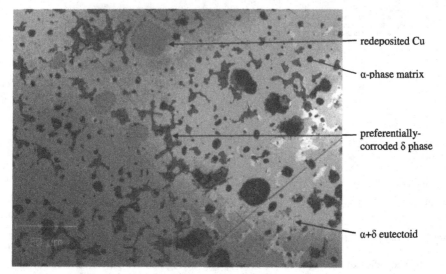

          — redeposited Cu

          — α-phase matrix

          — preferentially-
corroded δ phase

          — α+δ eutectoid

Figure 5. Optical micrograph of the body metal of *ting* (FSC-B-56). Again the alloy is two-phase with α+δ eutectoid. In this figure, selective corrosion of the δ phase can be seen along with areas of redeposited copper (darker round areas in top, left corner). (brightfield, unetched)

and the chaplet are heavily corroded, but the chaplet is more corroded. (Without additional investigation, it is difficult to determine if the corrosion is intergranular, transgranular or both.) Most of the corrosion appears light gray in bright field illumination and red with polarized light, which is characteristic of cuprite (Cu_2O) corrosion. This identification was confirmed with x-ray diffraction (XRD) and EDS analyses. Almost no remnant δ phase was found in either the body or the chaplet; this could be indicative of low tin content and/or selective corrosion of the δ phase. Inclusions of copper sulfide were also found and can be clearly seen as rounded, gray regions in the chaplet. Copper sulfide inclusions are due to the ore source. These inclusions were shown by XRD to be Cu_2S, chalcocite.

DISCUSSION

The coefficients of expansion for the reference bronze alloys and the *ting* were similar to literature values for Cu-Sn-Pb bronzes: COE (84-12-4) = 18.3 x10^{-6} /°C, COE (78-11-11) = 18.6 x10^{-6} /°C [4]. In the Chinese bronzes, the differences seen between the COEs of the body metal and the previously cast parts are associated with differences in both composition and corrosion.

For the *ting*, EDS compositions of the body metal and the leg were similar but these were made on relatively small areas of a heterogeneous casting. Quantitative bulk chemical analysis of the castings by Gettens found the leg to contain 5% more lead than the body metal [2]. Lead has a higher COE (29.3 x10^{-6} /°C [7]) than the bronze and the leg with more lead would be expected to have a higher COE. This difference is exacerbated by corrosion with the formation of lead carbonate, which has an even higher COE (50 x10^{-6} /°C [7]). In addition, the redeposited copper seen in the body metal serves to increase the difference between the two cast parts because the COE of copper at 16.8 x10^{-6} /°C [8] is lower than that of the bronze alloy.

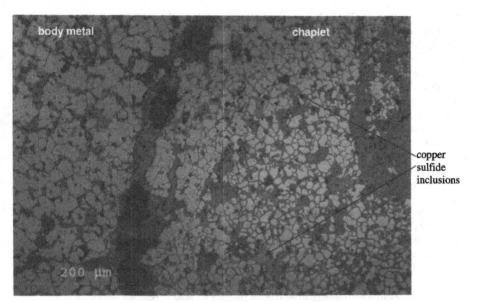

Figure 6. Optical micrograph of the body and chaplet of *p'an* (FSC-B-9). The chaplet is on the right side of the image. The entire sample is heavily corroded, but the chaplet can be seen to be more corroded than the body. The lighter gray regions visible in the chaplet metal are copper sulfide inclusions (see arrows), remnants of the ore source. (brightfield, unetched)

The COEs determined for the *p'an* as a whole were lower than those of the reference alloys or *ting*. This vessel is heavily corroded and the primary corrosion product, cuprite (Cu_2O), has a COE lower (5.7×10^{-6} /°C [9]) than that of metallic bronze. The presence of copper sulfide inclusions from the original ore may also reduce the measured COEs of the *p'an*. The COE of Cu_2S in the temperature range 110 - 327.5°C is lower than bronze (0.96×10^{-4} [7]). The corroded state of the *p'an* certainly plays a role in the low value of the measured COEs and may also explain the difference seen between the values for the chaplet and the body metal because the chaplet contains more cuprite than the body.

The microstructure of cast bronze typically consists of two copper-tin phases: α and δ. Both of these phases have cubic crystal structures [6,10,11]. Due to its low solubility in bronze, lead is found heterogeneously distributed throughout the structure. Assuming that residual stress formation begins when the last of the melt solidifies, the temperature of interest is the melting point of lead: $T_{m[Pb]} = 327.5$°C. The residual stress (σ) formed during cooling due to mismatch in coefficients of expansion (assuming no porosity or corrosion) can be calculated for the reference alloys using the equation:

$$\sigma = \frac{E \cdot \Delta COE \cdot (T_{m[Pb]} - 25°C)}{3(1 - 2v)} \tag{1}$$

The elastic modulus (E) was determined from the slope in the elastic region of the stress/strain curves obtained from the compression tests and was found to be on the order of 1.04×10^7 N/m^2

for the three reference alloys. Poisson's ratio (ν) was estimated to be 0.35 using literature values for copper alloys [8]. The maximum difference in coefficients of expansion measured for the reference alloys was used in the calculation. For COE differences of 1.5×10^{-6} /°C, the magnitude of the residual stress calculated is 5250 N/m^2. However, if one assumes cooling, and thus contraction, for only one piece of the casting, the calculated stress equals 66500 N/m^2 (ΔCOE = 19×10^{-6} /°C). In either case, the value is much lower than literature values for the ultimate tensile strength of bronze ($2.3 - 2.8 \times 10^8$ N/m^2 [4]) and it is evident that there are wide tolerances in composition that allow for a successful casting. Plastic deformation was seen to occur at stresses below the calculated residual stresses in the compression tests (approximately 2500 N/m^2 on a 75-15-10 alloy) and deformation lines are visible in Figure 4 in the corrosion of the bronze *ting* near the interface of the separately cast leg and body.

CONCLUSIONS

1. The coefficients of expansion measured for the reference alloys containing low lead and tin compositions were all approximately 18 to 19×10^{-6} /°C.
2. Significant differences in COE were measured in the ancient samples between the body metal and the previously-cast pieces. These differences are related to variations in composition and corrosion among the pieces.
3. It is unlikely that differences in composition alone would lead to failure of a bronze due to differential thermal expansion, however they do lead to residual stresses and deformation.
4. Inclusions such as sulfides in a bronze casting may affect local stress distributions. Their role in casting failures needs further examination.

ACKNOWLEDGMENTS

We would like to express our gratitude to Paul Jett, the Department of Conservation and Scientific Research at the Freer Gallery of Art and the Conservation Department of The National Museum of Denmark for their assistance and encouragement on this project. In addition, W. Thomas Chase and Keith Wilson provided valuable insights.

REFERENCES

1. W. T. Chase, *Ancient Chinese Bronze Art: Casting the precious sacral vessel* (China Institute in America, New York 1991).
2. R. J. Gettens, *The Freer Chinese Bronzes, Vol. II, Technical Studies* (Smithsonian Institution, Washington, D.C. 1969).
3. *Materials Characterization by Thermomechanical Analysis, ASTM STP 1136,* edited by A. T. Riga and C. M. Neag (American Society for Testing and Materials, Philadelphia, 1991).
4. D. Hanson and W. T. Pell-Walpole, *Chill-cast Tin Bronzes* (Edward Arnold & Co., London 1951).
5. W.T. Chase in *Ancient & Historic Metals: Conservation and Scientific Research,* edited by D.A. Scott, J. Podany and B.B. Considine, (Getty Conservation Institute, Marina del Rey, CA, 1994) pp. 85-117.

6.   *Metals Handbook, Vol. 8*, 8th ed., edited by T. Lyman (Metals Park, OH: American Society for Metals, 1973).
7.   *Gmelin Handbook of Inorganic and Organometallic Chemistry, 8th edition: Copper (System No. 60, B1) and Lead (System No. 47, B1, C1),* (GMELIN Institute for Inorganic Chemistry of the Max-Planck-Society for the Advancement of Science, Frankfurt, Germany).
8.   *Metals Reference Book, Vol. III,* 4th ed., edited by C. J. Smithells (Plenum Press, New York, 1967).
9.   *CRC Handbook of Physical Properties of Rocks,* edited by Y. Sumino and D. L. Anderson (CRC Press, Boca Raton, FL, **93,** 1984).
10. E. Sidot, A. Kahn-Harari, E. Cesari and L. Robbiola, *Mat. Sci. Engr. A* **393,** 147-156 (2005).
11. M. Taube, A. H. King and W. T. Chase, *Phase Transitions* **81**(2-3), 217-232 (2008).

Mater. Res. Soc. Symp. Proc. Vol. 1047 © 2008 Materials Research Society 1047-Y01-01

Variability in Copper and Bronze Casting Technology as Seen at Bronze Age Godin Tepe, Iran

Lesley D. Frame, and Pamela B. Vandiver
Materials Science and Engineering, University of Arizona, 1235 East James E. Rogers Way, Tucson, AZ, 85721

ABSTRACT

Excavations at Godin Tepe—a Bronze Age site in the Kangavar Valley of the west-central region of Iran—yielded a metal assemblage of 202 artifacts of which 91 are curated at the Royal Ontario Museum, Toronto, Canada. The assemblage consists of decorative objects (figurines, vessels, bracelets, rings, needles, pins) as well as weapons and tools (chisels, blades, daggers, and projectile points). Secondary dendrite arm spacing was measured on polished and etched metallographic sections of the eight samples that display cast structures. Cooling rates were calculated base on these measurements along with the average composition of the metal. Comparison to reference data shows that these cooling rates group into ranges typical of quenched and furnace cooled environments. In addition, the maximum temperatures reached during smelting and casting were estimated based on the microstructure and composition of technical ceramics and slag fragments.

Composition and microstructure information was obtained for these artifacts with the use of scanning electron microscopy and electron beam microprobe.

INTRODUCTION

The archaeological site of Godin Tepe is located in the highlands of west-central Iran. It lies along the so-called High Road, which leads from the Mesopotamian lowlands to the northern Iranian Plateau and beyond. Ideally located about half way between Kermanshah and Hamadan near modern day Kangavar, Godin Tepe acted as a center for the exchange of goods, transmission of ideas, and spread of technology. Its excavation during the late 1960s and early 1970s led by T. Cuyler Young, Jr. of the Royal Ontario Museum in Toronto, yielded burials, a long occupation sequence from Chalcolithic through Iron Age, and a ceramic sequence that is linked typologically with other highland sites. The later levels (Periods VI:1, IV, III, and II) contained more than 200 metal artifacts and some metallurgical processing debris [1]. Due to the nature of Godin Tepe as a trade center, it is probable that the metal objects represent the material culture of many sites along the trade routes. However, the production debris found at Godin Tepe also indicates the on-site processing of metals.

Metallurgical material from Godin Tepe is sparse, but it includes crucibles, furnace and tuyere fragments, ore, and metal artifacts dating to the late fourth through early first millennia B.C.E. The production materials were concentrated in only a few locations throughout the site, and they are indicative of local small scale production. The analysis of 70 metal artifacts and fragments of metallurgical debris [1] (over one-third of the two-hundred and two that were excavated) have contributed to understanding the variability in manufacture methods practiced during this time period, reported in Part I below.

Eight of these artifacts exhibited casting structures, which enabled the measurement of the secondary dendrite arm spacing (SDAS) for each artifact in order to determine the cooling method employed during casting (Part II of results). Faster cooling rates result in a microstructure with finer texture and closer spacing of secondary dendrites (Figure 1); therefore the cooling rates can be calculated based on the SDAS following the general relationship shown in Equation 1 [2].

$$d_2 = k \cdot \dot{T}^{-n} \qquad (1)$$

where d_2 is the SDAS (μm), $\dot{T}$ is cooling rate ($^{\circ}$C/s) and k and n are constants. Experimental values of n vary between 0.33 and 0.5, and k depends on the composition of the melt [3]. This relationship has been heavily explored for Al-Cu alloys [4, 5, 6, 7], but these aluminum alloys typically show a different set of SDAS cooling rate curves than Cu-base alloys.

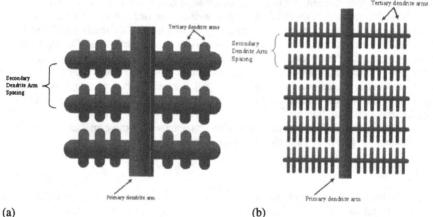

(a)   (b)

Figure 1. Schematic drawings of (a) the coarse dendritic structure typical of slow-cooling compared to (b) a finer structure from faster cooling.

The third part of this investigation reports the maximum temperatures reached during smelting and melting operations based on the composition and microstructure of the technical ceramics. This establishes the capacity of Godin craftsmen for a high temperature metallurgical technology. Each aspect of this study emphasizes the variability of the methods of manufacture employed during the Bronze Age in highland Iran.

METHODS

Seventy artifacts from Godin Tepe were made available for analysis by the Royal Ontario Museum, Toronto, Canada. These artifacts included pins, needles, bracelets, chisels, blades, projectile points, a figurine, processing lumps, furnace and crucible fragments, ore, and a single tuyere (Period III:2). The metal artifacts were sampled at the ROM using a jeweler's saw, and all technical ceramics and ore fragments were sampled using a diamond blade on a Buehler Isomet saw. Sampled sections were mounted in Buehler EpoThin epoxy resin and polished to a 0.5μm finish on a variable speed Buehler polishing wheel. The samples were carbon-coated for compositional analysis using wavelength dispersive spectroscopy electron beam microprobe on a Cameca SX50 SEM-equipped Electron Microprobe using 15-25keV, 20nA and a 1μm spot size. The carbon coating was removed from the samples, and they were polished to a submicron finish on a Buehler Vibramet using colloidal silica. Etching with ferric chloride solution revealed the microstructure for metallographic examination on a Nikon Eclipse Microscope equipped with a MicroFire CCD camera.

The methods of investigation employed here provide "method of manufacture" information and have become the standard for most archaeometallurgical investigations. However, when combined with published theoretical and experimental material properties data, microstructural and compositional information can be pushed to provide further insight into manufacturing processes. The eight artifacts from Godin Tepe exhibiting cast structures have been measured to determine the cooling rates achieved in antiquity. The cooling rates reported here are based on the equation from Miettinen [8]:

$$d_2 = 180 \cdot \dot{T}^{-0.33} \cdot \exp(-C^{0.3}) \tag{2}$$

C is a compositional term that varies with each alloy. This term has been calculated using a simplified version of an equation from Jyrki Miettinen [9]. Miettinen's equation is based on experimental and theoretical copper-base alloy solidification data. In the simplification (Equation 3), C_i is the composition (in wt%) of each element in solid solution with copper.

$$C = 0.17 \cdot (C_{Fe} + C_{Ni}) + 0.3 \cdot (C_{Pb} + C_{As} + C_{Sb}) + C_{Ag} + 0.37 \cdot C_{Sn} - 0.02 \cdot C_{Fe} \cdot C_{Ni} + 0.008 \cdot C_{Pb} \cdot C_{Sn} \tag{3}$$

Values for d_2 were measured on digital micrographs of the artifacts using the Adobe Photoshop measure tool, and pixels were converted to microns according to the scale for each image. Only the secondary dendrite arms making a perpendicular angle with the primary arm were included in these measurements. Equations 2 and 3 were combined to solve for $\dot{T}$ for each object.

RESULTS

Part I: The variability in manufacture of Godin Tepe metal finds

The non-utilitarian items, including decorative pins, needles, and bracelets, from Godin Tepe show remarkable variability in manufacture, especially when compared to the utilitarian items such as tools and weapons. The bracelets, even within the same period, show various cast, worked, and annealed microstructures (Figure 2), whereas tools, such as chisels, are all produced in a similar manner to achieve a limited set of performance criteria (Figure 3).

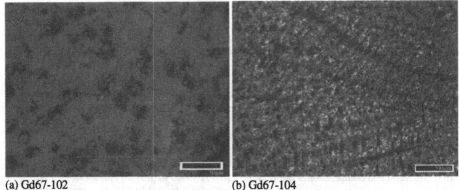

(a) Gd67-102   (b) Gd67-104

Figure 2. Metallographic sections of (a) Gd67-102 and (b) Gd67-104, Period III bracelets. These bracelets are similar in style with overlapping narrowed ends, but despite this typological similarity, they show considerable variability in microstructure. (a) Gd67-102 contains large, equiaxed grains with annealing twins that are typical of a metal that has been annealed, slightly cold-worked and annealed again, whereas (b) Gd67-104 reveals an as-cast structure with little evidence of subsequent working and annealing.

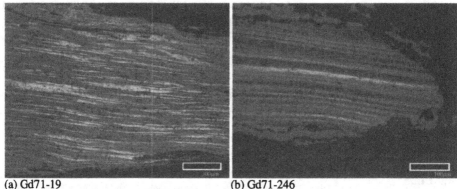

(a) Gd71-19   (b) Gd71-246

Figure 3. Metallographic sections of (a) Gd71-19 and (b) Gd71-246, early Period III chisels. The chisels from Godin Tepe each show signs of considerable cold-working to achieve a flat and hardened tip. Etching away the solute-rich regions of these chisels reveals the lamellar microstructure.

Part II: Cooling rates for cast metal artifacts

Due to this high degree of variability, the artifacts exhibiting cast structures are few and show a great range in artifact type. These include eight samples (Table I), and the secondary dendrite arm spacing (SDAS) was measured for each. Compositions were determined by microprobe analyses, and values reported in Table I are based on the average of "n" analyses, where "n" is reported in parentheses after the description.

Table I. Metal artifacts with dendritic microstructures (bdl=below detection limit).

Object #	Description	(n)	Period	Cu	As	Sn	Fe	Ni	Ag	Sb	Total
Gd73-312	Figurine fragment	(10)	VI	97.40	1.92	bdl	0.09	0.83	bdl	0.10	100.34
Gd71-261	Blade	(16)	IV	96.43	2.51	0.05	0.53	0.15	bdl	0.06	99.73
A1-1109ss511	Crucible prill	(2)	IV	93.62	4.82	bdl	bdl	0.66	0.07	0.13	99.30
Gd67-104	Bracelet	(20)	III:6	93.70	3.02	1.33	1.08	0.76	0.27	0.07	100.23
A1-69ss2	Nugget*	(4)	III:4	97.06	2.20	0.09	0.16	0.17	bdl	0.06	99.74
B1-38	Nugget*	(10)	III:4	90.97	0.93	7.63	0.12	0.15	bdl	bdl	98.92
Gd67-231	Bracelet	(15)	III:2	93.32	1.27	5.04	0.07	0.30	0.06	0.05	100.11
Gd73-419	Pin	(4)	II	97.11	1.59	bdl	0.44	0.27	bdl	bdl	99.41

* Probable raw material lump that was cast prior to forming.

Three of these artifacts are shown here (Figures 3b, 4a and 4b), and the values for d_2 are based on the average spacing between secondary arms for the sample (Table II).

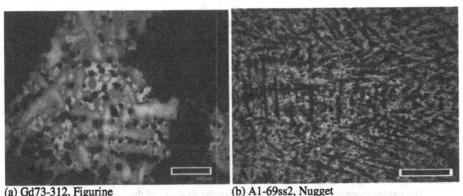

(a) Gd73-312, Figurine   (b) A1-69ss2, Nugget

Figure 4. Metallographic sections of two cast microstructures. (a) Gd73-312 is a figurine fragment, and (b) A1-69ss2 is a processing lump that is rectangular in cross-section; the slight deformation of the dendrites indicates that it was cold-worked into this shape after being cast.

Table II. Calculated cooling rates for Godin Tepe cast artifacts

Object #	Period	Description	Ave d_2 (μm)	C	$\dot{T}$ (°C/s)
Gd73-312	VI	Figurine fragment	29.72	0.76	14.05
A1-69ss2	III:4	Nugget	81.61	0.81	0.64
Gd71-261	IV	Blade	81.69	0.95	0.56
Gd67-104	III:6	Bracelet	64.90	2.03	0.52
Gd67-231	III:2	Bracelet	84.76	2.72	0.17
B1-38	III:4	Nugget	82.71	3.17	0.15
Gd73-419	II	Pin	145.30	0.64	0.14
A1-1109ss511	IV	Crucible prill	112.59	1.67	0.12

The rates indicate that there are three methods of cooling, or casting and annealing, which yield cooling rates of 14°C/s, ~0.55°C/s, and ~0.15°C/s, regardless of artifact type. Table III presents examples of cooling rates for quenching in water or a water-oil mixture as well as cooling in a furnace. It should be noted that the secondary dendrite arm spacing is indicative of the entire cooling process given that ripening occurs while the secondary arms are in contact with the melt [10,12]. Therefore, the calculation of a slow cooling rate (low $\dot{r}$), may suggest furnace cooling as well as possible annealing. Based on the values reported in Table III, most of the artifacts could be described as 'furnace-cooled;' this matches well with the low cooling rate estimated for the crucible prill (A1-1109ss511) since it is likely that the crucible containing this entrapped prill would have cooled slowly in situ after the smelting event was finished. 'Furnace-cooled' suggests slow and relatively controlled cooling, but the phrase should not carry the same precision as it does in modern industrial practice. These artifacts actually would have been cast into open, closed, or lost wax molds; however, the cooling rates for these presumed ancient practices have not yet been determined.

Table III. Typical cooling rates

Cooling Method	Cooling Rate (°C/s)	Reference
Quenched in water at 25°C	110	[10]
Quenched in 90% water : 10% oil at 50°C	16	[10]
Air-cooled in a furnace	~0.5	[11]

Part III: Maximum temperatures achieved in crucibles

In addition to the measurement of SDAS, the crucible slag was analyzed to identify the constituent silicate phases and determine the maximum temperatures reached during the smelting and melting operations. The compositions of technical ceramics and copper alloy prills are summarized in Tables IV and V, respectively.

The values reported in Table IV are microprobe analyses of the glass phase of each silicate melt. These compositions are all relatively high in SiO_2 (35.8 – 57%) and Al_2O_3 (8.3 – 17%), which corresponds to high melting temperatures. The limestone constituent with some magnesium oxide, in some cases a dolomite, is quite variable (CaO 6 – 37%, MgO 2.6 – 6.6%). The compositions of entrapped prills and the glassy phases of the slags were plotted on their respective phase diagrams [1,13], and the processing temperature ranges were approximated. Since the glass matrix phase composition is used here, the T_m values represent the minimum temperature to initiate melting for these systems. It was determined that all technical metallurgical ceramics reached at least 1250 to 1300°C.

Table IV: Composition of slag in Godin Tepe metallurgical ceramics

Godin #	Description	Period	Na₂O	K₂O	SiO₂	MgO	Al₂O₃	CaO	MnO	FeO	CuO	Total
A01-50ss40	Furnace Fragment	VI	0.55	1.35	35.84	6.56	9.55	35.97	0.21	5.01	0.67	95.72
B1-510ss28	Melting Crucible	VI	1.05	2.14	54.43	6.31	12.47	15.25	0.13	6.66	0.18	98.62
A1-1123ss521	Smelting Crucible	IV	0.13	0.55	43.09	3.63	8.31	10.71	0.20	31.27	0.22	98.09
A1-1109ss511	Smelting Crucible	IV	0.76	2.01	39.08	2.55	12.37	14.77	0.11	14.93	0.36	86.96
A2-1184ss63	Melting Crucible	IV	0.77	2.11	49.74	6.15	14.83	13.53	0.11	8.81	bdl	96.06
AA2-193ss9	Melting Crucible	III:4	2.30	6.17	56.98	2.57	16.96	5.94	0.17	7.64	0.22	98.96

Table V: Entrapped prills in Godin Tepe metallurgical ceramics: composition, T_m, and corresponding phase diagram reference.

Godin #	Location of analysis	P	S	Pb	Sb	Sn	Ag	Cu	Fe	As	Ni	Total	T_m (°C)	Ref.
A01-50ss40	Prill #1	bdl	bdl	bdl	bdl	bdl	bdl	85.48	bdl	bdl	bdl	85.48	1083	
	Prill #2	bdl	bdl	bdl	bdl	bdl	bdl	87.30	0.24	bdl	bdl	87.54	1083	
	Non-metallic inclusion	bdl	20.19	bdl	bdl	bdl	bdl	80.06	bdl	bdl	bdl	100.25	1130	[14]
B1-510ss28	Prill #1	bdl	bdl	bdl	bdl	bdl	bdl	99.92	0.09	bdl	bdl	100.01	1083	
	Non-metallic inclusion	bdl	9.12	bdl	bdl	bdl	bdl	93.21	0.42	bdl	0.03	102.78	1125	[15]
A1-1123ss521	Prill #1	0.07	0.08	bdl	bdl	bdl	bdl	93.83	3.88	0.77	0.04	98.67	1115	[16]
	Prill #2	bdl	1.37	bdl	bdl	bdl	0.09	89.77	7.43	1.65	bdl	100.30	1100	[17]
A2-1184ss63	Prill #1 (mineralized)	0.47	0.07	0.18	bdl	bdl	bdl	43.12	0.52	3.19	bdl	47.56		
A1-1109ss511	Prill #1	bdl	bdl	bdl	0.69	bdl	bdl	89.33	1.18	6.85	1.44	99.49	1000	[17]
	Non-metallic Inclusion	bdl	1.18	20.55	3.56	bdl	0.90	50.36	1.44	15.50	6.37	99.86		
AA2-193ss9	Prill #1	bdl	0.25	bdl	0.13	10.88	bdl	83.80	0.04	3.91	0.22	99.23	980	[15]

CONCLUSIONS

This paper applies a new method of studying heat treatments and casting practices of ancient metal craftsmen to gain greater insight into technological variability during the Bronze Age in Iran. This investigation has brought to light the great variability in manufacturing methods employed for non-utilitarian items, such as bracelets and pins, as compared to the uniform production methods employed for utilitarian objects (e.g., chisels and projectile points). This conclusion is based on (a) the differences seen in microstructure for the bracelets and pins including variations in grain size, presence of inclusions, and grain morphology; (b) the range of compositions for many similar looking bracelets and pins [1], and (c) the different cooling rates calculated for similar artifacts (e.g., Gd67-104 and Gd67-231). This conclusion—many typologically similar non-utilitarian items were produced using different technological methods—provides a strong argument for the presence of multiple producers of these goods. This is not surprising given the nature of Godin Tepe as a trade center during the Bronze Age. In fact, the material culture at Godin most likely represents a variety of regional production styles for the artifact types whose intended purposes do not constrain the method of manufacture.

Further, it is clear that Godin Tepe, though not a *main* producer of copper and bronze implements in antiquity, was in fact contributing to this variability in production style. The metallurgical processing debris from Godin Tepe indicates small-scale high-temperature processing of copper-base ore and metal as early as the late 4th Millennium B.C.E. (Godin Period VI:1). The temperature estimates reported in the results will be corroborated with differential thermal analysis of the crucible slags in a future work, and this is a necessary next step to fully understand the thermal history of these technical ceramic vessels

The next phase of this project will involve the experimental replication of the alloys and measurement of the dendritic microstructure resulting from various cooling rates and casting conditions.

ACKNOWLEDGMENTS

These artifacts were provided for analysis by the Royal Ontario Museum, and without the support of Mitchell Rothman, Bob Henrickson, Hilary Gopnik, Susan Stock, and Bill Pratt, this project would not have been possible. David Killick, Aaron Shugar, Vince Pigott, and Christopher Thornton provided advice regarding the archaeometallurgical portion of this study, and Jyrki Miettinen, Carlos Levi, and Robert Mehrabian provided valuable advice concerning the solidification equations and interpretation of the calculated cooling rates.

REFERENCES

1. L. D. Frame, M.S. Thesis, University of Arizona (2007).
2. D. R. Askeland, *The Science and Engineering of Materials* (PWS Publishers, Boston, 1984).
3. J. Miettinen, *Computational Materials Science* 22, 240-260 (2001).
4. B. P. Bardes, M. C. Flemings, *Modern Casting* 50, 100-106 (1966).

5. M. C. Flemings, T. Z. Kattamis, B. P. Bardes, *Transactions of the American Foundrymen's Society* **99**, 501-506 (1991).
6. M. E. Glicksman, R. N. Smith, S. P. Marsh, R. Kuklinski, *Metallurgical Transactions A* **23A**, 659-667 (1992).
7. T. Z. Kattamis, J. C. Coughlin, M. C. Flemings, *Metallurgical Society of AIME – Transactions* **239**, 1504-1511 (1967).
8. J. Miettinen, *Computational Materials Science* **36**, 367-380 (2006).
9. J. Miettinen, (Personal Communication, September 2007).
10. A.K. Sinha, *Physical Metallurgy Handbook* (McGraw-Hill, New York, 2003), p.13.11.
11. G. P. Anastasiadi, R. V. Kolchina, L. N. Smirnova, *Metallovedenie i Termicheskaya Obrabotka Metaliov* **9**, 35-37 (1985).
12. W. Kurz, D. J. Fisher, *Fundamentals of Solidification* (Trans Tech Publications, Rockport, MA, 1984), pp.65-96.
13. F. Kongoli, in *Yazawa International Symposium: Metallurgical and Materials Processing: Principles and Technologies*, Florian Kongoli, K. Itagaki, C. Yamauchi, H.Y.Sohn, Eds. (The Minerals, Metals, and Materials Society, Warrendale, PA, 2003), pp.199-210.
14. D.J. Chakrabarti, D.E. Laughlin, in *Binary Alloy Phase Diagrams Vol. 2*, T.B. Massalski, H. Okamoto, Eds. (ASM International, Materials Park, OH, 1990), pp.1467-1471.
15. A. Prince, H. Okamoto, *Ternary Alloy Phase Diagrams* **7** 4577, 9408 (1995).
16. L.J. Swartzendruber, in *Binary Alloy Phase Diagrams Volume 2*, T.B. Massalski, H. Okamoto, Eds. (ASM International, Materials Park, OH, 1990). pp.1408-1410.
17. P.R. Subramanian, D.E. Laughlin, in *Binary Alloy Phase Diagrams Volume 1*, T.B. Massalski, H. Okamoto, Eds. (ASM International, Materials Park, OH, 1990). pp.271-275.

Mater. Res. Soc. Symp. Proc. Vol. 1047 © 2008 Materials Research Society   1047-Y06-02

The Technology That Began Steuben Glass

Jacob Israel Favela, and Pamela Vandiver
Department of Materials Science and Engineering, University of Arizona, Tucson, AZ, 85716

ABSTRACT

Frederick Carder popularized art glass in America and is remembered as the founder and head of Steuben Glass Works. Carder, a designer and glass technologist trained in England, established the factory in Corning, New York, in 1903. The factory produced colored and highly decorated glass vessels that competed with but were less expensive than those of Tiffany Studios. To understand the differences in technology between the competing products of Carder and Tiffany, especially the type called "aurene," we analyzed and compared opalescent white glass formulations, iridized with thin-film, golden luster decoration and some examples decorated with combed trails containing silver. The methods of analysis are electron beam microprobe analysis and scanning-electron microscopy with simultaneous energy dispersive x-ray analysis. Analytical results show that Carder produced an affordable product by standardized processing that included opalescent compositions in a narrow range of soda-lime-silicate and lead-alkali-silicate glasses with calcium phosphate or boneash as an inexpensive but reliable opacifier, quite thick flashed "golden" lustrous coatings made from tin oxide or tin and silver, and relatively rough velvet- to satin-textured, iridescent, thin-film coatings that were formed during multiple rapid heat treatments.

INTRODUCTION

The Art Nouveau movement of the late nineteenth and early twentieth centuries combined sophistication, elegance, and novelty in art with the technological progress of the Industrial Revolution. The use of stylized natural motifs and unique materials, such as glass, enamels, shell, coral and ivory, was encouraged among artisans beginning in the 1890's. Two of the leading American glass manufacturers, Louis C. Tiffany and beginning in 1903 Frederick C. Carder, produced in their factories luxury glass goods that were similar in style; both used opalescent glass formulations, trailed and combed threaded decoration and fumed lusters for brilliant iridescent effects on the surfaces of various blown objects, such as lamps, vases and goblets [1, 2]. Tiffany called the production "favrile" and Carder patented the name "aurene."

Frederick Carder (1863-1963) was born and educated in the Stourbridge, the center of England's glass industry. He was educated as an artist, potter, sculptor and designer. He won several art competitions and studied with the cameo glass master, John Northwood, who replicated the Portland vase. Carder worked for the Stevens and Williams glass factory during the day and taught art in the evenings at the Wordsley School of Art, until in 1903 he accepted an offer from Thomas G. Hawkes, owner of a glass decorating business, to establish the Steuben Glass Works in Corning, New York. He was sufficiently successful in the production, particularly of electric and gas shades and vessels, that in 1913 Tiffany threatened to sue because the original could not be differentiated the Steuben imitations. The Steuben factory ceased production of colored glass in the mid-1930s, and Carder became less active at Steuben and pursued his own art projects until his death in 1963.

In general, glass shapes from Tiffany Studios are considered more asymmetrical and fluid, and the visual effect is more subtle and experimental; whereas, Carder's Steuben Glass Factory production is geometrically more regular and rigid, and the surfaces are well defined and easily readable from a distance. To what extent do these differences imply that Steuben employed more common initial use of molds, more rapid forming sequences and less cost and variation in raw materials? In the marketplace, Carder's glass often sold for less than that of Tiffany and developed a middle class market for art glass. While stylistic similarities are easily seen and understood, our task was to investigate how the technological differences and range of variation in composition and fabrication.

The glass technology of Tiffany studios is well characterized and understood [2]. To date, no technical studies have included works by the British-American glassmaker, Frederick Carder. To help understand production technology of Carder art glass, scientific investigation was carried out with energy dispersive X-ray microanalysis in a scanning electron microscope (SEM-EDS) and electron probe microanalysis (EPMA) on three fragments of Carder art glass.

EXPERIMENTAL

Morphological and compositional properties of bulk glass, inlays and iridescent surfaces were obtained simultaneously with energy dispersive X-ray microanalysis and scanning electron microscopy (SEM-EDS). The microanalysis in the SEM was carried out using a Hitachi S-2500 with an energy dispersive system, LINK eXL, on samples that were coated with only carbon. High-resolution images of iridescent surfaces and liquid-liquid phase separation of the bulk glass were obtained using a Hitachi S-4500 field emission SEM. These samples were coated with a thin conductive layer of Au/Pd alloy to facilitate high magnification imaging in the S-4500. To reveal the morphology of crystals and phase separation and prior to coating, etching by hydrofluoric acid was carried out using a 10% solution for 10 seconds on freshly fractured, cross section surfaces of the inlays and bulk glasses. The iridized surface coatings were imaged without an acid treatment.

A Cameca SX50 electron microprobe analyzer was used for quantitative analysis of the bulk composition and surface decorations using ground and polished cross sectional samples that were coated with about 200 Angstroms of carbon. Diffraction crystals TAP, PET, and LIF were used for quantitative analysis, and PC2 for qualitatively investigating the possible presence of boron. The beam conditions were 15 kV, 20 nA with a one-micron beam diameter. A dwell count of 20 seconds for peak and background for all elements was used with the exception of Na; for Na a 10 second dwell time was employed, and sodium was counted first to avoid degradation of signal due to beam effects. Well-characterized natural and synthetic materials similar in concentration to the unknowns were used as standards.

SAMPLE INVENTORY

The objects chosen for this study were obtained in 2001 and 2002 from a part of the Steuben Collection of the Rockwell Department Store that was not accessioned by the Corning Museum of Glass. The sample inventory and descriptions are as follows:

1. Carder lamp shade: The upper edge of a lamp cover with green and golden trailed and combed decoration was sampled with diamond tools. The inner surface is iridized with a gold color and

the outer white body is an even dense solid white in reflectance that is translucent in when illuminated. Probably made about 1921 and called "Steuben Gold Calcite," the shade is described in catalogs as having green decoration with gold aurene and an alabaster-like appearance. At this time August Jansson and Charlie Matthews were Steuben gaffers known to make decorated specialty shades. A similar shade is shown at left as Fig. 59, p. 33 of reference 1.

2. Carder moil from upper part of a shade: This piece was once part of a larger lamp cover, but was separated and removed in a finishing operation while the larger, lower portion was retained for sale. The body was made with multiple gathers, probably three, and the layers have varying opacity, indicating a "striking" opacifier that formed during multiple reheatings. The range of opacity in cross-section is dense milky white to an opalescent translucency. The exterior was then decorated with brown, iridescent blue/tan, and light tan combed trailings. Both the interior and exterior were fumed with a gold luster prior to annealing. A similar shade is is found in the center of Fig. 59, p. 33 of reference 1.

3. Carder saltcellar, formerly the base of a vase: This piece is cylindrical in shape with a flat bottom and has striations of an irregular, iridescent gold and orange coating in zigzag or zebra-like pattern. It once served as a base for a tall probably molded vase that subsequently was broken. The cup-like base was retained and reused. The rim has been ground and polished to allow use as a salt dish. The body consists of layered glass with of with a range of opacity of dense white to milky translucence. The base of the saltcellar, exterior diameter 6.2 cm, is similar to the gold aurene vase made about 1915 now in the Smithsonian Institution collection and illustrated in colorplate 9A, ff. p. 90, of Gardner [1].

Figure 1. Samples of Carder art glass chosen for this study, displayed as follows: Carder saltcellar, Carder Lamp shade, and Carder moil.

RESULTS

Experimental results for the study include quantitative elemental bulk glass compositions (Table I) in addition to qualitative elemental compositions of the iridized surface layer for each sample (Table II). High-resolution FESEM images are shown in Figures 2-5 and help relate the nature of each glass microstructure to its compositional character. .

Table I: Results of EPMA quantitative analysis of the bulk glasses and glass overlays are given as the average of "n" analyses in the first column with standard deviation presented in the second column for each sample. BDL, below detection limit, is generally in a few parts per hundred, except Sn which is about 0.1%. We conducted a separate EPMA analysis for boron, but it was also below detection limit. EDS analysis on the Cameca did not reveal other elements that would explain the low totals in the overlays. Dashed lines indicate the element was not included in the analysis, but was also not found by previous EDS analysis with a detection limit of 0.5%.

Wt.%	Carder Lamp				Carder Moil						Carder Saltcellar	
	Bulk Glass (n=13)		Glass Overlay (n=4)		Bulk Glass (n=7)		Glass Overlay (n=4)		Colored Additions (n=6)		Bulk Glass (n=7)	
SiO_2	48.90	±1.13	49.58	±0.25	65.31	±0.67	56.48	±2.54	59.35	±2.07	64.53	±0.47
Al_2O_3	0.59	±0.02	0.60	±0.01	1.48	±0.04	1.12	±0.11	1.75	0.99	1.48	±0.03
PbO	31.94	±0.32	32.17	±0.52	BDL		22.34	±0.48	22.11	±0.37	BDL	
ZnO	0.17	±0.02	BDL		7.06	±0.11	4.14	±0.22	5.04	±2.02	7.22	±0.16
As_2O_3	0.91	±0.05	-		-		-		-		-	
Na_2O	3.73	±0.79	4.05	±0.10	14.83	±1.04	9.78	±0.99	11.12	±3.39	14.14	±1.20
K_2O	2.82	±0.10	2.99	±0.02	3.39	±0.04	0.74	±0.60	1.43	±1.13	3.32	±0.06
CaO	0.86	±0.43	0.04	±0.01	3.70	±0.08	0.02	±0.00	0.06	±0.06	3.73	±0.08
MgO	0.05	±0.01	0.06	±0.01	0.03	±0.01	BDL		0.02	±0.00	0.02	±0.00
Fe_2O_3	0.10	±0.01	BDL		0.06	±0.00	BDL		0.08	±0.00	BDL	
P_2O_5	1.05	±0.04	BDL		3.89	±0.07	BDL		BDL		3.92	±0.06
TiO_2	0.08	±0.00	0.12	±0.00	BDL		BDL		0.07	±v0.00	BDL	
Ag_2O	-		0.32	±0.03	-		BDL		BDL		-	
SnO	-		BDL		-		0.16	±0.03	BDL		-	
Total	91.20	±1.48	90.01	±0.58	99.75	±1.24	94.62	±2.85	101.03	4.72	98.36	±1.31

Table II. (Right) Results of the qualitative analysis of chemical elements in the iridized surface by means of SEM-EDS. The layers are so thin that quantification is not possible without use of another surface analytical technique such as PIXE.

Object	Major Elements
Lamp	Sn
Moil	Sn, Ag
Saltcellar	Sn, Ag

Figure 2. Electron micrographs obtained using FE-SEM of the Carder Lamp sample: (a) Preferential etching of an immiscible phase by 1% HF acid solution for 10 seconds reveals a somewhat interconnected network of submicron spherical pores, and (b) BSE image of the wrinkled surface layer with some prominent droplets that yield a satin matte surface texture.

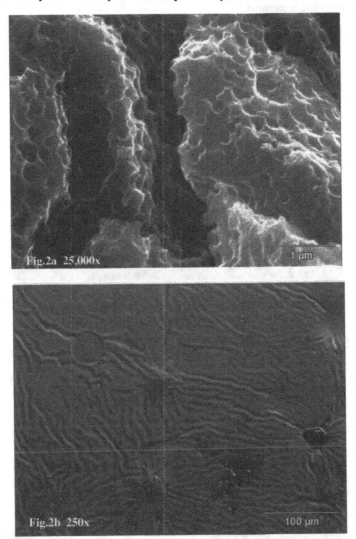

Figure 3. Electron micrographs obtained using FE-SEM of the Carder shade sample, (a) High magnification image of the 400-800 nm thick fumed, iridized coating and the rough interface with the underlying glass, (b) High magnification image of tin oxide particles precipitated as the surface iridescent layer that show evidence of agglomeration and sintering, evidence of further heat treatment beyond application.

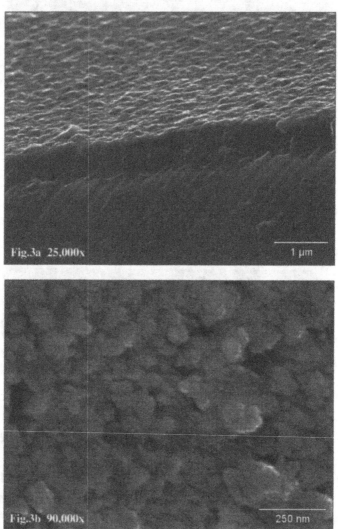

Figure 4. Electron micrographs obtained using FE-SEM of the Carder moil sample, (a) Preferential etching of the immiscible phase by 10% HF acid for 10 seconds reveals a somewhat interconnected network of micron and submicron pores, and; (b) Regions of nucleated crystal growth of a probable calcium phosphate phase that occurs in whiter areas of the cross-section. The larger calcium phosphate crystals from ref. 3 have a similar morphology, and; (c) micron-sized crystal growth arranged in opaque bands running parallel to the alternating layers of white opal and translucent phases of the Carder moil sample.

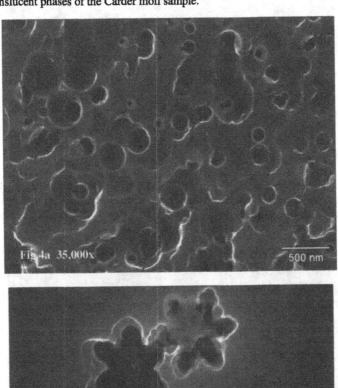

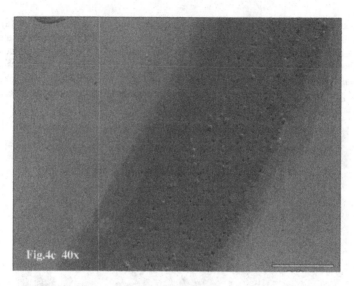

Fig.4c 40x

Figure 5. Electron micrographs obtained using FE-SEM of the Carder saltcellar sample. (a) Regions of variable size (0.1-0.7 microns), larger volume fraction and more interconnected, etched-out second white opal phase, and, (b) An area with a larger average, more dispersed, less interconnected band of opalescence.

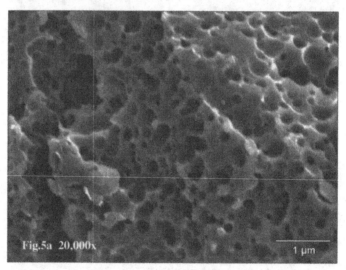

Fig.5a 20,000x

1 μm

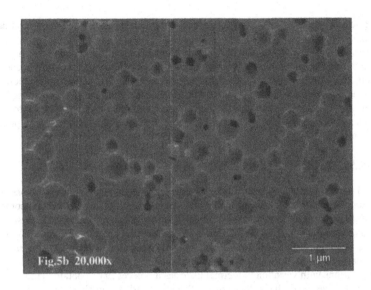

Fig.5b 20,000x   1 μm

DISCUSSION

Table III summarizes the distinctive features of Carder's aurene glass, and differentiates them from Tiffany's studio practices. These features are divided into three categories, as follows:

Bulk Glass Microstructure and Morphology

High-resolution images obtained with FE-SEM reveal a network of etched-out droplets of a second phase that now appear as micropores (Figs. 2a, 4a, 5a and b). This opalescence is caused by liquid-liquid phase separation during forming and reheating of the glass, essentially a process of undercooling such that two stable glass phases separate that are more stable than the single bulk phase. In addition, micron-sized crystal growth (Fig. 4b) was found arranged in opaque bands running parallel to the alternating layers of white opal and translucent phases of the Carder moil sample (Fig. 4c). The crystal features, probably calcium phosphate [3], are too large to cause light scattering and act to opacify the glass. Conversely, the etched-out droplets are on the order the wavelength of visible light (300-800nm) with a range in droplet size of approximately 0.2-0.5μm (200-500nm) and thus are responsible for the opalescent effects observed in Carder samples.

A range in of volume fraction of the micropore network was observed in bands parallel to the surfaces in the cross-sections of the saltcellar and moil samples that relate to opalescence. The saltcellar, for example, contains a network of micropores that appears dense and continuous (Fig. 5a). This network is irregular in other regions that have more dispersed droplets of comparable size and lesser opacity (Fig. 5b). These features are reasonable when we consider that the preferential etching of the immiscible phase by HF is more efficient, or appears "over-etched," in a dense network of higher volume fraction second phase and with a higher surface area [4]. This range of structural features also appears in the moil sample and is expected for samples containing several glass layers of varying opacity that indicate a variation in heat treatment with spatial depth.

However, in the shade sample the distribution of opalescence is homogeneous, that is, shows only a single morphological feature in cross-section. The opalescence gives only moderate opacity (Fig.2a), when compared to the dense micropore network observed on the more opaque samples. The homogenous feature is reasonable because the lamp sample is composed of one single visible layer of dense opal glass in contrast to several glass layers that characterize other samples. The homogeneity implies that the lamp received a uniform heat treatment, perhaps during annealing.

Bulk Glass Composition and Formulations

Seven to thirteen analyses were made from the interior to the exterior of each cross section. For each sample, traditional one-way analysis of variance (ANOVA) revealed that compositions did not differ significantly as a function of position across the cross-section of each sample: for the shade analyses $F(14, 120)=0.000225$ and $p=1$; for the moil analyses, $F(6,49)=0.000178$ and $p=1$; for the saltcellar, $F(6, 49)=0.000123$ and $p=1$. By extension, no statistical difference exists between compositions used for white opal and transparent phases of the same sample. Thus, two glass compositions with variation in thermal histories are necessary to obtain the range of opalescence observed in each Carder sample. The nucleated crystal observed in Figure 4b is a special case and supports 'striking' of a second phase by holding the glass melt at a lower temperature for longer periods of time with a subsequent reheating.

We have also considered the variability between bulk glass melts of samples of similar composition: the lead-silicate compositions of both the Carder saltcellar and moil pieces were also found to have little statistical difference between them. These results imply a standardization of Carder glass technology, especially when we consider the inherent variability that should be apparent between samples of similar appearance and composition produced over a period of years: This variability is not observed between the moil and saltcellar samples that appear to have similar color and opacity.

Two distinct glass systems –soda-lime-silica and lead-silicate–are represented by the compositions of the three glass samples with some variation in additives among them (Table I). The high content of lead (31.94 wt.%) in the Carder lamp sample is typical for decorative glass, as the so-called "crystal glass" ware typically contains greater than 20% wt. PbO. The high index of refraction of lead oxide serves to increase the light diffusing properties of the immiscible phase in the glass and intensifies the appearance of opalescence. Zinc oxide, about 7%, in the soda-lime-silicate glasses of the bulk moil and saltcellar serves much the sampe function but to a lesser degree.

The significant amount of $1-4\%P_2O_5$ suggests that it is the main opalizing agent in this glass formulation: P_2O_5 helps to nucleate phase separation, as well as serving as a glass former and intermediate flux in the melt. Calcium oxide and phosphorous pentoxide are present in an approximate ratio of 1:1 in the bulk glasses, totaling from about 2 to 8%. Further analyses would attempt to identify the likely crystalline phase or phases of interest, 2 CaO-P2O5 with about 45% CaO and 3 CaO-P2O5 with about 54% CaO [5]. Their eutectic is about 1275°C [5], with undercooling for nucleation being about 200°C lower, or about 1075°C, within the workshop glass working range. Remember that Carder was a student of John Northwood in Stourbridge at a time when Northwood was experimenting with white for cameo glasses and had just reproduced the Portland vase. Carder is know to have studied the role of calcium phosphate and phase separation in the formation of opalescence and opacity.

Fining agents, such as arsenic trioxide, are found only in the Carder lamp sample and help promote the removal of bubbles. Antimony and sulfur, used as the common fining agent sodium sulfate at a concentration of about 0.25%, were below detection limits.

No boron was detected using the PC2 crystal with EPMA; however, the low totals require that the possibility of the presence of boron be tested by other means.

Glass Overlays and Iridescent Surface Decoration

In the shade, the bulk and overlay glasses are the same, except for the 0.3% silver oxide that makes the threaded decoration appear a golden yellow. In the moil, the threaded overlay and colored additions are a similar soda-zinc-lead silicate, and the base glass is a soda-lime-silicate. Even though the expansion coefficients are sufficiently similar to one another that the two different glasses do not fracture or delaminate, the two glasses do not match in their temperature-viscosity profiles, as the additions are raised from the surface and are a stiffer, more viscous glass. This is an intentional visual effect that produces a slightly raised cameo-like texture and that required considerable experimentation for success. The saltcellar has no added glass trails.

Tin oxide, SnO_2, was found to be the main constituent of the iridized surface layer by SEM-EDS (Table II), but is known to give a silvery iridized surface when sprayed onto the hot surface of the glass just before the last reheat and prior to placement in the annealing oven. However, when silver is also present, the effect is golden. The thickness of the iridescent layer is shown for the Carder lamp sample (Fig. 3a) and measures approximately 0.5 to 0.8 μm, approximately five to ten times thicker than on Tiffany glass which is noted for a more translucent, variegated and phantasmal effect. Microprobe analysis revealed the glass overlay to be a lead-silicate glass statistically similar to compositions found in the bulk ($F(5, 36)=0.000157$ and $p=1$) with the exception of the SnO_2 additive (Table I). The glass overlay is also rich in about 0.3% silver oxide, a common yellow colorizing agent, or "silver stain," that forms a solid solution in the batch and is re-crystallized onto or near the surface when put in a reducing flame. Silver is not present in significant amounts in the glass overlay of the moil, but may be present as a surface precipitate in the iridized coating as seen by SEM-EDS (Table II). The presence of tin oxide in the glass overlay of the moil sample suggests that, at least for this glass piece, metallic salts were added into the lead-rich melt and co-precipitated onto the surface in a reducing atmosphere. The oxidation of lead oxide to lead tetroxide at the surface may aid in or be responsible for the reddish brown iridescent trails in the moil sample. These additions were trailed and combed onto the molten glass in the same manner as the copper oxide in combed green-leaf motif on the surface of the Carder lamp, but e found no increase in copper, iron or chromium oxides in the added red trails. The colorant in the moil is a subject for further study.

The buckling of the surface iridized layer observed in Figure 2b is due to multiple heating and cooling treatments at the last stage of making the glass article. The fine particle size – 100-250 nm (Fig. 3b)– of the precipitated particles on the surface allows them to deform plastically during the large compressive stresses introduced as the glass shrinks and expands during reheating and cooling cycles. The result gives an iridescent satin- to velvet-like appearance to the sheen of the glass article and is due to the increased diffusivity of light interplaying with the rough surface texture. Each surface sample showed a characteristic surface morphology of a multiple heat treatment regime when compared with replication studies reported in ref. 2.

Table III: Summary of differences in the technology of "aurene"-style art glass produced by Tiffany's and Carder's workshops based on results of this study and other references [2, 6, 7, 8].

	Tiffany	Carder
Bulk Material	• Opal glasses of variable compositions, mainly high-cost lead-borosilicate glasses • High PbO content (35-40 wt.%) • Contains up to 0.4 wt.% Ag_2O in bulk	• Opal glass of narrowly defined compositions: Soda-lime-silica and lead-silicate systems • High PbO content (~31 wt.%) • No Ag_2O detected in bulk material
Glass Overlay	• Typically the same composition as bulk glass material but with various coloring agents; • Contains up to 0.4 wt.% of Ag_2O	• Lead-rich glass overlay distinct from bulk, • Tin oxide containing composition • Contains up to 0.32% Ag_2O • Contains up to 0.16% SnO_2
Iridescent layer	• Precipitated SnO_2 and/or Fe_2O_3. • 0.05-0.1 µm, or 50-100 nm, in thickness	• Precipitated SnO_2, possibly minor Ag_2O • Up to 0.5-0.8 µm or 500-800 nm in thickness

CONCLUSIONS

Many of the design and processing techniques made famous by Tiffany appear in Carder art glass: the practice of layering glass with different optical properties, fumed "golden" lusters of tin oxide and "silver stains," and employing multiple heat treatments for increased diffusivity. Perhaps the most interesting feature of the Carder glass shade, however, is that a single glass formulation was employed to obtain two intermixed, phase-separated glasses with distinct optical features. Samples sharing the same degree of opacity, such as the moil and saltcellar pieces, are also similar, even near duplicates, in compositional character. These technological features imply standardization of raw materials and batch compositions in the glass technology of Frederick Carder-- an aspect antithetical to the elaborate experimental melting program conducted at Tiffany studios. Most Tiffany soda-lime-silica and lead-silicate compositions vary greatly in base elements, fluxes and additives [2]. A summary of the technological features that describe Tiffany and Carder art glass is given in Table III with information derived from this study and others [2, 6, 7, 8]. Substantial concentrations of up to 7% B_2O_3, an expensive but useful flux in the late nineteenth century that cost five times the price of lead compounds, are typical in Tiffany blown art glass compositions containing a range of lead borosilicates [2]. For instance, to make a glass opalescent one finds Tiffany glass containing in one glass more than one mechanism of producing the effect, as with opalescence being produced by phase separating borosilicates, potash lead compositions, cryolite and boneash. Less exotic glass formulations were used by Carder and with a more standardized production regime. This may have given Carder an edge that made possible lower production costs and thus the lower sale price that benefitted his patrons and made available to the middle class the art glass that had been exclusive, expensive and precious.

If samples can be found or provided, future studies should differentiate the blue, red and gold aurene compositions and the calcite, alabaster and cameo glass compositions of Carter and the cameo compositions of Northwood. A third important glass artist of the late nineteenth and early twentieth centuries who incorporated opalescent glass in stained and leaded window glass, John LaFarge, could be investigated to create a more complete appreciation of the variability in methods and materials.

ACKNOWLEDGMENTS

The authors gratefully acknowledge the assistance and collaboration of MSE students, Katelun May and Scott Cooper, Gary Chandler, MSE Microscopy Facility, and Kenneth Dominick of the Microprobe Laboratory, Lunar and Planetary Sciences Department, University of Arizona. Thanks are also expressed to the well-known collector of Carder glass, Robert Rockwell, for his help in acquiring samples and for his advice.

REFERENCES

1. P.V. Gardner, The Glass of Frederick Carder, Crown, New York, 1971, pp. 1-373.
2. W. D. Kingery and P. B. Vandiver, The Technology of Tiffany Art Glass, in Application of Science in Examination of Works of Art, edited by P. A. England and L.van Zelst, Museum of Fine Arts, Boston, 1985, pp. 100-116.
3. W. D. Kingery, Introduction to Ceramics, 1st ed., Wiley, New York, 1960, pp. 310.
4. W. D. Kingery, D. R. Uhlmann and H. K. Bowen, Introduction to Ceramics, 2nd ed., Wiley, New York, 1976, pp. 181-185.
5. E. M. Levin, C.R. Robbins and H. F. McMudrie, Phase Diagrams for Ceramists, American Ceramic Society, Columbus, OH, 1964, vol. I, Figs. 154-155, 246-247, 405-406.
6. D. Jembrih et al., Identification and Classification of Iridescent Glass Artifacts with XRF and SEM/EDX, Mikrochimica Acta 133 (2000) pp. 151-157.
7.D. Jembrih et al., Iridescent Art Nouveau Glass -IBA and XPS Characterization of Thin Iridescent Layers, Nuclear Instruments and Methods B 181 (2001) pp. 698-702.
8. M. Mäder et al., IBA of Iridescent Art Nouveau Glass- Comparative Studies, Nuclear Instruments and Methods in Physics Research Section B: Beam Interactions with Materials and Atoms, 239(1-2), (2005) pp. 107-113.

Mater. Res. Soc. Symp. Proc. Vol. 1047 © 2008 Materials Research Society   1047-Y07-03

Replication of Glazed Quartzite From Kerma, Capital of Ancient Kush (Sudan)

Lisa Ellis[1], Richard Newman[2], and Michael Barsanti[3]
[1]Conservation, Art Gallery of Ontario, 317 Dundas Street West, Toronto, M5T 1G4, Canada
[2]Scientific Research, Museum of Fine Arts, Boston, 465 Huntington Avenue, Boston, MA, 02115
[3]Ceramics, School of the Museum of Fine Arts, 230 The Fenway, Boston, MA, 02115

ABSTRACT

Glazes found on ancient Nubian quartzite sculpture were characterized in a previous study by scanning electron microscopy/energy-dispersive X-ray spectrometry (SEM/EDS). Now in the collection of the Museum of Fine Arts, Boston, these objects were excavated from Kerma, the capital of ancient Kush in the early 20th century by the joint Harvard University–Boston Museum of Fine Arts Expedition. The project presented here attempts to recreate the ancient technology used to glaze quartzite with compositions determined in the previous study. Raw and fritted experimental glazes were prepared, as well as an alkali paste mixed with a copper colorant. All of the samples were fired in modern electrical kilns. After firing, samples of the glazes and their quartzite substrates were examined with SEM/EDS to see which experimental glazes and firings most closely resembled the ancient samples.

INTRODUCTION

This study was designed to further understanding of an important collection of blue glazed, quartzite sculpture and fragments found at the Museum of Fine Arts, Boston (Figures 1 & 2). These unusual objects were excavated by the joint Harvard University–Boston Museum of Fine Arts Expedition at the site of Kerma, the capital of the ancient kingdom of Kush.

George Reisner, Director of the Harvard University–Boston Museum of Fine Arts Expedition, noted that the site of Ancient Kerma was strewn with green and blue glazed artifacts fashioned out of faience, steatite, quartzite, rock crystal and perhaps carnelian. Never in his extensive excavations in Egypt had Reisner ever seen blue glazed quartz and quartzite objects in such "number, size or workmanship" as at Kerma. These finds included surprisingly large sculpture as well as thousands of meticulously made smaller items: measuring around a quarter of a meter and now in Boston, a large fragment is all that remains from what is estimated to have been a meter long scorpion (Figure 2) along with approximately 3,800 blue glazed spherical-beads and a large quantity of small ring-beads that are speculated to have been sewn together into a net dress [1].

In "Egyptian Materials and Industries," first published in1934, Alfred Lucas attempted to link ancient manufacturing techniques used to produce glazes on steatite, faience, quartzite, quartz, and pottery. In his investigation of glazed stone from Egypt, he wrote simply that "Experiments…found that by strongly heating either potassium carbonate…or powdered natron, mixed with a small proportion of finely powdered malachite, on quartz pebbles, a beautiful blue glaze was obtained every time" [2].

Figure 1 (left), Fragment of Reclining Lion, 1700–1550 B.C., glazed quartzite, MFA 13.4229. Harvard University–Boston Museum of Fine Arts Expedition. **Figure 2.**(right), Scorpion fragment, 1700–1550 B.C., glazed quartzite, MFA 20.1666. Harvard University–Boston Museum of Fine Arts Expedition.

At the British Museum in the 1930s, Horace Beck and scientist Sir Herbert Jackson characterized outer layers of ancient blue glazed stone artifacts. Copper, sodium and a trace of calcium were identified spectroscopically in the composition of a bright blue glaze, which appeared black where it was thickly applied on a chert bead [3]. Like Lucas, Jackson and Beck also speculated that the silica in the glaze had come from the quartz on firing.

Beck's further examination of glazed quartz, milky quartz, quartzite and chert led him to distinguish between what he believed were two methods of glazing used in antiquity. He concluded that Egyptian specimens had a glaze applied in a slurry or as a mixed powder which was then fused to the surface with heat. Although some of the glazed stones studied from Mesopotamia looked like those from Egypt, there were others with an "extremely *high polish*" that Beck believed to have been prepared in a different manner. He speculated that these were first coated with an alkali and then heated until the alkali and silica had diffused into one another, or until the surface "flowed." Considerable amounts of soda were identified by Herbert on the surface of a similar object, a bead from Ur [4].

The more recent study by the current authors of blue glazed quartzite sculpture from Kerma at the Museum of Fine Arts, Boston [5] determined that the glazes found on these objects are largely inhomogeneous. Silica content ranged from ~ 43% to 85% of total normalized composition and copper levels were estimated to be from 2% to 24%. CaO was found to be a minor constituent from under 1% to about 2.5%. Assumed to be a contaminant in the copper oxide, small amounts of ZnO were found in many of the samples.

High alkali glazes in the samples were shown to have about 13% to 19% Na_2O and about 2% to 7% K_2O and low alkali glaze to have < 5% Na_2O and < 1% K_2O. Some cross-sections revealed blocky inclusions in what appeared to be an unweathered matrix of much lower alkali content. Analytical totals, however, for the matrix found it to be 7-9% lower than some of the inhomogeneous inclusions, indicating alteration had taken place in the matrix and suggesting that the analyses may not have accurately reflected the original composition. The compositional difference between the matrix and inclusions appeared to indicate that coarsely-ground, previously fired high alkali glaze had been mixed with a finely milled lower alkali mixture before application to the substrate and firing. Analysis of samples with high Na_2O contents also revealed that Na_2O and CuO content decreased from the top surface of the glaze to the substrate while SiO_2 content increased.

REPLICATION EXPERIMENTS

After characterizing the composition of the sculptural fragments, the authors of the 2004 study were interested in further investigating how these objects were made. It is clear that ancient manufacturing processes were not foolproof: many of the stone bodies appear to have been cracked or broken in firing, and the passage of time has resulted in losses of large amounts of glazes that did not fit well with the substrate. A small portion of the Kerma material, however, affirms that glazing large pieces of quartz and quartzite was not only possible but that the resulting artifacts could survive in good condition in the right burial environment.

Specific areas of interest to the authors included firing temperatures; glaze composition, particularly alkali content; adhesion of glazes to stone substrates; and the effect of high temperatures on quartzite and quartz. Replication experiments were undertaken with the help of the ceramic departments of both the School of the Museum of Fine Arts in Boston and the Ontario College of Art and Design in Toronto. Glaze ingredients were those regularly used in the modern ceramic studios. Potassium was supplied in the form of pearl ash, purchased from Sigma-Aldrich. Quartzite pebbles were picked up from the riverbanks of the Cheena River, Alaska, the beaches of Martha's Vineyard, Massachusetts, and the shore of Lake Ontario. These white and crystalline stones were examined with SEM/EDS and found to be composed of pure silica, according to the protocol described below [6].

Three types of glazes were created: the first, a raw glaze, was made up of pure, raw ingredients, mixed well with water and applied to the surfaces of the stone samples. Glazes were mixed according to recipes devised using Hyperglaze software [7] and were based on the compositions of two of the glazes characterized in the 2004 study, one with a high and the other with low alkali content (Table 1). The former had 17% sodium oxide and 3.2 % potassium oxide content, whereas the latter had 4.2% sodium oxide and 0.8% potassium oxide. These samples were fired at a range of temperatures from cone 019 to cone 10 [8] approximately 695°C to 1305°C respectively (Figure 3). Kilns turned off when maximum temperature was reached and the samples were, for the most part, allowed to cool slowly to room temperature in the unopened kilns without any special annealing or cooling cycle.

Figure 3. Experimental high and low alkali glazes on quartzite pebbles, fired between 695°C to 1305°C.

The second type, a fritted glaze, was prepared by firing the first two glaze formulations to cone 011 (around 894°C) and then grinding, with a mortar and pestle, the somewhat vitrified glaze. About 5% per weight alkali was added to the powdered material before firing for a second time to replenish alkali volatilized during the first firing. The powders were mixed with water to make slurries which were then applied to quartzite pebbles which were subsequently fired to cone 06 (about 999°C).

A third kind of glaze was devised from that described by Alfred Lucas, called here an "alkali paste." Two pastes were prepared. The first was made of a 4:1 mixture of potassium carbonate or pearl ash with copper carbonate. The second was a 2:2:1 mixture of sodium bicarbonate to sodium carbonate to copper carbonate. Provided on the internet by the Royal Ontario Museum [9], the combination of the soda carbonates is a pure substitute for the natron called for by Lucas, mixed here without halite. Water was added to the powders which were then applied to quartzite pebbles as well as quartz crystals. As the pastes dried, large crystals formed. After firing, the glazes were glossy and appeared to be well adhered to the quartzite substrate (Figures 4 & 5).

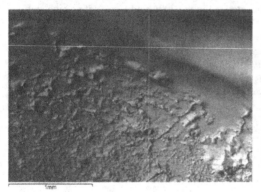

Figure 4 (above) showing soda alkali paste glaze fired at cone 011.
Figure 5, (left) back-scattered electron photomicrograph of same sample shows adhesion of glaze to quartzite and softening of surface topography.

At first glance, the sample made using the high alkali raw glaze formulation, fired to cone 010 or around 900°C, seemed to closely resemble the Kerma glazes. The glaze was glossy and appeared to be well adhered to the substrate. A cross-section of the glaze, examined by SEM/EDS [6], however, revealed that, unlike the Kerma glazes examined in 2004, it contained a significant amount of cristobalite, as, it turned out did all of the glaze samples fired at high temperature. A lower firing temperature of cone 019, about 700°C, was not sufficient to vitrify the high alkali glaze which looked only slightly glossier when fired at cone 016, about 790°C.

None of the samples made using the lower alkali raw glaze was successful. Not even a very high temperature firing at cone 10, about 1305°C, was sufficient to vitrify the glaze. Similarly, the fritted glazes, which appeared matte after firing, did not resemble Kerma glazes in the least. A cross section of the low alkali fritted glaze fired to 999°C examined under the SEM shows it to be unsatisfactory: many of the quartz particles have not vitrified, copper-rich regions seem to have separated from the glaze, and the glaze surface is very uneven.

Alternately, the soda-alkali paste glazes looked very much like the Kerma glazes after firing. Cross-sections of these glazes examined with SEM/EDS show a relatively homogenous glaze layer. Further EDS examination of the soda-alkali paste glaze demonstrated an inverse relationship between silica and copper content through the cross-section. Silica content was highest near the quartzite substrate at 67.1%, lessening as it approaches the surface copper oxide content where it was measured at 61.6%. At 17.3%, copper content was highest at the surface and diminished to 11.7% at the substrate. Sodium distribution seemed to remain constant throughout the sample, measuring between 20.9% to 21.2% without diffusion gradients. Cristobalite was also noted forming near the interface of the glaze with the stone substrate. When the surface of the quartzite was viewed under high magnification, it appeared that the interaction of the glaze and stone during firing had softened projecting features of the stone, almost dissolving them, as well as filling low areas.

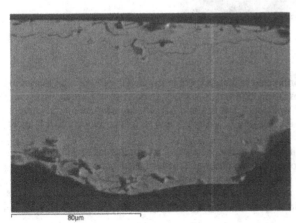

Figure 6, back-scattered electron photomicrograph showing cross-section of soda-alkali paste after firing at cone 011, around 894°C.

80µm

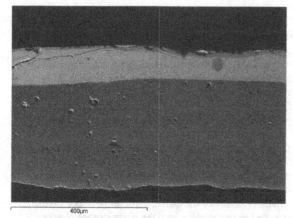

Figure 7, back-scattered electron photomicrograph of cross section from MFA 20.1184.

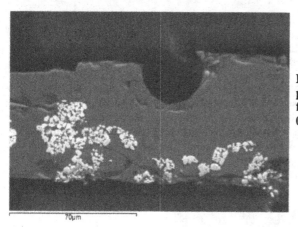

Figure 8, back-scattered electron photomicrograph of cross-section from soda-alkali paste, fired at cone 06 about 999°C.

DISCUSSION

Of the three glaze types formulated in these experiments, the glaze made with the soda-alkali paste was most like those sampled from the ancient artifacts from Kerma. Most surprising to the authors was the degree of diffusion of silica from the quartzite substrate throughout the glaze in the short firing to the relatively low temperature of cone 011, about 894°C.

The experimental results indicate that the low alkali glaze, prepared from either raw ingredients or as a frit, was not used in the glazing of the artifacts from Kerma. The high temperatures required to fuse such a glaze would probably not have been possible using ancient kilns [10]. Rather it appears that the low alkali content characterized in many of the ancient glazes is better explained by weathering and that a good portion of the material loss, as determined by analytical totals, was due to the loss of sodium and potassium oxides. Supporting this is the more or less consistent alkali content distribution throughout the thicknesses of glazes made using the soda-alkali pastes as compared to the distribution of alkali content noted in the ancient glaze samples. The latter was shown to increase from the stone substrate to the exterior of the glaze. While it was previously believed that this gradient developed during firing, it is now believed to be an effect of weathering.

Substantiated through the experiments was the alteration of the quartzite substrate during firing. The slightly melted surface noted on one of the samples in the 2004 study was no doubt caused by the interaction between the alkali in the glazing material and the quartzite substrate at high temperature.

Copper oxide crystals formed in glazes prepared with all three experimental methods. These are similar to those noted in the 2004 study in a sample from the large glazed lion in the MFA, Boston's collection. This indicates that the copper oxide crystals in the ancient samples were perhaps not the result of an intentionally added copper-based supplement, as suggested previously by the authors.

The replication experiments described here give some insight into the ancient manufacture of glazed quartzite sculpture in Kerma. Like the ancient glazes, the modern alkali paste samples share an excellent fit between stone substrate and glaze as well as softened or melted stone surfaces and the appearance of copper oxide crystals. Ancient phenomena not recreated in the modern samples include the blocky inclusions with higher alkali content as reported in 2004.

Table 1: Experimental Glazes

Glaze 1: High Alkali		Glaze 2: Low Alkali	
	% wt		% wt
Silica	53	Silica	74
Sodium Bicarbonate	20	Soda Ash	7.5
Pearl Ash	6	Pearl Ash	2
Copper Carbonate	21	Copper Oxide	16
Zinc Oxide	0.5	Alumina*	0.7
Total	100.5		100.2

*The alumina used in the experiment would in future experiments be replaced with sodium feldspar, which has a lower melting temperature.

ACKNOWLEDGMENTS

The authors would like to express their gratitude to Hannah Verlin, School of the Museum of Fine Arts, Boston; Pamela Vandiver, University of Arizona; Gordon Thompson, Ontario College of Art and Design; and Matthew Siegal, Lawrence Berman, Rita Freed, Susanne Gänsicke, Pamela Hatchfield and Abigail Hykin, all of the Museum of Fine Arts, Boston.

Members' comments at MRS Symposium Y, November 2007, especially those of Andrew Shortland are appreciated and acknowledged here.

1. George A. Reisner, *Excavations at Kerma*, Vol. VI, (Kraus Reprint Co., Millwood, N.Y., 1975) pp. 49-50, 153.
2. A. Lucas, *Ancient Egyptian Materials and Industries*, 4th Ed., revised and enlarged by J. R. Harris. (1964) pp. 172-4.
3. Horace C. Beck, *Ancient Egypt and the East: British School of Archaeology in Egypt*, Notes on Glazed Stones, Part II.-Glazed Quartz (June 1935) p. 23.
4. Beck, Ibid., 1935, p. 21.
5. Ellis and Newman (MRS 2004).
6. After firing, experimental glazed quartzite pebbles were sampled in one or more places, removing glaze and substrate where possible. Samples were mounted in Buehler Epothin epoxy resin and prepared as cross sections, which were coated with carbon for analysis. These were examined in a JEOL JSM-6460 LV scanning electron microscope with an Oxford Instruments 'INCA x-sight' energy-dispersive X-ray spectrometer (SEM/EDS) at 20 KeV. A variety of synthetic glass and pure element standards were used; matrix corrections were carried out using the XPP procedure. Two small glazed stone fragments were placed directly in the SEM in low-vacuum mode (chamber pressure 35 pascals). All samples were observed using the JEOL back-scattered electron (BSE) detector in 'shadow' mode, which combines the traditional BSE image with some topographical information.
7. HyperGlaze: Ceramics Software for Artists. Version 10.0.1. ©1989-2005 Richard Burket 6354 Lorca Dr, San Diego, CA 92115-5509. HyperGlaze@sbcglobal.net http//members.aol.com/hyperglaze Portions ©2000-2004 Runtime Revolution Limited. All Rights Reserved Worldwide.
8. Cone refers to the temperature reached in a kiln as gauged by the reaction of pyrometric cones manufactured by Orton, (see www.ortonceramics.com) when heated at 150°C/hour.
9. Recipe for natron is available on the Royal Ontario Museum's website, at http://www.rom.on.ca/schools/egypt/learn/mummyb.php.
10. See the following for maximum contemporaneous firing temperatures: Pamela B. Vandiver, C. Swann & D. Crammer, "A review of mid-second millennium BC Egyptian glass technology at Tell el-Amarna," *Materials Issues in Art and Archaeology II*, (MRS 1991), pp. 609-16; Paul Nicholson & Julian Henderson, *Ancient Egyptian Materials and Technology*, Ed. I. Shaw, (University of Cambridge, Cambridge, 2000) p. 200.

Methodology and Instrumentation

Mater. Res. Soc. Symp. Proc. Vol. 1047 © 2008 Materials Research Society   1047-Y04-04

Silver Nanoparticle Films as Sulfide Gas Sensors in Oddy Tests

Rui Chen[1], Laura Moussa[2], Hannah R. Morris[1], and Paul M. Whitmore[1]

[1]Art Conservation Research Center, Carnegie Mellon University, 700 Technology Drive, Pittsburgh, PA, 15219

[2]Department of Chemistry, Carnegie Mellon University, 4400 Fifth Avenue, Pittsburgh, PA, 15213

ABSTRACT

The preparation and performance of a silver nanoparticle-based sensor for use in Oddy tests are reported. A suspension of spherical silver nanoparticles (Ag NPs) (mean diameter of 30 nm, absorption of surface plasmon resonance (SPR) at 428 nm) in methanol was synthesized and the Ag NPs were assembled into monolayer films on glass slides, using polyethylenimine as a linking agent. UV-Vis spectrophotometry was employed to measure the SPR intensity of the Ag NP films in order to evaluate the extent of reaction. It was observed that the Ag NP films were quite stable under Oddy test conditions in a blank test, after a brief alteration of the spectrum due to particle dispersal, with no significant decrease in the SPR intensity after 1.5 months at 60°C and 100% RH. The sensitivity of Ag NP films to sulfide gases emitted from a test wool fabric in the Oddy test was investigated. UV-Vis spectra taken after the Oddy tests showed the disappearance of the Ag NP SPR peak and the growth of the UV absorption due to Ag_2S. Elemental analysis with energy dispersive x-ray spectroscopy confirmed that sulfur had been incorporated into the Ag NP film. Ag NP assemblies of lower NP density were created that indicated the presence of sulfide gases prior to significant tarnishing of a Ag foil. The results demonstrate that the Ag NP films can be used as sensitive, quantitative optical sensors to replace Ag foils in the Oddy test system.

INTRODUCTION

Air pollutants in a museum environment and gases emitted from storage and display materials can cause or accelerate the degradation of materials, particularly metals, used in works of art. For example, reduced sulfur gases such as hydrogen sulfide (H_2S) are known to be quite harmful to some artifacts because it creates tarnish on the surface of silver, black sulfides on copper alloy objects, and blackens white lead pigments [1]. A general and practical method widely applied in museums to evaluate the safety of storage and display materials is the Oddy test, which was first proposed by Andrew Oddy at the British Museum in 1973. The Oddy test is performed by enclosing a sample of the material of interest along with metal coupons (usually silver, copper, and lead) in a 100% RH environment. The test system is incubated at 60°C for 28 days, after which the coupons are visually examined for evidence of corrosion due to off-gassing emissions from the test material. The original Oddy test has undergone modification over the years. In order to decrease the variation between laboratories of assessments on identical materials and to reduce the test preparation time, standard procedures were devised and the experimental setup was simplified to the 3-in-1 test [2-4]. Meanwhile, because the visual evaluation on corrosion extent of metal coupons is subjective, other assessment methods such as measuring the weight gain of a lead coupon [5] and applying standard materials for assessment thresholds [6] have been proposed. More complex alternative tests have been suggested [7-9]; however, they have not yet been adopted in museum laboratories due to the specialized equipment and expertise required. Moreover, these tests are only sensitive to certain substances and not as useful as the Oddy test for indicating the presence and damage potential of a wide range of volatile compounds. Therefore, while very practical and useful, the Oddy test is time-consuming, and the extent of corrosion on the metal coupons is difficult to evaluate as well as not being easily quantifiable.

Developments in nanotechnology in recent years have inspired us to explore the potential of silver nanoparticle (Ag NP) assemblies as sensors to replace the silver coupon in the Oddy test. Nanometer-sized silver particles, having a high surface area, will react rapidly with the emitted pollutants (usually sulfur gases) and thus will have extraordinarily high sensitivity. Furthermore, Ag NPs exhibit a significant optical absorption in the visible region of the spectrum due to the surface plasmon resonance (SPR), which is a collective oscillation of the conduction electrons of Ag NPs in resonance with incident electromagnetic radiation. This property endows Ag NPs with a variety of colors depending on particle size and shape. The surface plasmon resonance is quite sensitive to the surface electrons of Ag NPs [10]. Its intensity will be attenuated as the surface electrons become bound electrons during a chemical reaction on Ag NPs. This change in the surface plasmon resonance will lead to color loss on Ag NPs and its assemblies, which is an indication that a chemical reaction has occurred. Thus, quantitative evaluation of the extent of reaction should be possible by measuring the changes in the visible absorption spectrum of the nanoparticles.

In this paper, we report the synthesis of a suspension of spherical Ag NPs in methanol and the fabrication of Ag NP assemblies. Subsequently, we used the Ag NP assemblies in the Oddy test with wool fabric as the test material, which is known to emit sulfide gases during its degradation. The performance of Ag NP assemblies as a quantitative, optical sensor for sulfide gases in the Oddy test system was evaluated.

EXPERIMENT

Materials

Water (HPLC) and sodium hydroxide (A.C.S. grade) were from Fisher Scientific. Absolute ethanol was from Pharmco Product Inc. Silver nitrate (99.9999%), sodium borohydride (99%), polyvinylpyrrolidone ($\overline{M_w}$ = 55,000), polyethylenimine (50 wt%) in H_2O ($\overline{M_w}$ = 750,000 by light-scattering), and methanol (HPLC) were from Sigma-Aldrich Co. Nitrogen (N_2) (low O_2 grade, concentration of $O_2 < 0.5$ppm) was supplied from Valley National Gases Inc.

Glass coverslips (No.2, 18 mm × 18 mm, from Corning) were used as a substrate. A staining jar (for 4 coverslips, 17 mm × 23 mm × 30 mm, from Ted Pella Inc.) was used for the cleaning procedure. Formvar/Carbon film copper grids (400 mesh) were from Electron Microscopy Science. Shell vials (4 ml), glass beakers (20 ml), and glass jars (125 ml) with a Teflon lined cap were from Fisher Scientific. Wool galoon fabric (N/500) was supplied from Testfabrics, Inc.

Synthesis of spherical Ag NP suspension in methanol

Under a flowing N_2 atmosphere, methanol (25 ml, HPLC) was injected into a 300 ml three-neck round bottom flask immersed into an ice bath. Polyvinylpyrrolidone (PVP) (1 ml, 66 mg) (0.595 mmol based on its unit) solution in methanol was quickly added to the flask while stirring (1200 rpm). The solution was allowed to mix for two minutes (min), and then AgNO₃ solution in methanol (2 ml, 6.8 mg, 20 mM) was added dropwise into the flask. After 10 min, freshly made NaBH₄ solution in ice-chilled methanol (20.2mg, 2.5ml, bubbles generated as soon as mixed) was injected dropwise into the system, and the solution immediately became light yellow, eventually turning brown-yellow. The solution was allowed to stir overnight in the dark.

A UV-Vis spectrum of the suspension usually showed a single peak, with peak wavelength around 420 nm. If a smaller peak wavelength was observed, the suspension was heated in an oil bath at 55 – 58°C for two days, while monitoring the peak wavelength of the absorption with a UV-Vis spectrophotometer. The suspension was centrifuged three times at 15300 rpm for 30 min each time in order to remove excess boric compounds and PVP. The particle sediment was isolated and dispersed into fresh methanol (20 ml), and the resulting Ag NP suspension was yellow-brown in color.

Assembly of Ag NP monolayer film on a glass substrate

A well-cleaned glass substrate is the key to successfully generate a monolayer assembly of Ag NPs. An alkaline washing procedure was adopted here in order to obtain reproducible assembled films. The alkaline solution (NaOH / ethanol) was made by adding NaOH pellets (7 g) into stirred water (28 ml, HPLC), adding ethanol (42 ml) after the NaOH had completely dissolved and allowing the mixture to stir for 30 min.

Glass coverslips were first cleaned using sparkleen™ soapy water with cotton swabs and rinsed thoroughly with distilled water. Then the glass coverslips were placed in a staining jar filled with NaOH / ethanol solution, followed by shaking for 2 hours in a wrist action shaker. The NaOH / ethanol was then poured out and the coverslips were rinsed twice with water, followed by ultrasonicating with HPLC water for 20 min at least four times. At the end, the used water tested neutral with pH paper. The cleaned glass coverslips had to be used within 2 weeks. For reproducible results, the coverslips were ultrasonicated in HPLC water for 20 min immediately before making a film assembly.

The Ag NP monolayer assembly was created by soaking a clean glass coverslip in 1% aqueous polyethylenimine (PEI) solution for 30 min, washing the PEI-coated coverslip in HPLC water for 30 min, rinsing it with water and subsequently with methanol to remove the water, then immersing the PEI-coated coverslip into the spherical Ag NP suspension in methanol for a certain time. After it was taken out of the suspension, the assembly was washed with copious amounts of methanol and then water. Excess water was removed from the Ag NP assembled film by gently touching it with a Kimwipe® or filter paper.

The Oddy test systems of wool fabric

A conventional Oddy test system was set up in a sealable glass jar (125 ml), with a piece of wool fabric (1 cm × 1 cm) in a glass shell vial (4 ml) and a glass beaker (20 ml) with a piece of Ag foil (5 mm × 15 mm) hung at the edge. The shell vial and the beaker were then placed side by side in the jar containing about 10 ml distilled water (Figure 1a). The jar was sealed and placed in an oven at 60°C.

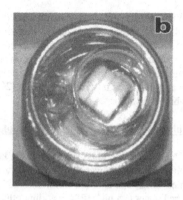

Figure 1. (a) The setup of a conventional Oddy test. (b) Top view of the setup of an Oddy test with a Ag NP assembly.

This same basic setup was used when Ag NP assembled films were substituted for the Ag foil in the Oddy test. A custom-made Teflon sample holder (a square block, 20 mm × 20 mm × 8 mm, with three slits cut across) was placed in the 20 ml glass beaker, and a Ag NP assembly was placed on the sample holder, standing vertically in the slit (Figure 1b). The Oddy test system was sealed and placed in the oven at 60°C.

<u>Instrumentation</u>

Optical spectra of the samples were recorded on a Perkin-Elmer Lambda 800 UV-Vis spectrophotometer operated in transmission mode. For the Ag NP suspension, methanol was used as background. For Ag NP assemblies, a PEI-treated glass coverslip was used for background correction of the optical spectra.

The transmission electron microscopy (TEM) analysis was performed on Ag NPs that had been assembled onto formvar/carbon films on copper grids using the same procedure as was used to make an assembly on glass coverclips. The samples were observed at 75 kV on a Hitachi H-7100 TEM. Digital images were obtained using an AMT Advantage 10 CCD Camera System and NIH Image software. Size and size distribution of Ag NPs were determined by measuring the Feret's diameter of particles on the TEM images.

Elemental analysis of the Ag NP assemblies on glass substrates was performed on a Phillips XL30 FEG scanning electron microscope equipped with an Oxford Instruments ISIS 300™ energy dispersive x-ray spectrometer, which was operated at 20 kV with a collection time of 100 s. The initial Ag NP assemblies and the fully reacted assemblies in the Oddy test of wool fabric were analyzed.

RESULTS AND DISCUSSION

<u>The Ag NP suspension and assembled Ag NP monolayer films</u>

As the size of Ag particles decreases to the nanometer level, the Ag NPs exhibit a significant optical absorption in the visible wavelengths due to surface plasmon resonance. The

SPR absorptions of the Ag NP suspension collected with a UV-Vis spectrophotometer is shown in Figure 2a. The spherical Ag NPs in methanol have a predominant SPR absorption at a peak wavelength of 428 nm. A slight shoulder at 350 nm indicates that the particles are faceted spheres, which is in agreement with the TEM images of the Ag NPs (Figure 2b). It is the predominant SPR peak at 428 nm that endows a yellow-brown color to the synthesized Ag NPs.

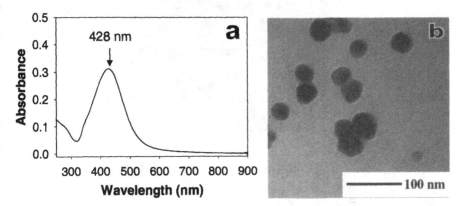

Figure 2. (a) UV-Vis spectrum of the Ag NP suspension in methanol. (b) TEM image of the Ag NPs.

Ag NPs were readily fabricated into films on glass substrates through the present assembling procedure. In order to examine the morphology of these assemblies, the Ag NPs were assembled on formvar/carbon film copper grids and characterized with TEM. Figure 3a, a TEM image, displays a relatively uniform particle distribution on the substrate and the assembly is predominantly a monolayer of isolated particles, with only a few clusters indicating occasional particle stacking.

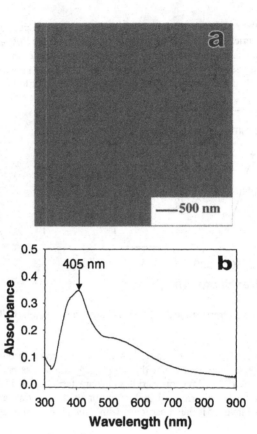

Figure 3. (a) TEM image and (b) UV-Vis spectrum of an assembled monolayer film of Ag NPs.

The Ag NP monolayer assemblies exhibited the SPR absorptions similar to those of the Ag NP suspension, with the peak wavelength blue-shifted by 23 nm relative to the suspension (Figure 3b). The shoulder absorption at around 520 nm may arise from the particle stacking as shown in the TEM image. The origin of shift in peak wavelength is not entirely understood, although it has been attributed to a change in refractive index of the surrounding medium of Ag NPs from methanol to air [11].

Stability study of Ag NP films under Oddy test conditions

The stability of the Ag NP assemblies when maintained under the Oddy test conditions (60°C, 100% RH) without the material of interest (for example, wool fabric) was investigated. The absorbance of their SPR absorptions was monitored with a UV-Vis spectrophotometer over a period of 46 days.

The Oddy test conditions seemed to alter the spectral profile in a way that suggests better particle dispersal over the Ag NP assemblies. Figure 4a shows the Ag NP assembly kept under the humidity and temperature treatment. The initial shoulder peak at 520 nm, which was assigned to particle stacking, vanished within 7 days, and the intensity of the predominant peak increased, with the peak wavelength shifting to 408 nm. The increase in the intensity of the predominant SPR and the disappearance of the shoulder at 520 nm suggest that the particle clusters may break down to create a more dispersed particle distribution during the thermal treatment. The humidity may also facilitate the movement of the Ag NPs, as water is absorbed on the particle surfaces probably on the PEI linker.

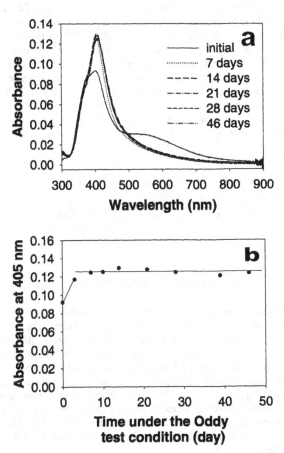

Figure 4. Stability study on an assembled monolayer film of Ag NPs under the Oddy test conditions: (a) its UV-Vis spectra for the varied time periods, and (b) plot of absorbance of the SPR peak at 405 nm versus the time periods.

Following the rapid increase of the SPR intensity during the first week of the thermal treatment, the optical spectrum of the assembly remained relatively constant for over a month (Figure 4b). This indicates that the Ag NP assemblies have good stability under the Oddy test conditions and can be used to detect sulfide gases in the Oddy tests. A decrease in intensity of the predominant SPR absorption should reveal that a reaction has taken place between Ag NPs and the sulfide gas emissions from the material of interest.

Ag NP assemblies as optical sensor in the Oddy test of wool fabric

To compare the performance of the new Ag NP sensor to that of the conventional Ag foil sensor, a conventional Oddy test was carried out with a piece of wool fabric (1 cm × 1 cm) enclosed with a piece of Ag foil. Wool fiber contains about 10% sulfur-containing amino acid cystine, in which the disulfide bridge can be dissociated and further react with water to emit H_2S [12]. The tarnishing of Ag by sulfide gases was visually evaluated and is shown in Figure 5. The Ag foil started to become visibly tarnished by sulfide gases emitted from the wool sample at day 3 of the Oddy test, showing a pinkish hue on the surface. Further reaction changed the pinkish hue gradually to a blue hue at days 4 and 5. However, further exposure of Ag foils did not significantly change their tarnished appearance. As has been commonly found, it was not easy to quantitatively evaluate the extent of tarnishing of the Ag foils during the test.

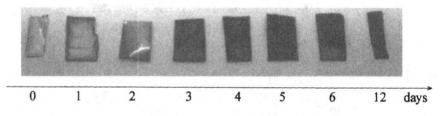

Figure 5. The tarnishing of Ag foils in the Oddy test of wool fabric.

The Ag NP monolayer films were substituted for the Ag foil and used to detect sulfide gas emissions in the Oddy test of wool fabric. The extent of the reaction between Ag NPs and sulfide gases can be quantified by measuring the optical absorption of SPR peaks of the assemblies with UV-Vis spectroscopy. As shown in Figure 6, a decrease in the absorbance of the SPR peaks at day 3 indicates that a reaction with sulfide gases had taken place on Ag NPs. With further reaction, the intensity of the SPR peaks of Ag NPs steadily decreased. At the completion of the reaction, the SPR peaks of Ag NPs in the visible region had disappeared and a shallow, broad absorption that monotonically increased into the ultraviolet region remained. The final optical spectrum is in good agreement with the optical spectrum of Ag_2S [13, 14], indicating that the free electrons of Ag NPs had turned into the bound electrons of Ag_2S NPs, which led to the disappearance of the SPR peaks.

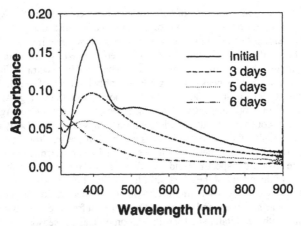

Figure 6. UV-Vis spectra of a Ag NP assembly during the exposure to sulfide gases in the Oddy test of a wool fabric sample.

To verify that this silver sulfidation had occurred concurrently with disappearance of the SPR peaks of the Ag NPs, elemental analysis of the exposed assemblies was performed with energy dispersive x-ray (EDX) spectrometer attached on a scanning electron microscope. Compared with the unexposed Ag assembly (Figure 7a), the EDX spectrum in Figure 7b clearly showed the sulfur (S) K_{α} peak, which confirmed that the sulfide gases had reacted with the Ag NPs during the exposure to the emissions from the wool sample of the Oddy test. The atomic ratios of Ag to S could not be accurately determined here due to the weak signals obtained from the nanometer-thick monolayer assemblies (the thickness of a monolayer is equal to the diameter of Ag NPs, about 30 nm).

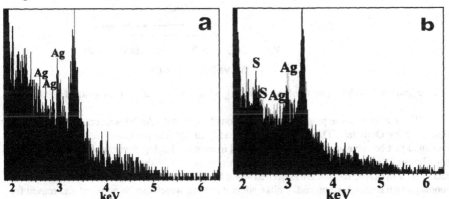

Figure 7. EDX spectra of (a) an unexposed Ag NP assembly and (b) an exposed Ag NP assembly to the Oddy test of wool fabric.

By varying the immersion time of a PEI-treated glass coverslip in the Ag NP suspension or changing the concentration of the Ag NP suspension, we are able to adjust the NP density of the assembly. Therefore, Ag NP films of lower NP density that could indicate sulfide gases in the Oddy test prior to significant tarnishing of a Ag foil, were created by immersing the PEI-treated glass coverslips in the Ag NP suspension for 10 min. The UV-Vis spectra in Figure 8 showed the complete extinction of the SPR peaks of a Ag NP assembly by day 3 of the Oddy test of the wool sample, when the Ag foil was just beginning to show perceptible tarnishing. Along with this significant change in the optical spectra of the assemblies, the assemblies of Ag NPs also lost their yellow color, becoming virtually colorless within 3 days during the Oddy test of the wool sample. This indicates that the Ag NP assemblies are much more sensitive to sulfide emissions than the Ag foil. Meanwhile, the thermally-induced particle rearrangement of Ag NPs, which rapidly altered the absorption spectrum and slightly increased the SPR intensity, did interfere with the detection of the extent of the silver-sulfide reaction. Future production of Ag NP assemblies for use as gas sensors should probably include a thermal conditioning step, in order to stabilize the particle arrangement and the optical spectrum before use. It is evident that the Ag NP assemblies are both quantitative and sensitive as an optical gas sensor for sulfide gases, and they should be considered alternatives for the detection of these gases in the Oddy tests.

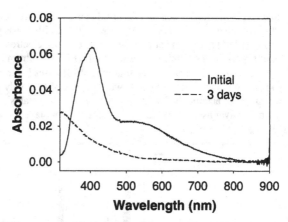

Figure 8. UV-Vis spectra of the initial Ag NP assembly and its fully reacted assembly.

There remain some practical considerations before these Ag NP sensors can replace silver coupons in the Oddy test. These devices react readily to ambient pollutants, so they must either be created just before use or they must be stored in inert packaging environments until they are used. Tests of the Ag NP sensors described in this study show a shelf life of at least four months when stored in plastic bags. Care is needed when handling the Ag NP films because the monolayer can be easily damaged. Unlike silver coupons, which can be polished and reused for successive tests, these Ag NP devices are single-use indicators. Consequently, the economics of production must be considered if the manufacture and quality control are to become commercially viable.

CONCLUSIONS

Ag NP monolayer films fabricated on glass substrates with polyethylenimine were used in the Oddy test of a wool fabric sample. The SPR absorptions of the Ag NP assemblies impart a yellow color to the assemblies from a predominant SPR absorption, and this feature can be monitored for quantitative evaluation of the extent of reaction. Compared with the conventional Ag foil, the performance of the Ag NP assemblies demonstrated their advantages as a sulfide gas sensor in the Oddy tests. The Ag NP assemblies can be made to be much more sensitive to sulfide gases, with the assemblies turning colorless within 3 days during an Oddy test, prior to significant tarnishing of the Ag foil. The extent of the reaction between Ag NPs and sulfide gases can be quantified by measuring the intensity of SPR absorption in the optical spectra of the Ag NP assemblies. For the fully reacted films, the SPR absorptions vanish and the optical absorption of Ag_2S NPs remains. In summary, the assembled Ag NP monolayer films can be used as a sensitive, quantitative optical sensor to substitute for Ag foils in the Oddy tests.

ACKNOWLEDGMENTS

This research work was supported by a grant from the Andrew W. Mellon Foundation.

REFERENCES

1. S. Bradley, J. Am. Inst. Conservat. 44(3), 159 (2005).
2. L.R. Green and D. Thickett, Stud. Conservat. 40(3), 145 (1995).
3. J.A. Bamberger, E.G. Howe and G. Wheeler, Stud. Conservat. 44(2), 86 (1999).
4. L. Robinet and D. Thickett, Stud. Conservat. 48(4), 263 (2003).
5. H. Berndt, J. Am. Inst. Conservat. 29(2), 207 (1990).
6. B. Pretzel and N. Shibayama, Conservation Journal (Victoria and Albert Museum, London), 43 (2003).
7. V. Daniels and S. Ward, Stud. Conservat. 27(2), 58 (1982).
8. J. Zhang, D. Thickett and L. Green, J. Am. Inst. Conservat. 33(1), 47 (1994).
9. C.L. Reedy, R.A. Corbett and M. Burke, Stud. Conservat. 43(3), 183 (1998).
10. J.-E. Park, T. Momma and T. Osaka, Electrochim. Acta 52(19), 5914 (2007).
11. C. Xue, Z. Li and C.A. Mirkin, Small 1(5), 513 (2005).
12. Á. Tímár-Balázsy and D. Eastop, Chemical Principles of Textile Conservation, (Butterworth-Heinemann, Oxford, 2002) p. 52.
13. J. Zhang, Z.L. Wang, J. Liu, S. Chen and G.Y. Liu, Self-Assembled Nanostructures, (Kluwer Academic, New York, 2002) p. 201.
14. K. Akamatsu, S. Takei, M. Mizuhata, A. Kajinami, S. Deki, S. Takeoka, M. Fuji, S. Hayashi and K. Yamamoto, Thin Solid Films 359(1), 55 (2000).

Mater. Res. Soc. Symp. Proc. Vol. 1047 © 2008 Materials Research Society   1047-Y06-07

The Grolier Codex: A Non Destructive Study of a Possible Maya Document Using Imaging and Ion Beam Techniques

Jose Luis Ruvalcaba[1], Sandra Zetina[2], Helena Calvo del Castillo[1], Elsa Arroyo[2], Eumelia Hernández[2], Marie Van der Meeren[3], and Laura Sotelo[4]

[1]Instituto de Física, Universidad Nacional Autónoma de México, Apdo. postal 20-364, Mexico DF, 01000, Mexico

[2]Instituto de Investigaciones Estéticas, Universidad Nacional Autónoma de México, Mexico DF, Mexico

[3]Coordinacion Nacional de Conservacion del Patrimonio Cultural, Instituto Nacional de Antropología e Historia, Mexico DF, Mexico

[4]Centro de Estudios Mayas, Instituto de Investigaciones Filológicas, Universidad Nacional Autónoma de México, Mexico DF, Mexico

ABSTRACT

The Grolier Codex has been a controversial document ever since its late discovery in 1965. Because of its rare iconographical content and its unknown origin, specialists are not keen to assure its authenticity that would set it amongst the other three known Maya codes in the world (Dresden, Paris Codex and Madrid Codex).

The document that has been kept in the Museo Nacional de Antropología in Mexico City, after its exposure in 1971 at the Grolier Club of New York, has been analyzed by a set of non-destructive techniques in order to characterize its materials including paper fibers, preparation layer and color compositions. The methodology included UV imaging, IR reflectography and optic microscopy examinations as well as Particle Induced X-ray Emission (PIXE) and Rutherford Backscattering Spectrometry (RBS) using an external beam setup for elemental analysis. All the measurements were carried out at 3MV Pelletron Accelerator of the Instituto de Física, UNAM. The aim of this work is to verify if the materials in the Grolier Codex match those found in other pre-Hispanic documents.

From the elemental composition we concluded that the preparation layer shows the presence of gypsum ($CaSO_4$), color red is due to red hematite (Fe_2O_3) and black is a carbon-based ink. These results agree with previous analyses carried out by Scanning Electron Microscopy (SEM-EDX) on few samples. However, the presence of Maya Blue in the blue pigment cannot be assured. The examination using UV and IR lights shows homogeneity in the inks and red color but dark areas that contain higher amounts of K in the preparation layer. This paper discusses the results obtained for the UV-IR examinations and the elemental analysis. A comparison with other studies on pre-Hispanic and early colonial codex is presented.

INTRODUCTION

There are sixteen codices from pre-Hispanic Mexico, only three of which come from the Maya area: the Madrid Codex, the Dresden Codex and the Paris Codex. If the controversial Grolier Codex is authentic, it would be the fourth Maya pre-Hispanic document known to this date.

The Grolier Codex discovery was strange; it is the only pre-Hispanic codex found in the

20th century, with the exception of a few archaeological fragments. The Mexican collector José Sáenz bought the manuscript in 1964. It was supposed to have been found in a dry cave, in Chiapas [1]. In 1971 Michael Coe presented the document at the Grolier Club in New York. Two years after its presentation, a facsimile was published with an iconographic study identifying the manuscript as a Post-classic Maya Venus calendar with Toltec features [2].

The painted section of the Grolier Codex (Figure 1), also known as Saenz Codex, consists in a 125 cm long strip of bark paper screen folded in 11 pages: both sides are prepared with a white layer, but only one is painted. Each page has a maximum of 19 cm height and 12.5 cm length, but the dimensions of the support vary substantially because of the losses. Three paper fragments, unpainted and without preparation, are associated with the manuscript, one of them has the remains of a red line.

Figure 1. The Grolier Codex

Over the white, thick and uniform preparation layer figures of glyphs, gods, priests and warriors with black and red lines are depicted. A few areas are filled with plain colors: brown, red and black, only page 11 has a pale blue-green color. A lot of preparatory drawings in brown, black or red washes that consistently differ in design consistently from the final outline can be seen.

There are many reasons to question the validity of this codex, apart from its recent discovery. Some scholars, [2] attributed the Aztec resemblance to a Maya-Toltec style, and noted that some renowned Maya codices were not completely painted, even though they were prepared, because it seems that the priests over-painted them. Others think that the combination of Central Mexican and Maya iconography in the context of a Venus calendar that repeats some of the Dresden Codex images in an inconsistent reading is probably the result of a falsification [3, 4].

Though radio-carbon dating of a free-standing sheet of paper placed it at AD (1230 ± 70), when Maya culture was receiving strong Toltec influences, detractors insist that despite the fact that the paper is antique, the painting might be the work of an experienced forger that has had access to the other three codices, particularly to the Dresden Codex. Although some scientific analyses (SEM-EDX, FTIR) have been practiced on some samples [5], a comprehensive study of the whole document has never been done. In this work, a general examination of this codex has been carried out using non destructive techniques such as ultraviolet (UV) imaging, infrared (IR) reflectography and optic microscopy examinations as well as Particle Induced X-ray Emission (PIXE) and Rutherford Backscattering Spectrometry (RBS) using an external beam setup for elemental analysis.

EXPERIMENT

The methodology proposed consists of a global analysis of the object without taking samples. The initial stage involved imaging techniques for a general examination of the codex, then characteristic UV and IR images can be registered and related to each part of the document. A lead sulfide Hamamatsu Vidicon tube camera was employed to perform IR reflectography with IR LED lighting (940 nm). The entire document was additionally registered with UV imaging of long wavelength (365nm). In a second stage, a detailed technical examination was made with a stereomicroscope. Finally for elemental analysis, about 60 spots in representative pages were analyzed with Particle Induced X-ray Emission (PIXE) and Rutherford Backscattering Spectrometry (RBS) using an external beam set-up. PIXE and RBS in-air measurements were carried out at the 3MV Pelletron Particle Accelerator of the Instituto de Física (UNAM) using a 3 MeV proton beam of 1 nA and 1 mm in diameter. The pages were placed on a PVC stand rotated 60° from the horizontal plane (Figure 2) and protected with PVC sheets on top that left only the analyzed areas uncovered.

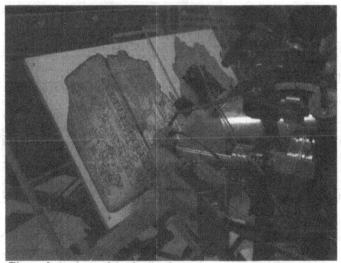

Figure 2. Analysis of the Grolier Codex by our external beam set-up.

Pages analyzed were 1, 2, 3, 5, 6, 8, 9 and 11. Measurements were taken in different points including fiber, preparation layer and pigments. Two detectors were used in the PIXE set-up; a Si-Pin for the detection of light elements and a LEGe for the trace elements detection. For RBS, a particle detector (Ortec R series) was used, placed at 45° to the beam direction. Reference materials of NIST SRM 2704, SRM 2711 and pure SiO_2 were used for calibration of PIXE and RBS detectors.

RESULTS AND DISCUSSION

Elemental composition by PIXE and RBS

The preparation layer is made of $CaSO_4$ (14.9% S, 49% Ca, 32.4% O) with some amount of strontium (2%). On the other hand, the PIXE spectra of red pigments showed a great amount of iron (average 23.9 %), together with the elements present in the preparation layer (Ca, S, O, Sr). Also, some magnesium, aluminum, silicon, titanium and manganese could be found. Mexican pre-Hispanic red pigments were made from either an organic colorant (carminic acid) or from inorganic pigments. No mercury or lead has been found; the presence of iron and elements that are commonly found in soils, indicate the use of natural red ochre, also called hematite red (Fe_2O_3 plus soil material such as clay) [Table I]. The black lines showed high amounts of carbon in the RBS spectra.

The traces of preparatory drawing, a thin red, brown or black wash does not contain an important iron presence. From Al/Fe ratios, the iron concentration in them resembles more that of the preparation layer than the one in red pigments [10].

Table I. Elemental concentrations determined by PIXE (%) for the pigments present in the Grolier Codex. Uncertainties are ±10%.

Color/Inks		Elements in preparation layer				Elements attributed to the pigments and inks									
		O	Ca	S	Sr	Mg	Al	Si	K	Cl	Ti	Mn	Fe	Cu	Zr
Black	P1	33.3	49.3	13.3	1.14	0.32	0.25	0.59	1.02	-	-	-	0.67	0.082	-
	P2	30.3	51.0	14.7	1.97	-	0.18	0.59	0.65	-	-	-	0.53	0.084	-
	P6	33.5	48.2	15.1	1.33	-	0.27	0.48	0.56	-	-	-	0.35	0.190	-
	P11	31.7	49.5	15.0	1.36	0.12	0.20	0.50	0.98	-	-	-	0.54	0.064	-
Red	P1	36.3	37.8	9.3	1.10	-	0.65	0.87	1.09	-	0.084	0.085	12.7	0.067	-
	P2	30.6	26.9	3.62	1.19	0.17	0.54	1.15	0.67	-	-	0.129	35.0	0.048	-
	P6	49.3	26.8	9.7	0.86	0.16	0.55	0.97	1.28	0.005	0.067	0.030	10.1	0.120	-
	P8	55.5	12.0	2.11	0.65	-	0.43	0.88	0.21	-	0.089	0.089	28.0	0.049	-
	P11	27.1	22.8	3.66	0.76	0.05	1.07	2.23	0.69	0.034	0.187	0.061	33.8	0.041	-
Previous drawing	P2	36.6	31.9	26.0	2.39	-	0.49	0.98	0.79	-	-	-	0.76	0.120	-
	P6	40.8	27.9	5.24	0.76	-	1.12	1.71	0.40	-	0.146	0.08	21.7	0.095	-
	P8	27.9	51.7	14.6	2.40	0.07	0.19	0.64	1.42	0.018	-	-	0.86	0.243	-
Blue	P11	29.0	54.6	8.15	1.38	0.87	0.52	2.70	0.98	0.049	-	0.014	1.52	0.26	0.012

Blue pigments in the Maya culture are usually "Maya Blue" pigments [6, 7, 8]. Maya blue consists of colorant (indigo) fixed on palygorskite or other similar clay. Indigo is an organic colorant extracted from the *Indigofera suffruticosa* plant and palygorskite is an ino-phyllosilicate that belongs to the sepiolite family. The Maya Blue appeared around the 8th century and was used up to 1580 [9].

While Indigo – being an organic colorant – cannot be identified with PIXE, elements present in palygorskite are possible to detect with this technique. The PIXE spectra of the blue shade show in fact a composition that would match that of palygorskite: Mg, Al, Si, K, Cl, Mn, Fe, Cu, Zn. However, as PIXE does not provide information of the compound but only of the elements present in it, and there are other materials also found in Mexican artifacts that should be considered as well, it is not possible to assert the presence of Maya Blue. A comparison with the analysis of two Maya blue mural painting fragments from Calakmul and Tulum sites did not show a good correspondence with the expected elemental profile of palygorskite and sepiolite clays. Though the presence of these clays cannot be certain we are able to conclude that no modern synthetic pigments have been found in the blue paint.

Infrared reflectography

The red color has a gray (middle absorbance) in IR reflectography, a common behavior of iron oxide earth pigments, and the black has a strong absorbance, usually seen in carbon black (Figure 3). It was observed with this technique that all the preparatory drawing lines are not IR absorbent so they are not seen (even though they are drawn in three different colors: red, black and brown) which means that the materials used either were organic, or were applied in a very low concentration and do not contain carbon black. In the IR reflectography imaging the brown degradation stains in the edges totally disappear.

UV imaging

The white preparation layer presents a slightly lilac tone under UV lighting, commonly seen in gypsum, which agrees the PIXE-RBS identification. All the painting lines have a strong purple response to UV (Figure 3).

Figure 3. Grolier Codex, page 7. UV lighting, visible light and IR reflectography images.

The brown stains on the edges had an unexpected behavior under UV: A strong fluorescence that turns from orange to dark violet. In a detailed examination, these stains do not permeate the surface, they have a halo effect as if two or three subsequent drops of dye or ink were carefully applied on top. Besides the borders are very well defined and in the spots where the preparation layer has been lost, the area beneath is unaffected. The degradation stains appear like coat of ink (Figure 4). PIXE analysis indicated that in these regions the amount of K increases and it can be related to the stains.

Figure 4. Halo effect, detail, UV lighting.

Microscopy technical examination of degradation

The fibers in the paper support are jointed in vertically oriented bundles, while the common direction observed in other Maya codices (Dresden and Madrid) is horizontal; the paper is also thinner. The pages without painting have a crossed pattern of bundles and are composed of only one layer of filaments, while in the painted sections, at least two superimposed layers can be detected, so it is possible that they are not the same kind of paper. There have not yet been any fiber tests to verify the species used in the manufacture of the paper.

The preservation of the manuscript is heterogeneous, all the edges are lost, eroded and stained, in contrast, the central parts are very well preserved, and the colors are unexpectedly bright. No single page is complete; all have lost between 10 to 50% of the paper. Some edges are too sharp, as if they were deliberately cut. In the microscope the fibers of the paper and the thick gypsum preparation layer have a clear and straight incision or a sudden disruption (Figure 5). In some cases the red lines used in the original design are painted over losses of gypsum preparation and painted the paper fibers.

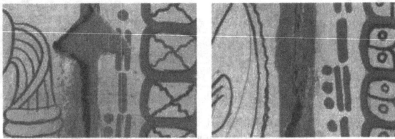

Figure 5. Sharp cut edges and red painting over losses.

The yellowish brown stains at first sight show degradation due to moisture appearance: pages 9, 10, 11 are entirely tainted. As it has been noted, under UV the stains look like an intentional alteration, showing a pattern of applied inks. It should also be considered that if the stains were indeed the product of moisture, the liquid should have affected more of the gypsum coat and the pictorial layers, because of the water reactivity of gypsum.

On the darker areas of the codex some regions contain clumps of dust and diverse materials adhered to the surface. Some sections present cracks and loses in the preparation layer, not always related to a natural cause of degradation (moisture, folds, etc).

CONCLUSIONS

Results obtained through external PIXE and RBS analyses indicate that no modern inorganic materials are present. The preparation layer consists of gypsum ($CaSO_4$). The presence of carbon-based black has been established by the use of RBS. Red pigments are made of red ochre (Fe_2O_3) and blue shades show some clay composition.

Although a gypsum composition preparation layer has been found in Colombino Codex, a Mixtec pre-Hispanic document [10], it is known that in the Madrid and Dresden codices the preparation composition is calcite ($CaCO_3$). This is the only material analysis that has been performed on those documents, so it is quite difficult to compare the results obtained with the results in this study. On the other hand, red ochre has not been observed in most of the codex already examined in our researches, (e.g. Colombino, de la Cruz Badiano, Azoyu [12]) a red organic colorant was used instead. More pre-Hispanic codices must be studied in order to establish the patterns of use of materials in codex manufacture and writing.

From what can be achieved with PIXE and RBS techniques, the Grolier Codex contains materials used in pre-Hispanic times, although the composition of the blue pigment could not be definitely established. Further analysis needs to be done for the identification of this blue pigment and also on organic materials.

The most unexpected features in the Grolier Codex are the degradation patterns. The stains and cut edges of the losses seem like an induced degradation, the UV lighting examination and the microscopy observation led to question the nature of the deterioration process. The irruption of the design painting lines over the degradation would not be easily explained if the document was indeed produced in the 13th century and then eroded and degraded after the moment of its production. The identification of the organic compounds in the codex seems crucial to establish the origin of the stains and in consequence the possibility to find out if they are a natural process or a forgery.

Although we are a bit closer to the determination of its authenticity, other factors must be considered, such as the iconographic content and the historical context. Materials analysis is just one of the methods that bring to light new questions.

ACKNOWLEDGMENTS

Authors would like to thank technicians K. López and F. Jaimes for their support at the Pelletron particle accelerator during PIXE-RBS measurements. Financial support was by projects MEC MAT2002-180, UNAM-DGAPA-PAPIIT IN403302, CYTED Proy.VIII.12, and CONACyT Mexico grant U49839-R.

REFERENCES

1. J.Alcina Franch, 1992. *Códices Mexicanos*, MAPFRE, Madrid, 219-220.
2. M.D. Coe, 1973. *The Maya Scribe and His World*, The Grolier Club, Nueva York.
3. E. Thomson, 1972. *A Commentary on the Dresden Codex: A Maya Hierogplyphic Book*, American Philosophical Society, Philadelphia.
4. C. F. Baudez, 2002. *Arqueología Mexicana*, Vol. X, num. 55, 70-79.
5. V. Rodríguez-Lugo, D. Mendoza-Anaya, L. E. Sotelo, Microstructural Study of the Grolier Codex by Means of LV-SEM, *Acta Microscópica*, October, (2001), 252-253,
6. M. Sánchez del Río, P. Martinetto, A.Somgyi, C.Reyes-Valerio, E. Dooryheé, N. Peltier, L Alianelli, B. Moignard, L. Pichon, T. Calligaro, J.C. Dran, *Spectrochimica Acta* Part B, (2004) 1619-1625.
7. M. Sánchez del Río, A. Sodo, S.G. Eeckhout, T. Neisius, P. Martinetto, E. Dooryhée, C. Reyes-Valerio; *Nuclear Instruments and Methods* B, 238 (2005) 50-54.
8. M. Sánchez del Río, P. Martinetto, C. Reyes-Valerio, E. Dooryhée, M. Suárez; *Archaeometry* (2006) 115-130.
9. M. Matteini, A. Moles, 2003. *La Chimica nel Restauro, I materiali dell'arte pittorica*, Nardini Editore, Firenze.
10. R. C. González Tirado, Masters Thesis, Monfort University, 1998.
11. C. López Binnqüist, PhD Thesis, University of Twente, Netherlands. Twente University press, Enschede. 2003.
12. J.L Ruvalcaba and C. González Tirado 2005. *Análisis in situ de documentos históricos mediante un sistema portátil de XRF* in La Ciencia de Materiales y su Impacto en la Arqueología. Vol II, Academia Mexicana de Ciencia de Materiales A.C. D. Mendoza, J. Arenas y V. Rodríguez coord., Ed. Lagares, México. p. 55-79.

Mater. Res. Soc. Symp. Proc. Vol. 1047 © 2008 Materials Research Society   1047-Y04-05

Thermal Volatilisation Analysis—The Development of a Novel Technique for the Analysis of Conservation Artifacts

James Pawel Lewicki, Deborah Todd, Perrine Redon, John Liggat, and Lorraine Gibson
Pure and Applied Chemistry, University of Strathclyde, Thomas Graham Building, 295 Cathedral St., Glasgow, G1 1XL, United Kingdom

ABSTRACT

Reported here is the development of a novel evolved gas analysis technique: Sub-Ambient Thermal Volatilization Analysis (SATVA) and its application in characterizing key analyte species from conservation artifacts. In this work SATVA has been applied to the study of volatiles evolution processes occurring in number of model conservation artifacts. The evolution of volatile species from cured formaldehyde resin, leather and metallic artifacts has been studied by SATVA. The specific analytes making up the total quantity of evolved material in each case have been separated and identified using sub-ambient differential distillation and a combination of online mass spectrometry, gas phase IR spectroscopy and GC-MS. The data gathered has been used to provide information on both the degradation processes occurring within the artifacts and the environmental history of the artifacts themselves.

INTRODUCTION

Sub-Ambient Thermal Volatilization Analysis (SATVA) is a versatile yet little known technique capable of: analyzing (in real time) the evolution of volatile species from an analyte, cryogenically collecting evolved volatiles; and characterizing the individual components of the total volatiles evolution process. Until recently SATVA has been almost exclusively confined to the field of polymer degradation [1,2,3] (where the roots of the technique lie) and had largely been forgotten. This work concerns the re-development of SATVA as a modern analytical technique and its application to a new field of research – conservation science.

In the context of conservation science, volatiles emissions are often linked with undesirable degradation processes occurring in the materials of an artifact as it ages. A detailed understanding of the nature of these volatiles evolution processes is necessary for conservators and conservation scientists so that the 'health' of important artifacts can be accurately monitored and remediation strategies are implemented correctly.

Volatiles evolution processes are also of importance in the context of archeological investigation of historical artifacts. Many materials, both natural and synthetic adsorb chemical species from their surroundings during their lifetime. These adsorbed species are often indicative of the uses that a particular artifact may have had or the environment that an artifact of interest has been exposed to.

The application of SATVA to conservation science is at an early stage and the work in this paper describes what is effectively an initial trial of SATVA. As such, the analyses described here are not truly non-destructive. The conditions employed (elevated temperatures and high vacuum) were selected in order to illustrate the ability of SATVA to collect, discriminate and identify volatile and semi-volatile analytes. It is the end goal of this work however, to replace the relatively harsh sampling regime employed here with an ambient temperature and pressure, inert gas/cold-trap cycling system. This modification is currently under development and SATVA in this form is potentially capable of extracting and characterizing adsorbed analytes non-destructively.

The volatile species evolved from conservation artifacts are often complex mixtures of differing chemical species; many of which are unknown initially to the analyst. The result of this is that many established analytical sampling techniques such as solid phase micro extraction (SPME) and related adsorption methods are rendered unsuitable due to their selective nature. When dealing with a large number of individual unknown analyte species there is an inherent risk in analytical adsorption techniques that important analytes may be missed due to a low affinity for the sorbent. Alternatively, an unnatural bias in the quantity of one analyte over another may be introduced due to differing affinities for the sorbent material. SATVA however, offers an alternative. Cryogenic trapping is *non*-selective and yields an unbiased assay of the total spectrum and relative quantities of volatiles emitted from an artifact. Detection of individual analyte species as a function of pressure under high vacuum is inherently sensitive and recent developments have allowed volatiles emissions from artifacts to be pre-concentrated in the cryogenic trap – further improving detection limits.

Other techniques more closely related to SATVA such as Headspace GC-MS have no such selectivity limitations and indeed, commercial Dynamic Headspace Analysis with its low detection thresholds and efficient analyte separation offer an attractive alternative to this rather unconventional technique. SATVA is however, extremely flexible and can accommodate virtually any sample size from complete artifacts to milligram quantities of sample. Additionally, SATVA has no MS low mass cut-off, unlike the majority of commercial headspace GC-MS instruments. Whereas many GC-MS systems cannot identify species below 50 amu without chemical derivatization, the gas-phase online MS of the SATVA system can detect and discriminate species from 1 to 300 amu effectively. This is particularly useful for the study of small molecules such as SO_2 and NH_3.

The SATVA technique is currently under development as an analytical tool for conservation science; however the initial results reported here demonstrate the potential of the technique for the analysis of conservation artifacts.

EXPERIMENTAL

Materials

For the purposes of this study, three model artifact types were chosen to be studied – each one representing a different aspect of volatiles evolution analysis. Model artifact one was a highly filled cast formaldehyde resin ball obtained from a commercial manufacturer. Model artifact two was a section of a leather book cover/binder of unknown age and origin. This was obtained from the author's personal collection. Model artifact three was the steel mouthpiece section of a large tobacco pipe. This item was obtained by author from a 'glory hole' junk shop in Glasgow. These items were selected as contrasting model materials for this work. Unfortunately, little is known about the history of these materials. However, as it was the intent of this initial work solely to demonstrate the potential of SATVA to detect and characterize analytes from such artifacts, this was deemed acceptable for an initial study. Figures 1 and 2 show the binder and mouthpiece respectively.

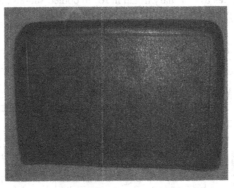

Figure 1. Photograph of artifact 2, the leather cover/binder

Figure 2. Photograph of artifact 3, the Tobacco pipe mouthpiece. A coin is shown also for scale.

Sub-ambient Thermal Volatilization Analysis

All SATVA analysis was carried out using a SATVA line which was built in-house, based upon the apparatus and techniques described by McNeill *et al.* [4,5]. The basic operating principle of the SATVA technique remains the same today as when it was first conceived during the early 1960s by McNeil and co-workers. The modern SATVA technique is based upon a continuously pumped, high vacuum system incorporating a series of cryogenic traps with sensitive pressure monitoring. The evolution of volatiles from an analyte is recorded as a function of pressure vs. temperature/time. Cryogenic trapping of volatiles is achieved by means of a liquid nitrogen cooled, sub-ambient trap. Controlled heating of the trap allows differential distillation of the captured volatiles to take place. Primary methods of separated component identification are online Mass Spectrometry, gas-phase IR Spectroscopy and GC-MS.

The SATVA apparatus consists of a sample chamber (heated by a programmable tube furnace) connected in series to a primary liquid nitrogen cooled sub-ambient trap and a set of four secondary liquid nitrogen cooled cold traps (see figure 3).

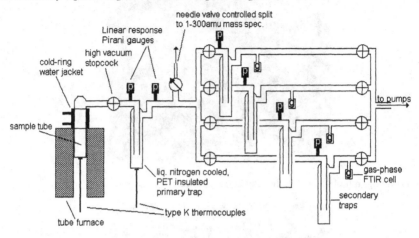

Figure 3. Schematic diagram of the SATVA apparatus showing the sampling stage and furnace linked in series to the primary sub-ambient trap, fractioning stage and pumping system.

The system is continuously pumped to a vacuum of 1×10^{-4} torr by means of a two stage rotary pump and oil diffusion pumping system. Volatile condensable products can be initially trapped at two stages: The water jacket cooled 'cold ring' (T~12 °C) immediately above the heated area of sample tube (this condenses high boiling point materials) and the primary liquid nitrogen cooled sub-ambient trap (T~ -196 °C) which collects all the lower boiling point species. Two linear response Pirani gauges are positioned at the entrance and exit of the primary sub-ambient trap to monitor the evolution of both condensable and non-condensable volatiles as a function of pressure vs. temperature/time from the sample. The use of linear response Pirani gauges allows valid pressure peak integrations to be carried out; where peak area corresponds to the quantity of evolved volatiles. Trapped, low-boiling species can be distilled into separate secondary cold traps by slowly heating the primary sub-ambient trap to ambient temperatures. These separated fractions can be subsequently removed into gas-phase cells for FTIR and GC-MS analysis. A series of non-linear Pirani gauges are placed at the entrance and exits of all secondary fraction traps to monitor the pressure changes as volatile species are distilled into separate traps and gas cells.

All SATVA runs were conducted under vacuum at temperatures ranging from 25 to 100°C. Each sample was held isothermally for a period of ~3 hours with continuous pressure monitoring and cryo-trapping of evolved volatiles. A 100 mg sample of the formaldehyde resin artifact was collected for analysis. Both the leather binder and the pipe mouthpiece were analyzed as complete items. A 1-300 amu VG single quadrapole Q300D mass spectrometer sampled a continuous product stream during both the isothermal and differential distillation runs. Sub-ambient differential distillation of collected volatiles was carried out by heating the primary sub-ambient trap at a rate of 4°C min^{-1} from -196 to 40°C. Volatiles were separated as a function of boiling point for subsequent IR and GC-MS analysis. A significant cold-ring fraction was collected during the analysis of the leather binder and pipe mouthpiece.

All FTIR analysis of the collected analytes was carried out using a Mattson 5000 FTIR Spectrometer used in transmission mode. High boiling 'cold ring' fractions were cast from a chloroform solution onto NaCl disks for analysis. Low-boiling volatiles were analyzed in the gas phase using gas phase cells with NaCl windows.

All GC-MS analysis of the collected analytes was carried out using a Finnigan ThermoQuest capillary column trace GC and Finnigan Polaris Quadrapole Mass Spectrometer. Suitable fractions were dissolved in chloroform and subsequently analyzed.

DISCUSSION

The work reported here has taken the form of three model case studies of three different artifact types; a cast formaldehyde resin ball, a leather binder of undetermined age and a steel mouthpiece section of a large ornamental tobacco pipe. Each item has been chosen to represent a different case for the study of a volatiles evolution process.

Case study 1: Analysis of off-gassing of degradation products from a formaldehyde resin ball

Many of the early 20th century 'plastic' artifacts in museums and historical collections today have been fabricated from polymers such as urea-formaldehyde or cellulose acetate/nitrate. These materials are subject to degradation over time and the conservation of such polymeric materials is a field in its own right. In many cases, the degradation products of the polymer are volatile and therefore amenable to study by evolved gas analysis techniques [6]. In this case study the off-gassing of free formaldehyde from a formaldehyde resin artifact has been investigated with SATVA. The volatiles evolution from a virgin (new) sample of the formaldehyde resin ball have been compared with a five year old sample in order to determine if SATVA can discriminate between the samples.

100 mg samples of a virgin and a 5 year old sample of the formaldehyde resin were held isothermally under vacuum at 70°C for 2.2 hours. Shown in figure 4 is the isothermal off-gassing trace obtained for the virgin resin sample.

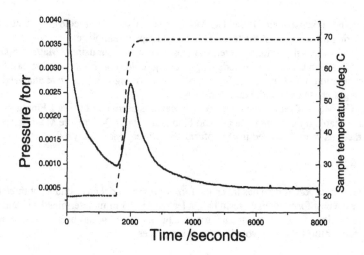

Figure 4. Isothermal off-gassing trace for the virgin formaldehyde resin held isothermally under vacuum for 2.2 hours. Solid line represents off-gassing pressure from the sample and dotted line represents the furnace temperature.

It can be observed from figure 4 that as the sample is heated moderately under vacuum significant off-gassing of volatiles occurs. The off-gassing rapidly reached a maximum rate and then tailed off gradually over the remainder of the isothermal holding period: This behavior is typical of many of the artifacts analyzed. A similar trace was obtained from the 5 year old sample of formaldehyde resin however the overall peak area was reduced indicating that the overall quantity of volatile species present in the sample had diminished.

The volatile species that off-gassed from both the virgin and aged samples were captured in the sub-ambient trap during the course of each isothermal run. At the conclusion of each isothermal run the sub-ambient trap was heated at a controlled rate; allowing the condensed analytes to distill differentially into separate components. The sub-ambient distillation traces for the virgin and aged formaldehyde resin samples are shown in figure 5.

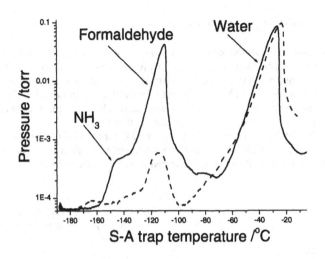

Figure 5. Sub-ambient distillation traces for the virgin and aged formaldehyde resin samples. The solid line represents the distillation trace for the virgin material and the dashed line represents the 5 year old sample.

During the course of each sub-ambient distillation run, the online mass spectrometer continuously sampled the analyte stream. Three primary analytes; ammonia, formaldehyde and water were identified from the virgin sample. However, it can clearly be observed from figure 5 that there is no ammonia present in the 5 year old sample and that the level of formaldehyde has been significantly reduced. Ammonia is thought to be a residue from the initial curing process which is lost to the environment over time. Water is present both from the initial cure and the surroundings therefore its level remains relatively constant. The level of formaldehyde is observed to reduce significantly between the virgin and aged material however it is still present at low levels in the aged material. The continued off-gassing of formaldehyde from the aged system is indicative of hydrolysis of the formaldehyde resin network as it degrades; indeed, it is well established that the hydrolytic ageing/degradtion of cured formaldehyde resins results in the production of free formaldehyde [7].

It is clear from this case study that SATVA can effectively monitor off-gassing processes from cured formaldehyde resin objects, can easily discriminate between levels of off gassing in a virgin and an aged sample and can detect the low levels of formaldehyde off-gassing during the hydrolytic degradation of a formaldehyde resin. Certainly in a real case – no such virgin material would be available, however the study demonstrates that the continuing evolution of formaldehyde from such an object could be characterized and quantified.

Case study 2: Analysis of a leather binder

The analysis of this leather cover/binder was intended to demonstrate if SATVA could identify any of the specific compounds responsible for the characteristic odor of the leather and to investigate if the technique could obtain any further information on the state and composition of the artifact. This artifact was chosen as it was effectively an unknown: visually the leather appeared to be relatively old; however nothing was known about its age. Additionally, the leather had a strong odor – characteristic of many old leather objects.

The leather artifact was held isothermally at 50°C for a period of 3 hours under vacuum with continual collection of all volatiles in the sub-ambient trap. Shown in figure 6 is the sub-ambient distillation trace for the volatiles collected from the leather artifact.

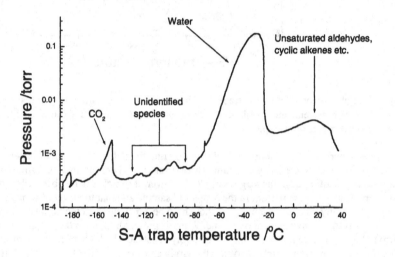

Figure 6. Sub-ambient differential distillation trace for the leather binder. A large quantity of water was evolved from the binder; however this has been effectively separated from the other analyte species. The CO_2 present is from residual atmosphere entrained within the sample.

The differential distillation of the collected volatiles from the leather has shown that it evolves a relatively complex mixture of volatile species. Several low boiling analytes were detected and have been labeled in figure 6. However, due to shortcomings in the sensitively of the mass spectrometer system, these analytes could not be identified. Water makes up a large proportion of the volatiles evolved which is in itself illustrative of the levels of water that natural materials such as leather retain from the environment. The final peak shown in figure 6 was identified as consisting of a mixture of higher molecular weight unsaturated and aldehydic species. This 'fraction' was extracted into chloroform and analyzed by GC-MS (see figure 7).

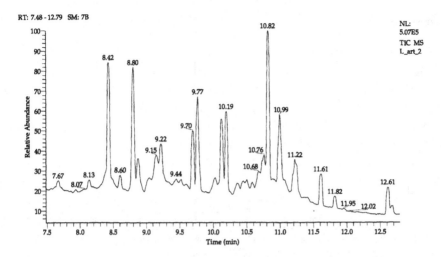

Figure 7. GC-MS total ion chromatogram (TIC) of the high-boiling volatile fraction obtained from the leather binder. Individual peaks correspond to separate chemical species.

Several of the peaks in the chromatogram (figure 7) are due to siloxane breakdown products from the column stationary phase, however the peaks at retention time 8.80, 9.15, 9.77 and 10.76 min have been identified as 2-isopropyl-5-methylhex-2-enal, a thiophene, 3-menthene and an unsaturated ketone respectively. The GC-MS analysis of the high boiling volatile fraction effectively indicates that many of the components of this fraction are unsaturated and aldehydic in nature. The compounds such as 3-menthene and 2-isopropyl-5-methylhex-2-enal have distinctive odors and it is thought that it is the combination of these various species that are responsible for the characteristic smell of the leather.

A significant 'cold-ring fraction' was also obtained during the analysis of the leather binder. This represented the highest boiling point material evolved from the leather which is not generally considered to be volatile under normal conditions of temperature and pressure. This fraction was analyzed by a combination of transmission FTIR spectroscopy and GC-MS. The GC-MS chromatogram obtained from this cold-ring fraction is shown below in figure 8.

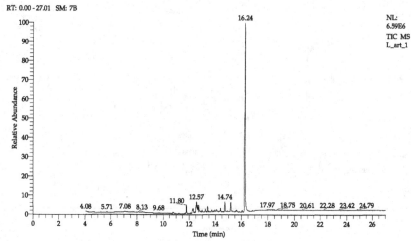

Figure 8. GC-MS total ion chromatogram of the leather binder cold-ring fraction. The single large peak at RT = 16.24 minutes has been positively identified as di-isooctylphthalate.

The cold-ring fraction was identified from the GC-MS and FTIR data as consisting almost entirely of di-isooctylphthalate (which is a modern plasticizer which was not used widely until the 1950s) together with low levels of related phthalate esters and derivatives. The presence of such large quantities of di-isooctylphthalate suggests that either the leather binder is contemporary in origin or that it has been treated at some point in its lifetime with a phthalate based agent.

<u>**Case study 3; Analysis of a tobacco pipe mouthpiece**</u>

In this final case study SATVA was employed to analyze trace volatile residues adsorbed onto the interior surface of a steel mouthpiece section of a large ornamental tobacco pipe. The trace volatiles of interest in this case were primarily; any residues of the tobacco that had once been used in this artifact.

The pipe mouthpiece was held isothermally at 100°C for a period of 3 hours under vacuum with continual collection of all evolved volatiles. A significant quantity of material was observed to off-gas during this isothermal hold period and once again; a significant cold-ring fraction was obtained. The volatile material evolved was identified as consisting primarily of water and a low level of higher boiling saturated hydrocarbon species.

An analysis of the cold-ring fraction revealed some unexpected results. Shown in figure 9 is the GC-MS total ion chromatogram of the cold-ring obtained from the pipe mouthpiece. Product identifications are presented in table 1.

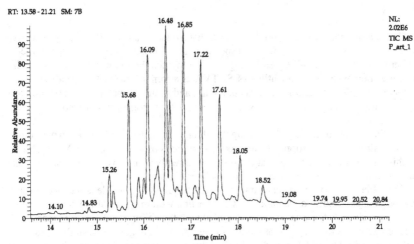

Figure 9. GC-MS TIC of the cold-ring fraction obtained from the analysis of the pipe mouthpiece. Each peak represents an individual chemical species. Product identifications are given in table 2 below.

Table 1. Identification of the components of the pipe mouthpiece cold-ring fraction.

Retention time /min	Identification
15.26	linear saturated hydrocarbon
15.34	Tetrahydrocannibol (THC)
15.68	linear saturated hydrocarbon
15.90	Resorcunol
15.95	cannicoumaronone
16.09	linear saturated hydrocarbon
16.31	Hexahydrocannabi-1,10-diol +Dronabinol
16.98	linear saturated hydrocarbon
16.56	cannibol species
16.70	cannibol species
16.85	linear saturated hydrocarbon
17.22	linear saturated hydrocarbon
17.61	linear saturated hydrocarbon
18.05	linear saturated hydrocarbon

It can clearly be observed from the data presented in figure 9 and table 1 that significant levels of THC (tetrahydrocannibol), a series of related canniboids and a series of hydrocarbon oils were present on the interior surface of the pipe mouthpiece. There is no evidence of any nicotine based compounds therefore it is reasonable to conclude that this pipe was used almost exclusively for the consumption of cannabis and not tobacco as it was initially assumed.

CONCLUSIONS

The data presented in this work has shown that SATVA can be utilized as an effective method for the analysis of volatile chemical species that are present in conservation artifacts. This technique is still at a relatively early stage of development however the three case studies presented here demonstrate the versatility of SATVA as a form of evolved gas analysis. Production of volatile degradation products in natural and synthetic materials can be monitored with this technique and adsorbed analytes such as the residues from cannabis smoke can be readily detected.

Certainly, some of the data for the high-boiling residues presented in this work could have been obtained by conventional solvent extraction methods – however SATVA is able to provide this data without having to treat the materials directly with solvent. The analysis can also yield an indication of the relative quantities of analytes evolved.

The unbiased nature of the cryo-trapping method provides tangible advantages over chemical or physical adsorption techniques when dealing with mixtures of unknowns. The wide range of sample sizes that can be analyzed, coupled with the ability to effectively separate individual components of a mixture of volatiles give SATVA a degree of flexibility and simplicity that is unmatched by related thermal analysis techniques such as thermal gravimetry-mass spectrometry (TG-MS).

The SATVA technique is currently in a developmental stage – however initial studies have demonstrated its potential. Further work is planned to refine the technique by improving the sensitively of both the mass spectrometer and the pressure monitoring. Finally it is the eventual goal of this work to apply a form of SATVA to the study of volatiles evolution processes in historical books and manuscripts. At this stage, the relatively harsh low pressure sampling regime will be replaced with an ambient temperature and pressure inert gas/cold-trap cycling system. This modification is currently under development.

REFERENCES

1. I.C McNeill in *Developments in Polymer Degradation -1*, edited by N. Grassie (Applied Science Publishers, 1984), p. 43.
2. W. McGill in *Developments in Polymer Degradation -5*, edited by N. Grassie (Applied Science Publishers, 1977), p. 1.
3. J.P. Lewicki, J.J. Liggat, M. Patel, R.A. Pethrick and I.R. Rhoney, Polymer Degradation and Stability, Article in Press.
4. I.C. McNeill, European Polymer Journal, 409 (1967).
5. I.C. McNeill, L. Ackerman, S.N. Gupta, M. Zulfiquar and S. Zulfiquar, Journal of Polymer Science Part A: Polymer Chemistry, 2381, (1977).
6. M. Watanabe, C. Nakata, W. Wu, K. Kawamoto and Y. Noma, Chemosphere, 2063 (2007).
7. J. Dutkiewicz, Journal of Applied Polymer Science, 3313 (1983).

Interdisciplinary or
Cross-Disciplinary Contributions

Mater. Res. Soc. Symp. Proc. Vol. 1047 © 2008 Materials Research Society 1047-Y01-04

Application of XRF and AMS Techniques to Textiles in the Mongol Empire

Tomoko Katayama, Ari Ide-Ektessabi, Kazuki Funahashi, and Ryoichi Nishimura
Graduate School of Engineering, Kyoto University, Yoshidahonmachi Sakyo-ku, Kyoto, 606-8501, Japan

ABSTRACT

X-ray fluorescence (XRF) and Accelerator Mass Spectrometry (AMS) techniques were applied to four pieces of ancient Mongolian textiles in order to assist cultural studies of the most significant era of medieval western Asian culture. Radiocarbon dating using Accelerator Mass Spectrometry (AMS) was performed in order to determine the historical age of these pieces. Then, X-ray fluorescence analysis using Synchrotron Radiation (SR-XRF) was carried out in order to obtain elemental maps as well as investigate their constituent elements. Results showed that the textiles were produced between12 th and 13 th century, and possessed elements such as Au, Cu, Fe and Ti were traced in these pieces whereas Au was used to make gold threads. Cu, Fe and Ti are well known as metallic mordant. In addition, high-resolution images were obtained using Scanning Electron Microscope (SEM) to observe the textile structure and their weaving conditions. The whole collected data can assist in bringing into light and facilitate a deeper understanding of the medieval Mongolian cultures, the textile technology, staining techniques, material process technology of the Mongolian Empire and their relations with the neighboring east and central Asian cultures, such as Persia, India and China.

INTRODUCTION

Advance techniques such as AMS and XRF were applied to ancient textiles collected in Mongolia. Mongolia is a country located in Eastern Asia. In 1206, Genghis Khan founded the Mongol Empire in Mongolia [1]. From 13 th to the 15 th century, it flourished as the largest contiguous empire in world history. Especially in the 13 th century, the power of the empire was tremendous and it conquered places from North China to Iran, throughout continental Eurasia. During this period, the culture of Mongolia was expanded throughout Asia where its influence lasted quite long. Therefore, the research of Mongolian culture is not only important for the historical science of Mongolia, but also for the whole of Asia. In this paper, four pieces of ancient Mongolian textiles were analyzed for the research of Mongolian culture as well as Asian culture. First, radiocarbon dating using Accelerator Mass Spectrometry (AMS) was performed to determine the historical age of the samples. AMS- ^{14}C dating is widely used for dating textiles such as silk [2]. Then X-ray fluorescence analysis using Synchrotron Radiation (SR-XRF) was carried out. SR-XRF analysis was applied for nondestructive trace element analysis and element distribution. Additionally, SR-XRF is capable of analyzing trace elements in the air with high-resolution and sensitivity because of the high brightness of synchrotron radiation [2][3]. After obtaining the spectrum, several main elements such as Fe, Au and Cu were selected, and elemental maps were obtained. High-resolution images using Scanning Electron Microscope (SEM) were also produced.

EXPERIMENT DETAILS

Four pieces of old Mongolian textiles, belonging to four different garments, supplied by the Center of National Heritage Mongolia were investigated using AMS-^{14}C dating in order to obtain their historical age. Appearance of the samples is shown in figure 1. As pretreatment, ultrasonic cleaning using hexane, isopropanol and acetone was performed in order to remove attached soil (60°C, 15min). Then, the modified cleaning procedure of the Acid-Alkali-Acid (AAA) is applied (HCl-NaOH-HCl, 85°C, 15 min) [3]. Halogens, and sulfur elements inside the samples were removed by CuO, Au and Sulfix. Sulfix, a kind of reagent for removing halogens and sulfur, is a heat-treated mixture of Cobalt oxide and silver oxide [4]. As Sample 2 was found to be plant fiber, Salfix was not used. Finally, carbon elements in the samples are graphitized. AMS-^{14}C dating was performed at Paleo Labo Co., Ltd., Japan, using compact AMS system called 1.5 SDH-1 Palletron (National Electrostatics Corporation (NEC), U.S.A.). The accelerating voltage was 500kV.

Sample 1 : reddish brown, dry

Sample 2 : dark brown, dry

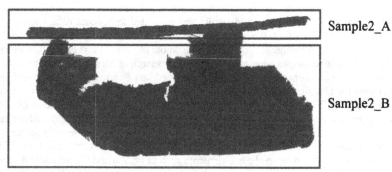

Sample2_A

Sample2_B

Sample 3 : brown, dry

Sample 4 : light brown and dark brown, dry

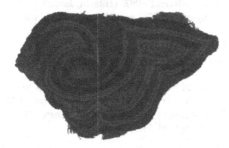

Figure 1. Appearance and condition of the Mongolian textiles

X-ray fluorescence (XRF) using synchrotron radiation was carried out at beam line 4A (BL4A) of Photon Factory (PF), High Energy Accelerator Research Organization (KEK), Tsukuba, Japan. Figure 2 shows the schematic drawing of the experimental set up at PF_BL4A. The electron beam energy in the storage ring was 2.5 GeV, with a maximum current of 400 mA. Incident X-ray energy was 15 keV. The cross-section of the beam was approximately 1(v) x 1(h) mm^2 on the sample. The synchrotron radiation was monochromated by a multilayered reflecting mirror. Precise beam size of monochromated X-rays was adjusted using slits. The incident and transmitted X-rays were monitored by ionization chambers that were set in front of and behind the sample. The fluorescent X-rays were collected by a solid-state detector at 90 degrees to the incident beam. Measurements were performed in air. Point spectra were measured for obtaining consistent elements of the samples. The spectra were obtained by using a multi-channel analyzer. The measurement time was 100 seconds for each spectrum. XRF imaging technique was applied in order to investigate the distributions of main elements. X-Y step pulse motors moved the sample stage. The measurement areas were divided into matrices of 20 x 20 pixels. At each pixel, the XRF yields for each element were integrated by single channel analyzers. The measurement time was three seconds for each pixel.

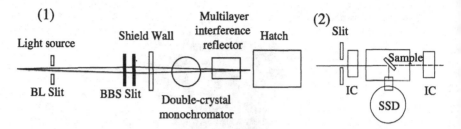

Figure 2. Schematic drawing of the experimental set up at Photon Factory, BL4A
(1) Overall view   (2) Inside of the hatch

The microstructures of the samples were investigated using a field-emission scanning electron microscope (FE–SEM) Model S-4500 (Hitachi, Japan), after coating the surfaces with Au in order to reduce charging effects. The accelerating voltage was 1.0 kV or 3.0 kV.

RESULTS AND DISCUSSION
The High Resolution Images of the Mongolian textiles

High-resolution images of Mongolian textiles were obtained by SEM (Figure 3). The textile structure and their weaving conditions were observed from SEM images. The diameter of the fibers was relatively thin: Sample 1 and 4 were approximately 12μm thick and that of Sample 2 and 3 were approximately 8 μm thick [5].

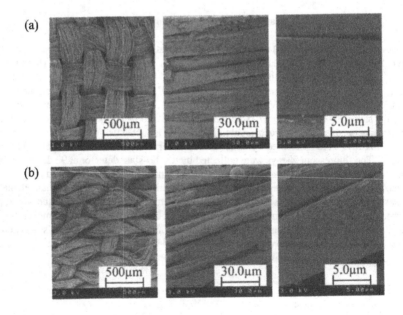

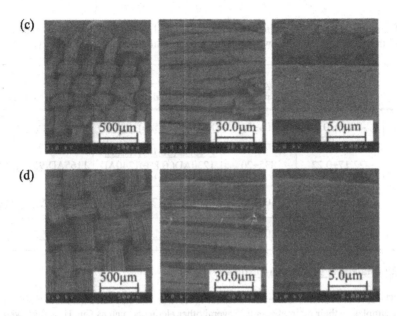

(c)

(d)

Figure 3. FE-SEM micrographs of the textiles with 1.0 kV (a_1 and a_2) and 3.0 kV (the others).
(a)Sample 1, the diameter is approximately 11 μm thick. (b) Sample 2 (9 μm thick)
(c) Sample 3 (8 μm thick) (d) Sample 4 (13 μm thick)

The Historical Age of the Mongolian extiles

The radiocarbon results were calibrated by accepting the OxCal 3.10 software and radiocarbon calibration curve (INTCAL04) [6-8]. Table I shows carbon isotope ratios (δ13 C), Conventional Ages, and calibrated age ranges. Sample 1, 3 and 4 were dated 13 th century, and Sample 2 was dated between 1165AD and 1260 AD. In the 13 th century, the Mongol Empire was flourishing under such powerful emperors as Genghis Khan and Kublai Khan [1]. Following this, it is thought that these textiles were produced during the peak of the Mongol Empire.

Table I. Results of AMS- ^{14}C dating: The date of Sample 1, 3 and 4 were probably produced during13 thcentury and Sample 2 was probably produced in the end of 14 th century.

Sample. No	carbon isotope ratios $\delta^{13}C$ (‰)	Conventional Ages (yrBP±1σ)	^{14}C calibrated age ranges	
			1σ calibrated age ranges	2σ calibrated age ranges
1	-20.89±0.2	704±20	1275AD(68.2%)1295AD	1260AD(90.0%)1300AD 1360AD(5.4%)1380AD
2	-22.17±0.22	835±20	1175AD(66.1%)1225AD 1230AD(0.8%)1240AD 1245AD(1.3%)1250AD	1165AD(95.4%)1260AD
3	-20.51±0.16	800±20	1220AD(68.2%)1260AD	1210AD(95.4%)1270AD
4	-21.91±0.17	805±20	1220AD(68.2%)1255AD	1205AD(95.4%)1270AD

The XRF Analysis of Mongolian Textiles

The results of XRF analyses are shown Figure 4. It is found that Fe and Au were detected from all samples as their main elements. Several other elements such as Cu, Ti, and Pb were also detected. The photo of Sample2 in Figure 1 shows that two different materials were used in Sample 2: Area A (Sample 2_A) and Area B (Sample 2_B). Figure 5 shows XRF imaging of Au in Sample 2. The XRF spectra and XRF imaging clearly show this difference. It is found from Figure 5 that Sample 2_A contains large amount of gold. Sample 2_A, Sample 3 and Sample 4 show similar element distribution. It can be said from these results that gold threads were often used in the time of Mongol Empire. Fe was detected in all samples. Iron oxide in the clay is a typical traditional metallic mordant or component for natural dying [9]. Several elements, such as Cu and Ti are also well known as metallic mordants [10]. This result suggests the possibility that dying using metallic mordants or pigment dying using iron oxide red were applied around that time.

(a)

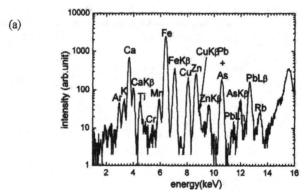

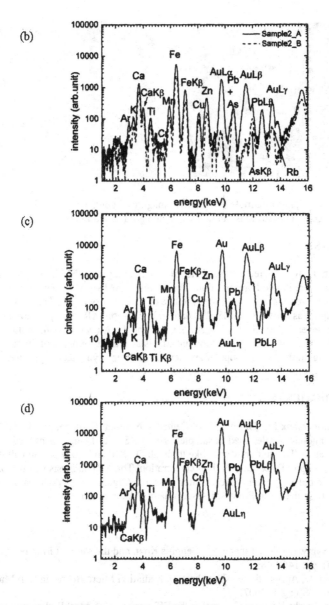

Figure 4. Representative XRF spectra of Mongolian textiles measured at Photon Factory BL-4A. The main element of Sample 1 (a) and bottom area of Sample 2 is iron (Fe). Sample 3, 4 and the other side of Sample 2 contains gold (Au) as well as iron.

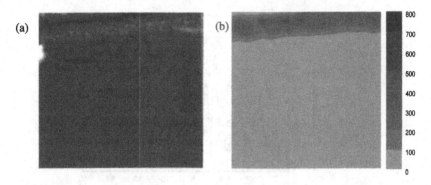

(a)   (b)

800
700
600
500
400
300
200
100
0

Figure 5. (a) Part of photo (Sample 2) (b) XRF imaging of gold. The "Sample 2_A" contains a large amount of gold. This string is thought to include gold constituents.

CONCLUSIONS

In this study, four ancient Mongolia textiles were analyzed. SEM images showed the Mongolian textiles are in good preservation. The results of AMS- ^{14}C dating showed that the Mongolian textiles were produced between the late of 12th century and 13th century, the height of the Mongol Empire, as shown by AMS- ^{14}C dating. In addition, XRF analysis revealed that these samples contain several elements such as Au, Fe, Cu, and Ti. Au was used for gold threads, salts of Cu, Fe and Ti are known as metallic mordants. Fe is also used for pigment dying. These results show that gold thread was popular and several different dying techniques were used in the Mongol Empire.

ACKNOWLEDGMENTS

The authors thank Mr. G. Enkhbat of Center of National Heritage, Mongolia who supplied the Mongolian textiles used as samples. The AMS- ^{14}C dating was carried out at Paleo Labo Co., Ltd. Japan. The authors would like to thank Dr. Koichi Kobayashi and all the member of AMS dating group in Paleo Lab for their cooperation. The XRF analysis was performed at Photon Factory in the High Energy Accelerator Research Organization, Tsukuba, Japan. The authors express their thanks to Professor Atsuo Iida of Photon Factory.

REFERENCES

1. Roux, J.P., Sugiyama, M., Tanabe, K., Genghis Khan and the Mongol Empire, (Sougen-sha, Japan, 2003), pp. 146-148.
2. Ide-Ektessabi A., Applications of Synchrotron Radiation Micro Beams in Cell Micro Biology and Medicine, (Spinger, 2007).
3. Turnbull1 J., Sparks R., Prior C., Testing the Effectiveness of AMS Radiocarbon Pretreatment and Rreparation on Archaeological Textiles, *Nuclear Instruments and Methods in Physics Research* Section B: Beam Interactions with Materials and Atoms, 172, 469 (2000).

4. Nakai, I., Kondo, N., Itabashi, M., Sasaki M., Terada Y., Nondestructive Characterization of Antarctic Micrometeorites Collected at the Dome Fuji Station by Synchrotron Radiation X-ray Fluorescence Analysis, *Antarctic meteorite research*, 13, 302 (2000).
5. Koseki, H., Natsume, Y., Iwata, Y., Evaluation of the Burning Characteristics of Vegetable Oils in Comparison with Fuel and Lubricating Oils, *J. Fire Sciences,*19, 31 (2001).
6. Hayashi, M., Sakai, T., Material for clothing, (Jikkyo, Japan, 1997), pp. 8.
7. Ramsey, C.B. Radiocarbon Calibration and Analysis of Stratigraphy: The OxCal Program. *Radiocarbon*, 37, 425 (1995).
8. Ramsey, C.B. Development of the Radiocarbon Program OxCal. *Radiocarbon*, 43, 355 (2001).
9. Reimer, P.J., Baillie, M.G.L., Bard, E., Bayliss, A., Beck, J.W., Bertrand, C.J.H., Blackwell, P.G., Buck, C.E., Burr, G.S., Cutler, K.B., Damon, P.E., Edwards, R.L., Fairbanks, R.G., Friedrich, M., Guilderson, T.P., Hoog, A.G., Hughen, K.A., Kromer, B., McCormac, G., Manning, S., Ramsey, C.B., Reimer, R.W., Remmele, S., Southon, J.R., Stuiver, M., Talamo, S., Taylor, F.W., van der Plicht, J. and Weyhenmeyer, C.E., IntCal04 Terrestrial Radiocarbon Age Calibration, 0-26 Cal Kyr BP. *Radiocarbon*, 46, 1029 (2004).
10. Kongkachuichay P., Shitangkoonb A., Chinwongamorna N., Thermodynamics of adsorption of laccaic acid on silk, *Dyes and Pigments*, 53 (2), 179 (2002).
11. Pizzolato P., Lillie R. D., The Staining of Mucopolysaccharides with Gallocyanin and Metal Mordants, *Histochemistry and Cell Biology*, 27, 335 (1971).

Mater. Res. Soc. Symp. Proc. Vol. 1047 © 2008 Materials Research Society　　　　1047-Y02-03

Preserving Intangible Aspects of Cultural Materials: Bonpo Ritual Crafts of Amdo, Eastern Tibet

Chandra L. Reedy
Center for Historic Architecture and Design, University of Delaware, 307 Alison Hall, Newark, DE, 19716

ABSTRACT

Ancient and historic products of past technologies exist in the form of material culture and archaeological finds, available for materials analysis. Technical studies and analytical work, coupled with the study of historical texts and archival documents, can help in reconstructing past technologies. But the act of making an object is, by its very nature, also an intangible part of human heritage. Production of material culture may be accompanied by specific rituals, social behaviors and relationships, music, knowledge gained from oral histories, meanings, intents, beliefs, and reasoning processes. For ancient objects, gaining access to these intangible aspects of cultural heritage may be extremely difficult, if not impossible. However, there are many societies where traditional crafts are produced within a context where the intangible aspects can still be recorded. Yet, these opportunities are disappearing at an alarming rate as development and globalization rapidly overtake more and more traditional communities. Documenting intangible data about craft processes can promote fuller understanding of the objects themselves, and aid long-term preservation of both the objects and the processes used to make them. Examples here are drawn from fieldwork conducted in 2007 at a Bonpo monastery (Serling) and nearby villages in the Amdo region of the eastern Tibetan culture area (in Sichuan Province, China). Bonpo practices, which pre-date the introduction of Buddhism into Tibet, incorporate a variety of ritual crafts that are strongly rooted in a complex web of intangible relationships, behaviors, meanings, purposes, and beliefs. This paper focuses on votive clay objects (tsha-tshas) and barley-dough offering sculptures (tormas). Processes encompassing intangible aspects that are explored include the decision to make an object, when to make it and in what form, selection of raw materials, methods for processing the raw materials, fabrication procedures, selection of who will be involved in fabrication steps, where to place the finished object, and whether it will be preserved for the long term or considered to be only a temporary object. Results are placed in the context of larger theoretical issues regarding documentation and preservation of intangible elements of cultural heritage as part of a study of materials and technological processes.

INTRODUCTION TO INTANGIBLE HERITAGE PRESERVATION

The UNESCO convention for the Safeguarding of Intangible Cultural Heritage dates to 2003 [1]. One of the domains that it explicitly incorporates within intangible cultural heritage is traditional craftsmanship [2]. The focus is not on the craft products themselves, but, instead, on the skills and knowledge that are crucial for the production of those crafts. The philosophy behind the establishment of this convention, and the preservation efforts that are organized around it, is that craft objects cannot be preserved in isolation from the knowledge and skills that created them. Another aspect to the UNESCO convention is the recognition that globalization, industrial efficiency, the reach of large multinational corporations, and other worldwide trends create barriers to continuation of traditional hand-made crafts. Documentation of technological knowledge and skills in action is needed before they can disappear. Other domains of intangible cultural heritage are often intertwined with craft production, including oral traditions, rituals and social practices, and knowledge and practices concerning nature and the universe, and these too need to be documented before they are lost.

Ancient and historic products of past technologies exist, but only as material culture and archaeological finds, sometimes supplemented with historical texts and archival documents. A study of written documentation plus materials analysis can help in reconstructing many aspects of past technologies. However, the act of making an object is, by its very nature, an intangible part of human heritage, so for past craft products we can never really know the whole story. We can identify the source of raw materials, reconstruct fabrication processes, and analyze distribution networks. But the rituals, social behaviors and relationships, music, oral history knowledge, meanings, intents, beliefs, and reasoning processes associated with that material culture may, in most cases, be beyond our reach.

However, there are many societies throughout the world where traditional crafts are produced within a context where the intangible aspects can still be recorded. Yet, these opportunities are disappearing at an alarming rate as development and globalization rapidly overtake more and more traditional communities. Documenting intangible data about craft processes can promote fuller understanding of the objects themselves, and aid long-term preservation of both the objects and the processes used to make them.

This paper focuses on two examples drawn from fieldwork conducted in 2007 at Serling, a Bonpo monastery, and nearby villages in the Amdo region of the eastern Tibetan cultural area (in Sichuan Province, China). Bon is the native, pre-Buddhist religion of Tibet. Although Buddhism is now the most widely-practiced religion in Tibet, Bonpo communities remain active, especially in eastern Tibet. The production and use of Bonpo ritual crafts are strongly rooted in a complex web of intangible relationships, behaviors, meanings, purposes, and beliefs. The discussion presented here highlights two ritual craft traditions, small votive clay objects known as tsha-tshas, and barley-dough offering sculptures called tormas. A variety of technological processes incorporating intangible data are explored, such as the original decision of when and why to make an object, what form to make it in, selection of raw materials, methods for processing the raw materials, fabrication procedures, selection of who will be involved in fabrication steps, where to place the finished object, and whether it will be preserved for the long term or is considered to be only a temporary object.

FIELDWORK SITE

Serling monastery (Tibetan *gser gling dgon pa*, or alternative spellings of *gsas ling dgon pa* or *sa dmar gser gling dgon pa*) is a Bonpo monastery [3]. Bon is the native religion of Tibet, although the number of practitioners is now dwarfed by the number of Tibetan Buddhists. Over the centuries, Bon has incorporated many elements of Tibetan Buddhism, while at the same time it has greatly influenced the development of Buddhism within Tibet. Bonpo practitioners preserve strong ties to nature and nature spirits, among other traditions [4].

The small monastery of Serling is located about 270 kilometers north of Chengdu and 77 kilometers southwest of Songpan, in Songpan District of Sichuan Province, China. The Tibetan name for Songpan District is zung chu dzong. This is the lower part of a region known as shar khog, on the borders of traditional eastern Tibetan areas of Khams and Gyarong. However, the villagers clearly identify themselves as being in the area known as Amdo, and they speak a version of a dialect of Tibetan called Amdokhe. The monastery itself is in the village of Zhang ngu kho, but it serves many surrounding villages. The village this paper focuses on is one of the closer ones to the monastery, Lhayul (also known as La yas).

Lhayul currently houses about 56 families. Homes are strung along the river valley (a tributary of the zhang ngu chu), with most of the land near the river devoted to fields, and the homes clustered along the hills above. A system of terraces support fields growing a variety of crops such as barley and beans. Green vegetables are mainly collected along the river or on the surrounding mountains. The mountains serve as pasture land for goats, yaks, and horses. The pigs that are also raised are fed with leftovers from family meals.

The "main street" of Lhayul consists of a small store, a post office, and the entrance to the village mill where barley is ground into tsampa, a main source of food. A typical home has wooden post-and-beam construction, with three stories. The lower one includes a barn for the animals, and fronts onto a courtyard where various work activities take place. A ladder leads up to the second floor, where the family resides. The kitchen/dining area is the center of the home, and can accommodate large gatherings. The upper floor is for storage. Since wood, food, and animal feed may be stored there, to keep it well aired there are often open areas or gaps along the rafters and walls of this upper floor. Traditional roofs are finished with wooden shingles, while modern ones may have clay tiles.

In the villages of this region, the home, family, animals, fields, mountains, and rivers are all crucial to survival, social relationships, and life satisfaction. As a result, they figure prominently in Bonpo rituals. While the monks study religious theory and philosophy, and participate in a variety of meditation practices within the monastery, they are also responsible for conducting rituals for the village and the householders. These rituals are often performed in homes rather than in the monastery.

TSHA-TSHA PRODUCTION

Tsha-tshas are small clay objects, stamped or made in a mold, having religious imagery and purposes. Tibetans have been making tsha-tshas since at least the seventh century, and probably earlier [5, 6] Both Buddhist and Bonpo practitioners make and use tsha-tshas, and a wide variety of deities, teachers, or religious structures have appeared on Tibetan tsha–tshas [7, 8]. These objects are still made today in the Tibetan Autonomous Region and in Tibetan cultural areas of many provinces of China [9], as well as in the Tibetan diaspora outside of China [10]. Although traditionally made of clay, a variety of recipes were used, including small additions of sacred ingredients intended to increase spiritual power; recently, modern molding materials have started to substitute for clay outside of Tibet [11]. This paper focuses on documentation and discussion of the intangible aspects of tsha-tsha production in Lhayul and in Serling monastery.

Decision to make a tsha-tsha

Periodically, villagers consult with monks in the monastery about what actions or religious practices might improve their circumstances. The monks may suggest a variety of things. One option that is sometimes prescribed is to commission the production of tsha-tshas.

Collection of clay

The villager arranging for tsha-tshas to be made will collect the necessary clay. Lhayul has only one suitable source, located on a hill above the village. The clay mine is just one small hole in the ground, providing very red, hard, dense clay that has to be hacked out in chunks using a pick (Fig. 1). The clay has few stones and little sand, so is considered good for transferring the fine details of the tsha-tsha molds.

Figure 1. The main clay mine above Lhayul is small and produces a hard, dense red clay.

Other surrounding villages utilize this source as well, for tsha-tshas and for small clay sculptures made at the monastery. No pottery is made here. Bricks used in some construction and for fencing are made from sandy clay mixed with stones, collected elsewhere near the streambed. There is another high-quality source of clay in the region, tan colored, located further away behind the regional school near the monastery. It is also sometimes collected and used for tsha-tshas and clay sculptures.

Typically, when collecting the clay, a large bag is filled even if only a few tsha-tshas are to be made. It is considered respectful to collect and prepare extra clay so that the monastery will have some material to use later if they need to make tsha-tshas or sculptures. The heavy bag of clay must be carried some distance down the hill and back to the home where it will be prepared.

Preparation of raw materials

The raw materials are prepared in the courtyard of the villager arranging for the tsha-tshas to be made. Barley grains, which will be added as a consecrated material, are placed in a basket to be winnowed and cleaned. Any stones are thrown out, and any chaff is removed and given to the pigs to eat (Fig. 2). The grains are then washed in water and set out to dry.

Figure 2. Barley grains for a consecrated additive are winnowed and carefully cleaned.

The clay itself also has to be prepared for use. Hunks are placed in a basket and pounded with a long-handled wooden mallet (Fig. 3). This produces finer clay, subsequently pounded again on top of a stone placed inside the basket. Any small stones are pulled out, and the processed clay is sieved through the holes in the basket so that the finer clay is separated. Remaining coarser clay continues to be pounded and sieved again. The material is eventually transferred to a finer-grained sieve basket for continued pounding and sieving until it reaches the consistency of a fine flour. The finer material will better take the shape of the mold design. Villagers also note that more careful preparation of materials results in better karma for everyone, leading to better rebirths.

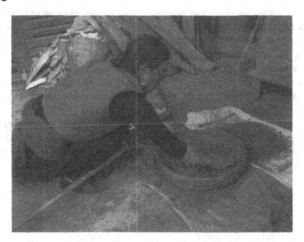

Figure 3. Clay is prepared by pounding with a wooden mallet and sieving to a fine consistency.

Figure 4. The only flat tsha-tsha mold available in Lhayul produces a multi-deity image.

The final step that must be accomplished by the villager is to secure the use of a tsha-tsha mold. These cast metal objects are expensive and few families in the village own one. In Lhayul, only one person (a metalworker) owns tsha-tsha molds, and he has two. For the tsha-tshas documented here, one with a three-dimensional design was borrowed, because the one for stamping out a flat tsha-tsha (Fig. 4) was already out on loan. As is customary, a variety of gifts are offered to the mold owner in exchange for the use of his object. Clearly, in this region the range of images depicted on tsha-tshas is limited due to limited availability of the necessary molds, not due to religious or ideological reasons.

The brass mold used in this case produces a solid clay object with a pyramid shape (Fig. 5). While the underside is plain, the four triangular-shaped sides of the pyramid are covered with rows of small raised dots, intended to represent deities. These symbolize 1,000 deities, so this form is called a thousand-god tsha-tsha form, or tong ku (Tibetan *stong sku*).

Figure 5. Pyrimid-shaped tsha-tsha mold produces a form that symbolizes 1,000 deities.

Consecration Ritual

The prepared materials and the borrowed mold are taken to the monastery. Before any tsha-tsha fabrication begins, the cleaned barley grains are blessed, or consecrated. This is done in a small shrine room in the monk's residential area. One monk reads from a consecration text in front of the barley seeds while another turns a wheel counterclockwise. A third enters from time to time and presents sacred juniper tree branches. The purpose of the ceremony is to make the barley into a more clean and pure offering for the tsha-tshas.

Tsha-tsha fabrication

Whereas the materials were prepared by laypersons, in this village only monks make the tsha-tshas. The villager who arranged for the tsha-tshas does not pay the monks for this service. Instead, modest gifts of food for the monastery or butter for the temple lamps are offered.

The monks first add water to the clay (Fig. 6), and let the mixture sit for about half a day to achieve the right consistency. The clay is then kneaded until it reaches a good working texture. Boards are placed next to the temple wall inside the main courtyard, to form the working surface. The inside of the mold is oiled prior to each use, using vegetable oil (Fig. 7).

Figure 6. Monks add water to the prepared clay and let the mixture sit for about half a day.

Figure 7. The brass mold is oiled before each use, in the courtyard of Serling Monastery.

Clay is pushed into the mold by hand. A space is hollowed out (Fig. 8), into which seven (a sacred number) of the blessed barley grains are dropped. The monk can either push the existing clay over the center hole to close it, or can add an additional chunk of clay if more is needed.

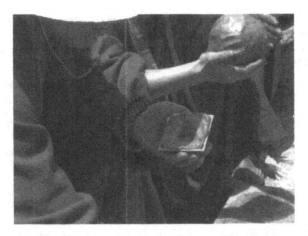

Figure 8. The blessed barley seeds are added to the tsha-tsha in the open underside.

The working board surface is then oiled, and the monk gently pushes the clay-filled bottom of the mold onto the board, trying to ensure that the mold gets fully filled by the clay (Fig. 9).

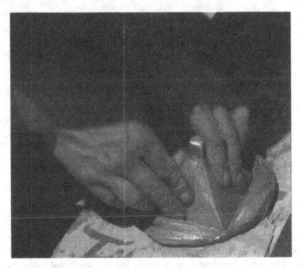

Figure 9. The monk carefully pushes the clay-filled mold onto an oiled board.

While pulling the mold gently off of the clay, the monk uses a knife to cut excess clay off from the edges of the mold, and, finally, to lift the mold off the board (Fig. 10). The finished tsha-tsha is placed on a drying board. It will be left out in the temple courtyard for several days, depending upon the weather (Fig. 11).

Figure 10. The finished tsha-tsha is lifted from the preparation board with a knife.

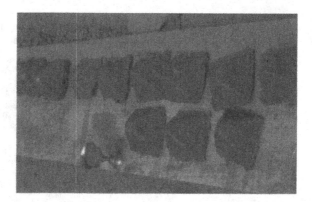

Figure 11. Finished tsha-tshas are left to dry in the monastery courtyard for several days.

Placement and storage of tsha-tshas

The main criterion for where the finished tsha-tshas are placed is that it should be a clean, safe, and respected place. Some are left in the monastery, while others are placed in the upper floor of a home. The third place they are typically left is a sheltered niche. Some of these are natural niches, such as recesses in a stone cliff face near the village. Most often, the sun-dried tsha-tshas quickly dissolve when moisture invades such natural niches, so many of those are no longer apparent in the cliffs today, although villagers claim to have left many there in the past.

Figure 12. A *tsaugau* is a building constructed as a shrine for completed tsha-tshas.

Constructed niches, or small shrines built to house tsha-tshas (called *tshagau* in this region), are also scattered throughout the village (Fig. 12). A site is chosen by a monk from the monastery who will walk around the village and decide where it would be most auspicious to build one. They can be built of either brick or stone masonry, with a wooden roof. A small opening is left, into which the tsha-tsha are placed. After construction, a monk returns to conduct a ceremony at the time that the tsha-tshas are placed inside.

TORMA PRODUCTION

Torma (Tibetan *gtor ma*) are offering sculptures usually formed out of a dough with a flour base. Most often, roasted barley flour is used, along with other additives [12, 13]. They are usually made and used only during a particular ritual, and then destroyed, although some may be kept in use for a longer period of time. Recently, more permanent tormas have been made out of clay, wood, resin modeling products, and even injection-molded plastics [14]. Tormas appear in both Buddhist and Bonpo practices. A wide variety of recipes and forms may be found, depending on local traditions and on the ritual for which the objects are to be used. Tormas range from very simple to extremely elaborate, and may have a plain brown color of the original dough mixture or may be brightly painted in complex color schemes. For special occasions, such as the

Tibetan New Year, butter tormas may be made instead of dough-based tormas. Those objects tend to be very complex and elaborate in color and design, and the butter is often mixed with candle wax to provide longevity.

The tormas discussed here were made during a two-day ritual called Sokha (Tibetan *gsol kha*) [15]. To conduct this ceremony, the monks from Serling monastery came to the home of the sponsoring household in Lhayul, where a room had been prepared for the ritual to take place. The tormas were made during the first day of the Sokha ritual; the monks stayed overnight in the home, and continued early the following morning. The primary purpose of Sokha is to bring good fortune and blessings to the household where the ceremony is held. Additionally, these blessings are sent out to all animals and to all people of the village and of the world. Each household in the village will try to have the ceremony performed in their home periodically. The discussions below focus on torma creation and use, rather than on the many other details regarding performance of the Sokha ceremony.

Preparation of raw materials

All of the necessary materials for making the tormas and for conducting the Sohka ceremony are first collected and set on or near the altars in the room set aside for the ceremony. The monks and the household members work together to assemble these materials. When this is done, the torma dough is prepared. One of the monks mixes roasted barley flour (tsampa) with clean water, using a large bowl. For most of the dough portion, a small amount of oil is added, and then the material is kneaded thoroughly. Dough for one special type of torma that will be eaten during the ceremony is mixed with fermented barley beer (chang) insted of water, and has sugar added as well. The monks note that depending on the type and purpose of tormas, they can have a variety of ingredients. Some of these are mixed in, and others are added onto the surface of the torma. They say that tsampa itself represents all sustenance, so its use is a prayer that in the next life we are asking to have much food and drink available to us.

The Sokha tormas will have certain materials added to their surfaces at various points in the ritual. These materials are also gathered at the outset and put aside on or near the altars to await their use. Included are butter, melted butter with or without a red colorant, red paper, grains, and candies. Also kept on the altar during the ceremony are special medicinal herbs, representing good health in the next life. While one monk mixes the ingredients for the torma dough, two other monks cut pieces of clean wood to serve as platforms for the assembly of tormas. These will also be used for moving the tormas as needed.

Torma fabrication

Bonpo practitioners in Amdo report that tormas are always made as part of a special ceremony, they are never made as separate objects to put on the altar, or as decoration. Although there are many different kinds here, all of them are made only by monks, not by laypersons. Fabrication of the Sokha tormas follows that pattern. The two monks who share the duty of fabricating the tormas start by oiling their hands before handling the barley dough. Then, they break off chunks of dough from the ball of material in the mixing bowl, and begin to form shapes (Fig. 13). Most of the forms are constructed of many separate pieces (Fig. 14) that are then joined, with complex and varied patterns. The full construction of all of the tormas that will be used takes several hours.

Figure 13. Pieces of dough material are torn off and tormas are formed by hand.

Figure 14. The torma assemblies require many smaller images that are joined together.

One of the monks made a set of tormas of a type that he described as embodying "the home's god." These tormas provide the patron deity of this household with a place of residence during the ritual. The other monk made another type, said to represent three different gods. Both sets of tormas include numerous separate pieces and are very complex in form.

Design elements are added to some of the torma surfaces using a long carved wooden block (Fig. 15). Each surface of the block contains a row of incised design elements that can be impressed onto the torma surface. The monks select the appropriate one for each particular torma, based on consultation with a text that accompanies the Sokha ritual. Since the household hosting the ceremony did not have these objects, they were borrowed from another villager.

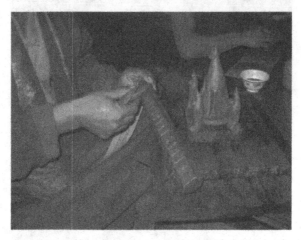

Figure 15. Design elements are added by pressing the torma dough onto a long wooden block with a series of images incised on each surface.

Yak butter is dotted onto the surfaces of many of the tormas (Fig. 16). The monks dip a piece of butter into cold water, then form the butter into shapes resembling flowers. They report that flowers made of butter represent the beauty that we want to fill our next life, symbolized by the prayer that after death we want to live in a place with many flowers. The butter material itself, they say, represents the milk that we pray will always be flowing like water in our next life.

Many of the tormas are painted with small feathers dipped into melted butter (Fig. 17). Different torma forms are said to represent different gods, which in turn represent different human emotions. White torma are called "soft" (peaceful) ones, and those are painted with just the melted yak butter. Others (representing wrathful aspects of nature) need red paint. For this paint, a small amount of red pigment powder is mixed into melted butter. The monks call this pigment "drimo" (Tibetan 'gri mo), and they prepare it themselves from a special plant (also called "drimo") that they collect on the nearby mountains. The leaves are boiled, and the floating residue is collected and dried, then ground into a fine powder.

Small ceramic bowls are lined with butter roundels, and added to the offerings. Candies are placed in the crevices of some of the torma forms, to represent power. The monks explain that this indicates the householders are praying that in the next life they hope to have many things and to be a powerful leader. Grains (of rice, barley, and corn) are also sprinkled around the tormas. These are said to represent the good crops one hopes to have in abundance in the next life. Oil, water, and butter lamps (representing the sun and moon) are also placed around the assembly, along with incense and juniper branches.

344

Figure 16. Yak butter flower-shapes are dotted onto the surface of some tormas.

Figure 17. A feather brush is used to paint melted yak butter onto the surface.

Strips are cut out of paper that is red on one side and white on the other. The strips are then cut into abstract flower shapes. Long wooden sticks are sliced, and the paper strips are glued onto them using dots of tsampa dough as the paste. The strips are then set along the edges of some of the tormas (Fig. 18) as a protective zone. The torma assembly is now complete.

Figure 18. Paper strips outline some of the tormas. The eating torma is in the right foreground.

Use and Placement of finished tormas

Over the next day and a half, the two-volume Sokha text is recited in front of the tormas. The recitation is accompanied by music from cymbals, horns, and a drum, at the appropriate passages (Fig. 19). Lamp and incense offerings are made to the altars. From time to time, some of the tormas are removed and either placed outside for the birds to eat, or are burned along with juniper branch incense offerings in front of the home.

Figure 19. After the tormas are fabricated, the Sokha ceremony begins.

Other activities during the ceremony involve the household members. They burn juniper branch incense and other offerings at an incense burner in front of the home while the mother recites prayers, a typical Bonpo purification ritual [16]. This fire is started as soon as the ceremony begins, and must be kept burning until it is concluded. A pile of items belonging to absent family members who had to miss the ceremony are left next to the altar, and are blessed along with the participants. Three days after the ceremony, they will be removed and given to those absent individuals to bring them luck and blessings. The mother of the house prepares a special food to place on the family altar in the adjacent room.

When it is time for the eating torma (in the shape of a turtle, as seen in Fig. 18) to be used as a feast offering, a bowl containing chang (fermented barley beer) is carried from person to person by one of the monks. Using a juniper-branch brush, some is put into each person's left hand. The person places some chang on their forehead, then in their mouth. The monk sprinkles some on the pile of objects next to the altar, for absent family members. The eating torma (believed to have the power to heal) is cut up with a wooden knife, and a small piece is given to each person. Some of the piece is eaten, but some is retained and thrown off as an offering.

For the final part of the Sokha ceremony, the family members, dressed in special robes, throw grain at the altar while reciting prayers. They sit in front of a flag representing wealth, tied to a wooden pole with a feather on top, and set into a wooden base filled with crop grains. The eldest son places the board holding some of the Sokha tormas onto the top of his head, while standing in front of an altar (Fig. 20). Other family members throw grain at the tormas.

Figure 20. Near the conclusion of the Sokha ceremony, the eldest son holds some of the tormas on top of his head while other family members throw grain offerings.

The family members form a procession following the son, as the tormas are carried out of the ceremony room. Everyone climbs a ladder up to the upper floor, where another ladder leads to the roof. The board and all of its tormas, with the specially-prepared food added, are placed on the roof as offerings for the birds (Fig. 21). This offering is a representation of generosity, kindness, and sharing.

Figure 21. The family procession ends at the rooftop, where the tray of torma offerings is left for the birds to enjoy, as a symbol of generosity, kindness, and sharing.

CONCLUSIONS

The two examples selected here illustrate the importance of documenting intangible aspects of craft production in traditional societies, while it is still possible to do so. The examples document how much richer and complex cultural knowledge associated with craft products is than is usually possible to deduce from materials analysis in isolation. Ethnographic fieldwork can inform laboratory analyses of ancient and historic products, so that even if intangible data cannot be fully recovered, it can still be taken into consideration in the development of experimental designs and hypotheses.

Although this paper focuses on the discussion of fieldwork, laboratory analyses of art and archaeological objects such as tsha-tshas, reported in detail elsewhere [11, 17], have benefited from insights provided by such fieldwork. For example, thin-section analyses of ancient and historic tsha-tshas typically reveal a non-clay component such as small linen fibers (Fig. 22) or hair. Ethnographic fieldwork clarifies that these components were almost certainly chosen primarily for their perceived spiritual power rather than for functional purposes. Thus it can be surmised that linen fibers might be clipped from the linen backings of thangka paintings, or may come from cloth owned or worn by a religious teacher.

We also know that these spiritual additives may be placed in just one spot in the center of a three-dimensional tsha-tsha form (as with barley grains), thus only visible by imaging techniques; or, in the case of flat tsha-tshas, may be identifiable from sampling anywhere on the object, but difficult to interpret. A craftsman in Dharamsala, India, adds animal fur and hair from clothing that Tibetan nomads have turned in as part of their pledge to the Dalai Lama to halt any unnecessary killing of animals. As a reflection of religious dedication and great compassion, small cuttings from these fur remnants are considered to carry great spiritual power.

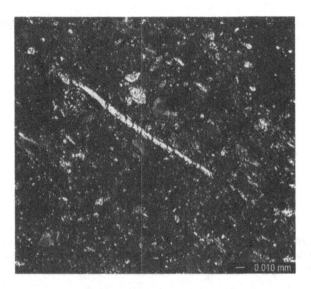

Figure 22. A thin section (crossed polarized light) of an eighteenth-century tsha-tsha shows a fine-grained sandy clay with tiny linen fibers scattered throughout the material.

Another tsha-tsha spiritual additive observed in ethnographic research is sand from a disassembled sand mandala. Knowing this type of material might be a possibility, one can look in thin section for bits of pigment powder in association with sand particles. In another workshop, paper fibers are added to flat tsha-tshas. These are not just any paper clippings; instead, only paper on which mantras or sacred texts appeared would be used. Knowing this, one can look for bits of ink in association with paper fibers. Powdered rosary beads are added in some modern workshops; finding small bits of wooden or stone pieces may indicate that is also what was used in an historic or archaeological tsha-tsha.

Thin-section analyses have helped to illuminate many steps in ancient and historic tsha-tsha production. The raw material selected for use, additives present in small amounts, how often the material is fired versus being dried in the sun, porosity, and surface decoration layers (Fig. 23) can all be discerned by laboratory studies. However, it is clear that not all information about the choices made in fabrication and use, and the reasons for those choices, can be gleaned without accompanying ethnographic fieldwork to record the more intangible elements of tsha-tsha production and use. Fieldwork is also helping identify where and why changes are occurring in technological processes.

Even for living craft traditions, ethnographic fieldwork is still critical. Many of the tangible and intangible aspects of the craft processes for making and using tsha-tshas and tormas in the Bonpo monastery and village in Amdo that was the subject of the fieldwork discussed here show variations from what has been reported in the scant literature on these two crafts. Most examples of these ritual crafts have been studied in a Tibetan Buddhist context, and from larger and more accessible monastic centers. There are often regional and cultural variations in choice of technical procedures, and especially in ideas about what the objects mean and represent and in how they should be interpreted and used. Understanding the range of that variability can help in making more sophisticated interpretations of past craft products.

Figure 23. A thin-section (plane polarized light) of an eighteenth-century tsha-tsha reveals a sandy clay material covered by a thin clay slip that serves an interface between the porous clay body and a paint layer (hematite).

A crucial point that was illustrated by the results of this fieldwork is that for both tsha-tshas and for tormas, the products that are made are not considered to be as important as the process of making them. In the case of the tsha-tshas, they are intended to be left in a protected area and so can survive for a very long time. In the case of tormas, most of them are destroyed immediately after use. Such objects represent what is probably a rather large category of "missing" ancient and historic objects, not recovered in the archaeological record and so unavailable for analysis today. They highlight the fact that no matter how good our analysis and interpretations are, there are always going to be major gaps in our understanding of past cultures.

Finally, traditional tormas were retained more permanently on an altar or in a monastery only occasionally. However, that situation is now changing, and more permanent materials are rapidly coming into use in Tibetan communities located outside of Tibet. New materials and fabrication techniques are emerging for both tsha-tshas and tormas in the new international Tibetan diaspora, and along with them, some new uses and contexts are appearing, too. Thus, while we rush to document disappearing traditional crafts, at the same time, the opportunity exists to document and explain the expansion of some craft traditions made possible by the introduction of new materials and technologies.

ACKNOWLEDGMENTS

This fieldwork was made possible by a grant from the Center for International Studies at the University of Delaware. I would like to thank the cooperation and assistance of monks from Serling Monastery, especially bstan pa and tshe dwang 'dud bdul, who made the tsha-tshas and tormas, carried out the Sokha ceremony, and graciously hosted me at the monastery. I am also grateful for the hospitality and assistance of the villagers of Lhayul, especially LuMen Namchu and her family, who let me stay with them during my visit and incorporated me into life in their village.

REFERENCES

1. UNESCO, Convention for the Safeguarding of Intangible Cultural Heritage, 2003, http://www.unesco.org/culture/ich/index.php?pg=00006 [accessed Nov. 16, 2007].
2. UNESCO, "What is Intangible Cultural Heritage?," 2007, http://www.unesco.org/culture/ich/index.php?pg=00002 [accessed Nov. 16, 2007].
3. S. G. Karmay and Y. Nagano (editors), *A Survey of Bonpo Monasteries and Temples in Tibet and the Himalaya* (National Museum of Ethnology, Osaka, 2003), pp. 622-624.
4. C. Baumer, *Tibetan's Ancient Religion: Bön* (Weatherhill, 2002).
5. Y. Bentor, in *Tibetan Studies: Proceedings of the 5th Seminar of the International Association of Tibetan Studies, Narita 1989*, edited by I. Shōren and Y. Zuihō. (Naritasan Shinshoji, Narita, 1992): Vol. 1, pp. 1-12.
6. G. Tucci, *Indo-Tibetica, Vol. 1: Stupa, Art, Archtectonics, and Symbolism (Mchod-rten and Tsha-tsha in India and Western Tibet)*, English version of 1938 Italian edition (New Delhi, 1988).
7. T. Huber, in *Tibetan Studies: Proceedings of the 5th Seminar of the International Association of Tibetan Studies, Narita 1989*, edited by I. Shōren and Y. Zuihō. (Naritasan Shinshoji, Narita, 1992): Vol. 1, pp. 493-496.
8. D. Liu, *Ca Ca: Zang chuan fo jiao mo zhi ni fo Xiang (Tsha Tsah: Tibetan Clay Molded Buddhist Images)* (Tianjin Shi, 2000).
9. J. Li, *Asian Arts* (1995) http://www.asianart.com/li/tsatsa.html [accessed Nov. 16, 2007].
10. Lama Zopa, Rinpoche, *Commentary on Tsa Tsa Practice* (n.d.) http://www.lamayeshe.com/lamazopa/tsa-tsa.shtml [accessed Nov. 16, 2007].
11. C. L. Reedy, in *The Object in Context: Crossing Conservation Boundaries*, edited by D. Saunders, J. H. Townsend, and S. Woodcock (International Institute for Conservation of Historic and Artistic Works, 2006), pp. 144-150.
12. C. Bütler, *Chö-yang* 7, 38-52 (1996).
13. R. Beer, *The Encyclopedia of Tibetan Symbols and Motifs* (Shambala, 2003), pp. 320-325.
14. Tormas, http://www.khandro.net/ritual_tormas.htm [accessed Nov. 16, 2007].
15. J. V. Bellezza, *Spirit-mediums, Sacred Mountains, and Related Bon Textual Traditions in Upper Tibet* (Brill, 2005).
16. S. G. Karmay, *The Arrow and the Spindle: Studies in History, Myths, Rituals and Beliefs in Tibet* (Mandala Book Point, 1998), pp. 380-412.
17. C. L. Reedy, in *Sino-Tibetan Art Studies*, edited by W. Luo (Third International Conference on Tibetan Archaeology and Arts, Beijing, in press).

Mater. Res. Soc. Symp. Proc. Vol. 1047 © 2008 Materials Research Society   1047-Y02-04

Replications of Critical Technological Processes and the Use of Replicates as Characterization Standards: An Experiment in Undergraduate Education

Pamela B. Vandiver, Heather Raftery, Stephanie Ratcliffe, Brian T. Moskalik, Michelle Andaloro, Katelyn Sandler, and Alicia Retamoza
Dept. of Materials Science and Engineering, Program in Heritage Conservation Science, University of Arizona, Tucson, AZ, 85721

ABSTRACT

The technology of artifacts is analyzed and reconstructed by comparison with known craft practices, the physical and chemical constraints imposed by the raw materials, and the sequence and steps for processing those materials to achieve certain optical and mechanical properties. Understanding of craft knowledge is best pursued by practice, coupled with technical analysis. Six case studies of hands-on, undergraduate student laboratory projects are presented. The studies include testing parameters for the making of stenciled hand images similar to those at caves such as Gargas from the Upper Paleolithic period in France, the variation in processing required to produce Egyptian blue pigments and objects, controlling composition to form either green or turquoise-blue colors in Islamic lead-containing glazes, optimizing the ratio of various pigments to gum Arabic medium in tomb paintings to evaluate application and durability, molding East Asian *gokok* beads in imitation of jade, and making and radiographing a mock-up of a damaged statue on the façade at the San Xavier Mission as a standard for comparison with the original. In each case, various parameters are varied to model the appearance, structure and composition of an object, and the students benefited from the experience of developing research questions and from their involvement in original research projects.

INTRODUCTION

Replicates are attempts to understand processing methods and craft knowledge [1]. To make a replicate requires the appropriate raw materials, an understanding of the geological context of the archaeological site where the technology was practiced, and knowledge of the presence and range of variability of resources at or near a site. The steps in a process are often found on a site and occasionally in a workshop or industrial context. Based on study of the variability of a particular artifact at a site, and the contexts of those finds, whether a living floor, whether redistributed or collected in infill, trash midden or waste pile, or whether mixed or degraded through another process of redeposition. By examining several examples, especially fragmentary ones, one may search for and find examples of errors in processing and use them to come to some understanding of the rate limiting steps in a process. Similarly, one may come to some understanding of what was desired or considered better quality by examining variability of artifacts on a site, with those found in a restricted or high status context, for instance a burial, special collection or elite storage area. In designing a replicate, one tries first to replicate the sequence of steps and the range of variability in those steps. However, one then proceeds to attempt modeling or replicating those steps that presented the most challenge or difficulty to the craftsmen. These are often the rate limiting steps. If one understands what was considered the best quality or which objects display the most highly prized appearance, then further experiments in replication may wish to discover the range of parameters necessary to produce an object of a particular, desired quality. In the sense that a replicate is attempting to understand a rate limiting step, a physical constraint of some material or process, or the processing required to produce a special appearance or optical property or to fulfill a special design function or performance criteria, then, if done correctly, the replicate can be considered a standard for the process or object. If the appearance or process depends on a particular step or if it requires some particular craft knowledge, then a critical technological process is being replicated. If the materials, processes, properties and performance characteristics are analyzed such that they are indistinguishable from the original fragments or object, then the replicates act as reproducible standards that can be used as a characterization references for that specific technological style of making a particular object type.

First, the educational context of the replication studies will be considered. Second, six student lab projects will be reported. Finally, the matter of finding problems and choosing parameters and ranges of variation will be addressed in the discussion.

INCORPORATING REPLICATION STUDIES IN MATERIALS SCIENCE COURSES

Hands-on projects enhance undergraduate-level science courses, but students rarely have the experience or understanding to undertake a research project without considerable mentoring. In the springs of 2006 and 2007 a one-unit lab component was attached to a general-education science course, "Materials Science of Art and Archaeological Objects," and six motivated students enrolled. The lab provided a forum during the first half of the semester for their introduction to specific test equipment and safe laboratory practices. During the second half of the semester students conducted a set of experiments to solve, or propose a probable solution to, a problem. At the end of the term, the lab studies were presented to the lecture class. Examples included replication of Paleolithic stenciled hands from Gargas in France, investigation of ways of varying the Egyptian blue color palette, modeling the variation in green to blue colors in the Islamic lead-potassia-silicate glazes, the use of gum Arabic media and its effect on various pigments and pigment overlays, experimentation with the use of raw glass batch or glass frit to mould Korean comma-shaped ornaments, called *gokok*, and making of a mock-up for comparison by radiography with a damaged eighteenth century CE statue at San Xavier Mission in Tucson. In the last example, possible interior armatures and layering of various plaster compositions were modeled in order to find out what may be below an unfortunate cement coating applied in the 1950s. In each of these cases and based on discussion of previous investigations and analyses, we tried to conjecture the possible processes or sequences of manufacture; each of the possible routes then are examined and tested.

In the lecture class, students undertake several short workshops or small group projects that give them some practical experience in assessing objects. At the end of the assessment, one member of the group shares their assessment with the group. Examples are: (1) museum object condition reports to evaluate materials, sequence of manufacture and damage and degradation of Chinese, Japanese, Tibetan and Korean scroll paintings, (2) evaluation of ancient pottery using 10x loupes for firing demonstration, temperature range using hardness and porosity to separate earthenware, stoneware and porcelain, for color application under, in or on top of a glaze, and reflectivity and brilliance to determine presence of lead, (3) written description of ancient jewelry with stones or imitations of precious materials to understand skeuomorphism and technology, (4) use of portable x-ray fluorescence, near-infrared light and Geiger counter to determine presence of uranium oxide in solution in eighteenth century pressed "vaseline" glass and in particulate form in a Fiesta-ware lead-glazed plate from about 1940. The students try individual projects lead by a craftsperson in which the aim is to make a finished object. Some examples are: (1) flaking of vitreous porcelain, flint and obsidian into blades, scrapers and burins using striking and sometimes finishing with pressure flaking, (2) egg-tempura panel painting on poplar panels, first undercoating with gesso and hide glue, then painting with pigments students have synthesized or ground in the lab, with optional gilding of surfaces and plaster decoration, (3) replication of a pre-glass ceramic, called Egyptian faience, made of crushed quartz, a natron flux and copper carbonate that forms a blue glaze through efflorescence of the salts during drying followed by firing-- all without the intentional application of a glaze. The copper carbonate used as a colorant in the Egyptian faience and panel painting is then smelted into copper, the iron oxide pigment is transformed into slag, and the black carbon pigment is used as part of the fuel. Some of the copper prills from the previous class are separated, cast in a small ingot mold and the pounded into strips, wire and spoons. This occurs as a co-operative class project near the end of the term. Each project emphasizes a transformation of materials. Students also visit on-campus museums and the conservation labs to examine Roman glass, obsidian tools, and panel paintings, and each student chooses a rare book or special collections object for examination, study and a short report.

During the first six-weeks of the term, the students taking the laboratory do experiments that involve the following: (1) breaking glass sheets using various medieval and modern methods combined with optical microscopy of surfaces, (2) grinding of poorly weathered clays and synthesis of Maya blue and determination of enamel melting temperatures by fusion button tests, (3) annealing, pickling and enameling on copper in cloisonné style and SEM-EDS of the results, (3) pottery and glazing and microprobe analysis of a Song dynasty Chinese glaze, (5) glass forming by pressing medallions, drawing rod, understanding annealing by making and breaking Prince Rupert drops, followed by blowing a glass, (6) brazing and welding of steel, and preparation of furnace and materials for copper smelting. Students have seen and understand the relationships of processing to composition, microstructure and properties.

SUMMARIES OF SIX STUDENT PROJECTS

1. Replication of Upper Paleolithic Period Stenciled Hands from Caves in Southern France: The earliest images on the walls of caves and rock shelters throughout the area that is now France and Spain occurred about 27,000 years ago in the Upper Paleolithic period. Some of the earliest rock art in this area consists of hundreds of hand stencils [2]. Many researchers have proposed that a fine spray of paint was blown, spat or flicked onto a hand placed against the wall of a cave (Fig. 1). Several caves contain hand stencils, among them are the caves of Pech-Merle, Gargas and Cosquer in southern France.

Michel Lorblanchet has proposed that the stencils were created by spitting pigment through a hollow bird bone tube onto cave walls, and using a pigment that was crushed to a powder and combined with water or saliva [3]. However, the flicking of a wad of hair or plant fibers dipped in wetted pigment also produces a splash pattern. The goal of this project is to reproduce these two techniques, to determine which technique produces splatter patterns that best resemble the splatter patterns depicted in photographs of the Paleolithic hand stencils, and finally to determine if a range of optimal conditions and parameters within the technology exists that produces the details and patterns that most resemble the originals.

Pre-wetted, heavy watercolor paper (Fig. 1), and later limestone tile (Fig. 2) was set vertically on an artist's easel as the substrate. Borosilicate glass tubes of various inside diameters were used as the "bird bones," and were cut to various lengths as needed. Several pieces of rabbit skin, with hair attached, were used as the wads of hair. The single pigment was a submicron to 3 micron powdered red iron oxide (Fe_2O_3, hematite, Sigma-Aldrich reagent grade) mixed in various proportions with distilled, de-ionized water. Table 1 gives a list of parameters for each factor. The resultant matrix of tests consisted of just over 2000 individual tests.

Table 1. Variation in Tested Parameters (* indicates best fit of parameter with photograph of ancient splattered hand stencil patterns, and +, the easiest to execute, and ^ indicates did not work). (Note that the first four parameters have a range of values that are equally consistent with images of the original.)

Tube inside diameter, mm	2^	4	6*	7*	11	
Tube length, cm	5	7.5	10*	15*	17.5*	20
Distance of tube end from wall, tile or paper, cm	2.5	5	10*	15*	20*	30
Wetness of pigment, 2.0 g Fe2O3 pigment mixed in	1 mL water	1.0 mL	1.5 mL*	2.0 mL*	2.5 mL	
Wetness of wall	dry	damp*	semi-wet	wet		
Strength of breath	soft	medium	hard*			
Length of breath	1 sec+	>1 sec				
Directionality of breath	straight on*	slightly up*	down	from side		

Splatter patterns similar to ones shown in photographs of the stenciled hands are characterized by having a circular central splatter pattern with very little substrate showing through, a uniform spray of droplets radiating outward on all sides from the central splatter pattern with average droplet size decreasing as the distance from the central splatter pattern increases, and a relatively small overall droplet size throughout the splatter pattern. The 'wad of fur" experimentation was abandoned after the first set of trials because elongated, linear patterns, or trails, resulted that had no resemblance to splatter patterns in the hand stencil photographs.

During the "bird bone" testing, a small amount of pigment was put into the tube by swiping one end of the tube along the side of a pigment-covered spatula. A sheet of watercolor paper, moistened with a sponge, was fastened with tape onto the easel. The pigment was blown from the end containing the pigment onto the paper at an angle perpendicular to the surface. The distance from the tube to the easel was kept constant by a stiff metal wire attached to the easel that acted as a jig, being horizontal, perpendicular to the substrate, and ending at the forehead of the blower. The still wet paper was removed from the easel, laid flat to dry while weighted on all four corners. When dry, it was sprayed with a clear fixative.

Several factors were controlled in the experiments, including tube diameter (measured as inside diameter), tube length, distance from wall (measured from the surface of the paper to the closest end of the tube), "wetness" of pigment, wetness of the wall, strength of breath, length of breath and directionality of breath. A chosen measurement for each factor was used as a standard and remained constant during the testing of the other factors. Some of these were changed during the experimentation process, if determined to be less efficient than another measurement. Due to the lack of familiarity with this method of painting, an iterative series of experimental trials was deemed necessary. Once each parameter was optimized for the desired visual effect, a final set of trials was conducted with the following parameters: 6 mm tube diameter, 10 cm tube length, 10 cm distance from wall, wetness of pigment a ratio of 1.5 mL of distilled $H2O$ per 2.0 g unit of dry pigment, wetness of wall about semi-wet (a light sheen of water on the surface of the paper), strength of breath fairly hard, length of breath about one second, and directionality of breath straight onto the paper, at a perpendicular angle to the surface.

The order of experimentation was chosen based on what had the greatest effect on the other factors. The order for the final trial was: pigment wetness, strength of breath, wetness of wall, tube length, tube diameter, distance from wall, directionality of breath and length of breath. A detailed discussion of results and measurements is given in reference [4]. Tube length and diameter are more indicative of a hollow reed or large bird feather than a bird bone. Two to three mm diameter tubes are difficult to discharge and clog easily. Pigments without sufficient water lead to small central patterns raised in relief and fine dusting of the surface. Insufficient wetness causes the tubes to clog, but when blown out, the pigment scatters without a central splatter. Overly hydrated samples ran out of the tube and distorted in a downward direction on the paper. A soft blow is inadequate to get most of the pigment out of the tube and results in a few large splotches. Medium blows result in larger droplets, many of which are below the center of the splatter. Hard blows produce uniform-sized, radial alignment of droplets with central dense splatter pattern. If the wall was too wet, vertical flow in the splatter pattern occurs and, if too dry, pigment droplets are raised in relief. Blowing for more than one second did not change the pattern, but blowing at an angle resulted droplets elongated in the direction of the source.

Another interesting feature of the blown pigment is the flow instabilities (Fig. 1) that result in somewhat elongated droplets, averaging 2-4 mm in diameter and elongated 2 to 4 times the width, that are outside the central splatter and that are interspersed with smaller very elongated droplets with an aspect ratios of 9 to 20 in length to 1 unit in diameter. These characteristics are common in patterns made with aqueous blown pigment. Another feature that is indicative of hand stencils is the penetration of elongated lines of pigment around the raised parts of the hand, as shown in Fig. 2. Finally this study did not consider blowing directly from the mouth, another of Lorblanchet's conjectures. Further study will investigate the alignment of the small platy iron oxide particles and particle agglomerations using scanning electron microscopy. Preliminary results at optimal wetness show local agglomerates, but no preferential alignment parallel to the cave wall surface. Presumably such alignment may result from the cyclic moisture variation in the interior of the caves.

Fig. 1. Splatter pattern in red iron oxide with central solid area surrounded by radially elongated droplets. Larger droplets are oval with a 1:2 to 1:4 aspect ratios. Small droplets are elongated with aspect ratios of 1:9 to 1:20. These instabilities are a trait of blown aqueous pigment. Pigment was blown using parameters marked optimal in Table 1 on a substrate of heavy watercolor paper.

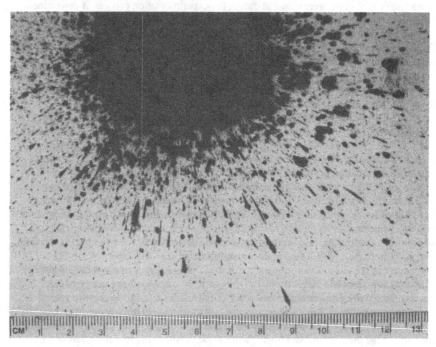

Fig. 2. Hand stencil with elongated droplets of hematite scattered around the raised areas between fingers and around knuckles characterize a stencil made by blowing on a limestone substrate. One central splatter pattern covers the upper palm and lower part of the fingers and another covers the ends of the fingers.

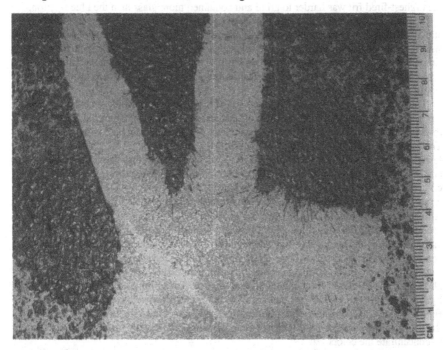

2. Ways of Varying the Egyptian Blue Color Palette and the Production of Objects of Egyptian Blue: Egyptian blue is the first synthetic pigment made in Egypt during the mid-third millennium B.C.E. Many studies have reviewed and characterized the compositions and microstructures of the pigments [5,6,7] and made comparisons to similar Roman and Chinese pigments [8,9] and to objects with similar properties [10]. Egyptian blue is fired as a frit made from 60-75% silica, 10-15% alkaline materials and 10-15% copper carbonate. Properly fired with slightly more copper carbonate than alkali results in crystallization of a copper-calcium-tetrasilicate($CuCaSi4O10$ or $CuO \bullet CaO \bullet Si4O8$) that is chemically identical to the rare, naturally occurring mineral cuprorivaite. The formulation of A.P. Laurie from 1914, replicated and modified by W.T. Chase, was used (62.71% $SiO2$, 12.04% $CaCO3$, 14.95% $CuCO3$ or malachite, and 10.3% natron, a naturally occurring alkali) [11]. The natron composition was 33.02% sodium bicarbonate, 32.99% sodium carbonate, 14.17% sodium chloride and 19.82% sodium sulfate. The mixed powder was placed in crucibles in a Lindberg box furnace and fired for 22 hours to three temperatures, 750°, 850° and 950°C. At 750°C very little of the crystal, cuprorivaite, forms. A gray friable material resulted (Fig. 3, upper right). The firing to 850° produced a brilliant blue frit (Fig. 3, upper middle), but the firing to

950° forms a significant glassy phase, resulting in a variegated areas of blue and green frit with some black glass particularly on the top surface (Fig. 3, upper left). Significant glass stops the crystallization process. The high-fired sample is a bit harder than the intermediate fired sample but both have a similar range of porosity. The green fraction of the highest-fired frit was harder to grind and contained more glass than the blue fraction.

A streak test on an alumina plate was conducted to test suitability for use as a chalk stick. The low-fired gray crumbled; the medium-fired bright blue easily made a streak when wetted but required several passes to deposit blue powder when dry. The high-fired blue areas of the sample behaved similarly to the medium-fired sample, but the green high-fired fraction left no color when used wet or dry. Only the medium-fired sample is suitable as a chalk stick.

The effect of particle size on color was studied. The three pigments were ground with a mortar and pestle and sieved to coarse (150 microns to 1 mm), medium (38-150 microns) and fine (less than 38 microns) fractions. Each color lightened with grinding. The low-fired fraction becoming a yellowish-gray when fine (Fig. 3, left column, and lower left), the medium-fired one producing a strong light blue (Fig. 3, middle column), and the high-fired one producing a lighter powder blue with a greenish cast (Fig. 3, right column). A small fragment of an Egyptian blue bowl from Tell Yi'nan, Israel, was ground finely and produced a gray blue color, indicating a probable intermediate firing temperature of about 800°C [12]. Another test of grinding overfired Egyptian faience produced a light to medium blue. Therefore, the tints of blue, green and gray may be varied at will by changing the firing temperature of the raw material and the shade can be controlled by the particle size of the pigment. In addition, recycled Egyptian blue and Egyptian faience can be used to make new objects or added to change the color of an already existing Egyptian blue pigment. In this particular example, the change is to a lighter gray with the Tell Yi'nan bowl or a lighter blue with an addition of the ground Egyptian faience. To make a strong green or strong blue of the high-fired raw material, the two colors were separated after the firing and during the initial crushing stage in order to concentrate the colors.

Spreadability was tested using an aqueous medium by taking samples of the fine and medium particle ranges of each of the three fired samples, mixing them with about one third their volume in de-ionized, distilled water, and drawing the pigment along with a brush (a #2 round brush) on glass and on paper substrates. The medium fraction was still too coarse to spread, but the fine fraction produced a reasonable paint (Fig. 4). The fraction fired to 850°C produced the brightest, intense blue. The high-fired fine fraction produced a more greenish blue. A binder stronger than water is required to maintain the pigments on the glass slide or paper. A binder such as egg yolk turns the blue pigments to a greenish color and darkens them, but gum Arabic maintains the blue color.

Lastly, each of the nine powders, shown in Fig. 3, were formed into beads and small objects first by adding water. These remained friable after firing to 850° for 6 hours. Next a boiled saturated solution of sodium carbonate in water was added in an approximate volumetric quantity, 1 part of water mixed with 4 parts powder, and beads or cubes were formed, dried and fired to 850°C for 6 hours. After firing, these objects were stable, durable and nonfriable, and their color, hardness and texture could be controlled. They were gray-blue and heterogeneous when starting with the low-fired raw material. With the medium-fired raw material, they were bright blue when coarse and lighter blue with finer particle size. Greenish-blues resulted from the higher-fired raw material, with lighter colors with finer particle size. Firing the fabricated beads to 950°C, however, produces overfired, dark blue to black misshapen lumps (Fig. 4, left, only the fine and medium particled examples are shown).

Egyptian blue was tested in the three ways this synthetic pigment was used by Egyptians, as a raw material, as a pigment, and as a fabricated and fired object. The experiments produced a range of color and texture by varying the temperature of heating the mixed raw materials, by controlling the pigment particle size, and by designating the object starting materials, their particle size and the use of alkaline sintering aids. Finally, the colors and textures were compared with an Egyptian blue vessel fragment from the archaeological excavations at Tell Yi'nan, Israel [12] and with colors produced by grinding useless, overfired Egyptian faience.

Fig. 3. Egyptian blue fired to 750°C is gray (left column), to 850°C is blue (center column on left) and to 950°C is blue and green (right column on left). Ground pigments shown in the top row have a coarse particle size; the middle row has medium ones, and the lower row fine particle sizes. Overfired and ground Egyptian faience is shown in the upper right and a ground sample from an Egyptian blue bowl from Tell Yi'nan, Israel, shown in the lower right.

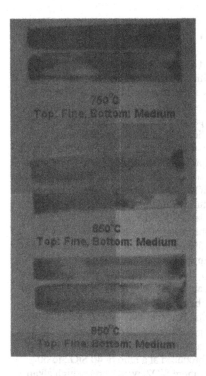

Fig. 4 (left). Ground pigments, fired to three temperatures, were mixed with about 25% by volume water and painted onto a glass substrate. The finer particles show better flow and adhesion.

Fig. 5 (below). Left side shows range of color and texture in initial raw material firing to 850°C (blue, upper left) and 950°C (lower left with green above blue lump). Second column shows blue objects made from powders fired to 850° with coarse particle size in upper row, medium in center row and fine in lower row. The third column shows green objects made from powders fired to 950° with coarse particle precursor in the upper row, medium in the middle and fine in the lower row in a turquoise blue-green. The fourth column, dark green to black and overfired, is made from powders fired to 850° with the final object fired to 950° with the same particle sizes as columns 2 and 3. The last column was fired both times to 950°C from the same coarse, medium and fine particle ranges, and is overfired green to black in color.

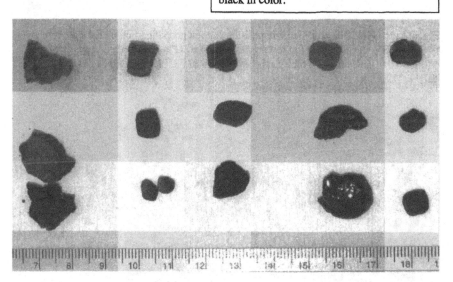

3. The Shift from Green to Turquoise Blue in Islamic Lead-Potassia-Silicate Glazes Colored with Copper Carbonate as Analyzed in Line-blend Tests: Some Iznik tiles made in the sixteenth century C.E. near Istanbul, Turkey, were examined. These either have a brilliant copper turquoise blue in a low-lead glaze or they are an olive green, also caused by copper, in a high lead glaze [13]. Traditional Uzbek potters say that they observe a color change when 15% or more PbO is added to a mixed-alkali silicate glaze that is fired to 1000°C [14]. We know that lead-containing glazes require an oxidizing atmosphere so the lead will not precipitate as an opaque black metal film on the surface. The problem involved determining the ratios of silica to lead and potassium oxides at which the color change occurs. Line-blend tests commonly used by potters offers a rapid combinatorial approach but have messy final chemical compositions (Fig. 6). Mixtures of SiO_2 and PbO were compounded in increments of 10 wt%: for instance, 10:90, 20:80, 30:70, 40:60, 50:50, 60:40, 70:30, 80:20, 90:10 SiO_2 to PbO. The same amount of copper carbonate, 3 wt.%, was added to each mixture. Three percent is the minimum amount needed for an intense blue-green color, based on Vandiver's unpublished results; adding more than 3% results in the color staying the same intense blue. One set of variable ratios of SiO_2 to PbO was prepared with 5% K_2CO_3, another set with 10% and a third set with 15%. At first, the mixtures were applied to red- and white-bodied tiles, but the red-bodied tiles tended to obscure the color change. Thus, only white-bodied tiles were used. Lead-rich glazes were fired to 825°C, and silica-rich glazes with a ratio of 50 SiO_2: 50 PbO or more SiO_2 were fired to 1050°C by consulting the SiO_2-PbO0K2O phase diagram and considering Islamic pottery firing practices that are commonly in the range of 850 to 1050°C. The color changed from green to turquoise blue between 20 SiO_2: 80 PbO, and 30 SiO_2 to 70 PbO. The change in the amount of K_2O had no effect on the color change. The brightest turquoise was produced at a ratio of 40 SiO_2 to 60 PbO. Glazes with a 60:40 ratio of SiO_2 to PbO, or more SiO_2, were a crusty high alkali fluxes.

Fig. 6. Line-blend with lead-rich compositions on left and silica-rich on the right; both rows have 3% CuCO% with top row on having 5% K_2O in glaze on red body and lower row has 10% K_2O on white tiles. Left tiles fired to 825°C and right tiles fired to 1025°C.

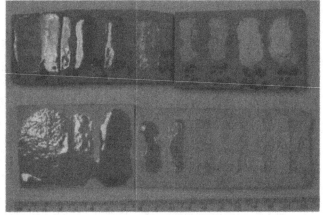

4. The Effect of Varying the Amount of Gum Arabic Media on the Application and Durability of Various Pigments and Pigment Overlays: Between 1987 and 1992, a large-scale conservation project was conducted in the Tomb of Nefertari under the direction of the Egyptian Antiquities Organization and the Getty Conservation Institute [15]. The tomb of Nefertari is located in Luxor, Egypt, in the Valley of the Queens and contains about 520 square meters of relief paintings that were created in 1255 B.C.E. The organizations recognized that preservation of this ancient treasure was, and continues to be, essential to the culture of humanity. To stem ongoing degradation, a "virtual experience museum," similar in concept to Lascaux II, has been proposed [15]. Analyses of the diversity of pigments, binders, plaster substrate and damaging salts have been published [15,16]. The pigment binding medium is gum Arabic from the native acacia tree, and the pigments consist of ferric oxide (hematite), burnt umber, yellow ochre (goethite), Egyptian blue and green and manganese dioxide black [16,17].

In these studies, no estimates of the pigment to binder ratios were given, but before a tomb replica could be created this unknown should be researched. Binders can be as low as 0.25 wt% in red chalk to 60% in some oil painting pigments. To test how little binder is required, three amounts of gum Arabic were chosen, 10, 15 and 20wt% mixed in a 1:5 ratio of gum Arabic to de-ionized, distilled water. Pigments included burnt umber, yellow ochre and Egyptian blue. The Egyptian blue was too coarse with particles up to 50 microns, and the pigment would not flow. A synthesized cobalt blue pigment of 1-5 microns was substituted and a fine charcoal black pigment was tested, even though neither of these is found in the Nefertari tomb. The charcoal black absorbed considerable medium and tended to crack off the surface when dry. This result demonstrated likely reasons why this pigment was not used. For the three pigments with suitable flow characteristics, the 10% binding medium mixture provided the smoothest surface and suffered the least amount of flaking. Mixtures with 20% binder were the roughest and had the most visible flaking paint, contrary to initial expectation. Areas where one pigment-binder mixture was applied on top of another, perhaps before the bottom layer of pigment was completely dry, exhibited the worst flaking and delamination, demonstrating that moisture is indeed detrimental to the stability of the paintings. Further tests should be conducted using less binder.

Figure 7 (next page). Limestone panel on next page painted with yellow ochre and burnt sienna (center), Egyptian blue (left), stabilized cobalt blue (right) and charcoal black pigments (columns to right and left) in a gum Arabic media of 20% (left), 15% (middle) and 10% (right). Note flaking due to overlays (upper register) and pigments with coarse particle size.

5. The Use of Frit or Glass Batch to Mould Korean Comma-Shaped Ornaments or *Gokok*: *Gokok* ornaments and beads were analyzed by X-ray fluorescence spectroscopy and contrary to expectation, scientists at the Freer Gallery of Art discovered that some of the beads that resembled jade, in fact, were made of glass [18]. *Gokok* are comma-shaped beads common to ancient Korea, China and Japan that are found in first millennium C.E. burial sites as part of other larger artifacts, including gilded gold crowns. The presence of glass *gokok*, or *magatama* in Japanese, requires a different stabilization and conservation plan than the preservation of jade. Knowing the condition of the surface would help to inform this plan. Molding followed by grinding and polishing to final form has been suggested as the most likely method of manufacture [18]. This finishing operation would make the glass more likely to weather than a fire-polished surface. In addition, the glass skeuomorph of jade imitates the heterogeneities in the jade, such as subtle color variation, translucency to opacity and vein-like or striated textures.

Since no trace of hot working appears on the surfaces of the glass beads, two methods of molding were investigated. Four glass batch compositions (two soda-lime-silicates and two lead-alkali-silicates) that are similar to the analyses of the glass *gokok*, were supplied by Blythe McCarthy of the Freer Gallery of Art. These compositions are similar to those reported for other glass beads and ornaments [19,20,21]. Even though some experts believe that the beads can be molded from raw glass batch materials, this is not physically possible because the raw materials outgas and react strongly with the refractory molds. Our first experiment demonstrated that this was the case as the raw batch melted into and stuck to the four types of molds that were made of fireclay, steatite, fireclay coated with powdered bauxite or alumina wash and pressed, carved mica. The glasses were heated to 1050°C for 24 hours. The glass beads could not be removed from the molds without fracturing because of thermal expansion incompatibility. The next experiment involved pre-melting the glass batches in fireclay crucibles at 1050°C for 24 hours where only the melted glass was used as the raw material. After melting, the crucibles were chiseled and cut away. The glass was ground to a range of particle sizes from 0.1-2 mm in maximum diameter in order to promote sintering of the particles. Some of the glass was ground in an iron mortar and pestle to introduce an iron impurity that would promote nucleation of crystals in the final firing; some of the glass was discolored with brown or gray surfaces and particles. This contaminated frit was separated as another sample to test whether this impurity might add color or textural variegation to the final product. Only the two soda-lime-silicate glasses, and not the lead formulations were tested, and each of the two batches was divided into two parts, one that was pure and another with iron contamination. The fusing temperature was tested with a fusion button test. Samples fired to 760°C or above for 12 hours were too transparent. Firing to 640°C for 10 hours and to 620°C for 12 hours followed by a 6 hour annealing produced the required translucency. The second period of heating at a cooler temperature was intended to promote nucleation and growth of secondary crystalline phases. The iron-contaminated sample had significant textural heterogeneities. Optical microscopy showed the presence of spherical bubbles and green striations in the bluish-green glass. Grinding and polishing produced a satin-matte surface texture. Further testing by x-ray diffraction and SEM-EDS is recommended.

Figure 8. Steps in the process of making a Korean *gokok* ornament of the Three Kingdoms and Unified Shilla periods; from left, the steps are melting and crushing the glass, filling the mold, fusing the glass frit in the mold, removing residual mold material, and surface polishing.

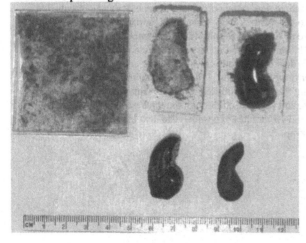

6. Manufacture of a Replicate Section of an Eighteenth Century C.E. Statue from the Upper Façade at the San Xavier Mission: The San Xavier Mission is located in the Santa Cruz Valley south of Tucson, Arizona [22]. The Tohono O'odham settled Bac, first with Jesuits and later with Franciscans from Spain. The mission was built between 1783 and 1797. No documentation of the original construction has been found, but evidence of many repairs is present and deterioration of some of these repairs, mainly by salt efflorescence, is evident.

Some of the repairs are affecting the integrity of the original structure. In 1906 and in the 1950s, layers of Portland cement were added to the roof and dome to make them impermeable to rain that is particularly heavy during the late summer monsoon season. In 1984 a whitewash of Portland cement was applied to the exterior walls as well as paint to help seal the surfaces. In the 1950s cement was applied to the statue above the main façade. The weight of this added cement was considerable, and drainage was also a problem. In 2000, a plan to remove the cement and replace it with a lime and sand plaster was accepted, and work has continued to the present.

The outer layers of cement were removed and replaced by a calcite-based lime plaster. The treatment consists of two layers of lime and sand with a mucilage additive extracted from *nopal*, or prickley pear cactus. This sugar-rich additive may slow the recrystallization of the lime allowing agglomeration and liquid-phase sintering of the crystals, thus increasing the effective cross-sectional area of the plaster, according to preliminary results of scanning electron microscopy. In summary, this treatment allows the dome and walls to "breathe," and allows the load to self-adjust during periods of monsoon and earthquake.

Our task was to evaluate nondestructively whether a similar treatment might be applicable for the cement-encased statue, presumably of a Franciscan monk that has broken and is missing a section above the waist (Fig. 9). It is feasible to place a portable x-ray on the roof and to mount film on a frame beyond the statue. Exposure would have to occur at night when there are no visitors, and calm conditions often prevail. Radiography might determine the variation in materials and presence of armatures or voids. Ultrasound is problematical due to the porosity in the plasters and cements that scatters the signal. In order to test various x-ray techniques, or parameters, a mock-up of the statue was made with several structural support materials embedded. Various density materials, such as wood, brick, limestone and metal, were selected to form inner supports and armatures (Fig. 10). Two different mixtures of lime plaster and sand were used, one in the lower half, and another for the upper half. The completed mock-up was dried for four weeks, then x-rayed using 50, 75 and 100 KV, 5 mA, 1-15 minutes of exposure [23]. The mock-up is 25x16x40cm, and penetration through the thickest section requires a technique of 100 KV, 5 mA and 15 minutes exposure. The next task is to x-ray the statue and compare it with the mock-up.

Fig. 9. The conically shaped sculpture is centered at the top of the baroque façade in the center of the upper "U"-shaped architrave. Photo taken in 1897 and preserved at the Arizona Historical Society, Tucson, AZ.

Fig. 10 Making the mock-up of a cross section of the statue using 10% by volume sand to lime plaster in the bottom half and 30% by volume sand to lime plaster in the top half. Brick reinforcement at left, wood in center and metal armatures to right inside metal mold.

DISCUSSION

Each case study involved finding a problem, determining useful experiments and varying parameters. In most cases the problem was in the published literature or was discussed by professionals in the popular press, and the students found no reports that a process had been tested by replication. The experiments with blowing or flicking red iron oxide pigments to produce stenciled hands answer a first order question, what is the detailed process and what can we measure about the process that might better define it or negate its possible use. Blowing within certain parameters produced patterns of the right size and shape as photographs of images in three caves, and flicking did not. Another possible process of daubing pigment with a sponge of liken or dried grass was not sufficiently specific and a preliminary trial had pigment dripping and out of control and so was not tested. The raw material was not part of the problem because they have been measured and characterized by several authors including Vandiver.

Egyptian blue pigment has been synthesized, analyzed and reported many times, but the making of Egyptian blue objects of different colors and textures has not been investigated. To do the second required a source of raw material from the pigment synthesis, and the student, Ratcliffe, needed to become sufficiently familiar with powder processing that observation would lead to prediction. The parameters of pigment preparation, particle size, media and final firing temperature were varied. Further work would analyze the results and further vary forming processes.

In the case of testing the amount of gum medium required to make an on-site virtual reality reconstruction of an Egyptian painted tomb for tourists, the student, Andalaro, put herself in the role of an artist trying to make a successful replication. Media and pigments were identified, but the concentration or volume fraction of dried medium was not. The experiment was to vary concentration of medium with five pigments and observe qualitatively the variability in flow and adhesion as the process was varied with single and double layers of application. In the other three case studies, the questions were brought to the students by professionals because very little is recorded or known about the craft process or the understanding of craft process required to practice a technology. The tests function to gain a basic understanding of a problem and are preliminary to setting up more detailed experiments.

CONCLUSIONS

In the case studies summarized here, non-science-major undergraduate students fulfilled a second-year science requirement by each conducting a lab experiment to help them to understand a complex technological process. In studying ancient technologies, we can analyze composition, microstructure and measure properties of an artifact precisely, but processing and performance (or function) are much more difficult to reconstruct. Our knowledge of processing technologies is based first on the physical and chemical constraints of materials and processes. Beyond these constraints, we employ conjecture, replication and interpretation, as in these case studies. The solutions are probabilistic in the sense that they often involve a narrative where multiple solutions are possible, with the one fulfilling the necessary and sufficient conditions being the more likely. These experiments are self-contained steps on a path to understanding and explanation. The students said that they had fun, that they felt like they were involved in a pioneering hands-on experience and that they learned something of value they will not forget.

ACKNOWLEDGMENTS

We gratefully acknowledge help and advice from Jean Clottes, Ian Freestone, Blythe McCarthy, and Joseph Simmons and Robert Vint. The Department of Anthropology,

University of Arizona, undergraduate research fellowship program is thanked for support of H. Raftery's travel in France to study several Paleolithic cave images.

REFERENCES

1. P.B. Vandiver, "Reconstructing and Interpreting the Technologies of Ancient Ceramics," in Materials Issues in Art and Archaeology, vol. 1, eds. E.V. Sayre, P.B. Vandiver, J. Druzik and C. Stevenson, MRS Symp. Proc. vol. 123, 1988, 89-102.
2. Leroi-Gourham, A., et al., L'Art des Cavernes: Atlas des Grottes Ornees Paleolithiques Francaises, Ministere de la Culture, Paris, 1984, pp. 467ff, 514ff, and J. Clottes and J. Courtin, La Grotte Cosquer, Editions du Seuil, Paris, 1994.
3. Lorblanchet, M., personal communication, 1988, and demonstrated in National Geographic Society video on Paleolithic Cave Art, 1994.
4. Raftery, H, "Paleolithic Hand Stencil Technology," Department of Anthropology 399, and Dept. of Materials Science and Engineering 258 course records, fall 2006 and spring 2005, respectively, University of Arizona, Tucson, AZ, unpublished ms.
5. Delamare, F., "Le Bleu Egyptien, Essai de Bibliographie Critique," in La Couleur dans la Peinture et L'Emaillage de L'Egypt Ancienne, eds. S. Colinart and M. Menu, Centro Universitario Europeo per i Beni Culturali, Ravello, Italy, Edipuglia, 1998, 143-162.
6. Pages-Camagna, S., "Pigments Blue et Vert Egyptiens en Question," op. cit., 163-176.
7. Delamare, F., "De la Composition Bleu Egyptian Utilise en Peinture Murale Gallo-Romaine," op.cit., 177-194.
8. Schiegl, S., and A. El Goresy, Archaeometry 45/4 (2003) 637-658.
9. Wiedemann, H.G., G. Bayer and A. Reller, "Egyptian Blue and Chinese Blue," op.cit., 195-203.
10. McCarthy, B. and P. Vandiver, "Ancient High-Strength Ceramics: Fritted Faience Bracelet Manufacture at Harappa (Pakistan), ca 2300-1800 B.C.," in Materials Issues in Art and Archaeology II, eds., P.B. Vandiver, J. Druzik and G.S. Wheeler, MRS Symp. Proc., vol 185, 1991, pp. 495-510.
11. Chase, W.T., "Egyptian Blue as a Pigment and Ceramic Material," in R.H. Brill, ed., Science and Archaeology, M.I.T Press, 1971, pp. 80-90.
12. Vandiver, P.V., "An Egyptian Blue Bowl from Tell Yi'nan," in H. Leibowitz, ed., Excavations at Tell Yi'nan, U. of Texas Press, Austin, 2003, pp. 75-88.
13. Kingery, W.D., and P.B. Vandiver, Ceramic Masterpieces, Free Press, 1986, pp. 123-134.
14. Vandiver, P.B., "Craft Knowledge as an Intangible Cultural Property," Materials Issues in Art and Archaeology VII, eds. P. Vandiver, J.L. Mass and A. Murray, MRS Symp. Vol. 852, 2005, pp. 331-352.
15. McDonald, J.K., House of Eternity: The Tomb of Nefertari, J. Paul Getty Trust, Los Angeles, CA, 1996, 6, 42, 113-4.
16. Saleh, S. A., "Pigments, Plaster and Salts Analyses," in Wall Painting of the Tomb of Nefertari: Scientific Studies for Their Conservation, ed. M.A. Corzo, J. Paul Getty Trust, Century City, CA, 1987, 96.

17. Palet, A., E. Porta and D. Stulik, " Analyses of Pigments, Binding Media, and Varnishes," and S. Rickerby, "Original Painting Techniques and Materials Used in the Tomb of Nefertari," in Art and Eternity, in eds. M.A. Corzo and M. Afshar, J. Paul Gettty Trust, Los Angeles, CA, 1993, 57, 63.
18. Douglas, J, B. McCarthy and I. Lee, "Gokak: Korean Glass and Stone Comma-Shaped Beads at the Freer Gallery of Art," Ornament 25/4 (2002) 34-39.
19. Jo, Kyung-mi, H.S. Yu, and H.T. Kang, "Scientific Analysis of Glass from Hwangnam-daech'ong Tomb no. 98," Conservation Science in Museum, National Museum of Korea, 1 (1999) 61-74.
20. Kim, Gyu-Ho, "Scientific Analysis and Interpretation of Old Glass Beads in Korea," Jour. of Ho-Am Museum 5 (2000) 69-77; and G.H. Kim and W.Y Huh, "SEM and X-Ray Microanalysis of Ancient Glassy Materials: An Analysis of Glass Beads Excavated at Maha-ri, Pongtam-myun, Hwasung-gun, Kyunggi Province," Jour. of Ho-Am Art Museum 3 (1998) 65-74.
21. Lee, Insook, "Ancient Glass Trade in Korea," Papers of the Brit. Assoc. for Korean Studies 5 (1994) 65-82; and Ibid., "Early Glass in Korean Archaeological Sites, Korean and Korean-American Stud. Bull. 8.1/2 (1997) 14-23.
22. Robert Vint, personal communication, April 2006, and historical records and photographic documentation of the San Xavier Mission at the Arizona Historical Association, Tucson, AZ.
23. Radiographs on disk available from A. Retamoza, Department of Materials Science and Engineering, University of Arizona, Tucson, AZ.

AUTHOR INDEX

SUBJECT INDEX

Printed in the United States
By Bookmasters